Bone
Histomorphometry:
Techniques
and
Interpretation

Editor

Robert R. Recker, M.D.

Associate Professor
Director
Metabolic Research Unit
Creighton University
School of Medicine
Omaha, Nebraska

CRC Press, Inc.
Boca Raton, Florida

Library of Congress Cataloging in Publication Data
Main entry under title:

Bone histomorphometry.

  Bibliography: p.
  Includes index.
  1. Bone—Measurement. 2. Histology.
3. Morphology. I. Recker, Robert R.
[DNLM: 1. Biometry—Methods. 2. Bone and
bones—Anatomy and histology. WE 101 B712]
QM569.B72   611'.0184   81-4101
ISBN 0-8493-5373-4

Direct all inquiries to CRC Press, Inc., 2000 Corporate Blvd., N.W., Boca Raton, Florida, 33431.

© 1983 by CRC Press, Inc.

International Standard Book Number 0-8493-5373-4

Library of Congress Card Number 81-4104
Printed in the United States

THE EDITOR

Dr. Robert R. Recker is Associate Professor of Medicine at Creighton University School of Medicine and Director of Metabolic Research there. He received his M.D. degree from Creighton University in 1963, interned at Wilford Hall U.S.A.F. Hospital in San Antonio, Texas and took residency training in general internal medicine at Creighton University. He finished fellowship in endocrinology at Creighton University in 1971 with research interest in metabolic bone disease and calcium metabolism. He has been director of Metabolic Research at that institution since 1974 and has been chief of clinical endocrinology teaching and patient care services. Dr. Recker is a member of five professional and/or scientific organizations and has published over 50 research papers in the area of calcium metabolism and metabolic bone disease. He serves as journal referee for the *Journal of Laboratory and Clinical Medicine, Metabolic Bone Disease and Related Research, Journal of Clinical Endocrinology and Metabolism, Calcified Tissue International* and *Clinical Orthopedics and Related Research.* His current research interest includes the pathogenesis and treatment of postmenopausal osteoporosis, the mechanism of age-related bone loss in postmenopausal women, the effect of calcium intake and female hormones on bone metabolism, and refinement techniques in bone histomorphometry.

CONTRIBUTORS

Roland Baron, D.D.S., Ph.D.
Associate Professor
Department of Medicine
Yale University School of Medicine
New Haven, Connecticut

H. M. Frost, M.D.
Southern Colorado Clinic
Pueblo, Colorado

Z. F. G. Jaworski, M.D.
Professor of Medicine
University of Ottawa
Ottawa General Hospital
Ontario, Canada

Webster S. S. Jee, Ph.D.
Professor of Anatomy
Department of Pharmacology
University of Utah
Salt Lake City, Utah

Donald B. Kimmel, D.D.S., Ph.D.
Research Assistant Professor
Department of Pharmacology
University of Utah
Salt Lake City, Utah

J. Kragstrup, M.D.
University Institute of Pathology
Aarhus Amtssygehus
Aarhus C, Denmark

F. Melsen, M.D.
University Institute of Pathology and
  Medicine
Aarhus Amtssygehus
Aarhus C, Denmark

L. Mosekilde, M.D., Ph.D.
Department of Medicine III
Aarhus Amtssygehus
Aarhus C, Denmark

Lynn Neff
Assistant in Research
Department of Medicine
Yale University School of Medicine
New Haven, Connecticut

A. M. Parfitt, M.B., B.Chir.
Director
Bone and Mineral Research Laboratory
Henry Ford Hospital
Detroit, Michigan

D. Sudhaker Rao, M.B., B.S.
Research Associate
Bone and Mineral Research Laboratory
Henry Ford Hospital
Detroit, Michigan

Angelica Santa Maria
Assistant in Research
Department of Medicine
Yale University School of Medicine
New Haven, Connecticut

Ann Silverglate, M.S.
Assistant in Research
Department of Medicine
Yale University School of Medicine
New Haven, Connecticut

James M. Smith, Ph.D.
Research Associate Professor
Department of Pharmacology
University of Utah School of Medicine
Salt Lake City, Utah

Agnes Vignery, D.D.S., Ph.D.
Research Associate
Department of Medicine
Lecturer in Cell Biology
Yale University School of Medicine
New Haven, Connecticut

TABLE OF CONTENTS

Chapter 1

INTRODUCTION

Robert R. Recker

The study of bone tissue in health and disease made a remarkable leap forward during the 1960s and 1970s. Prior to 1960, microanatomists and pathologists examined bone relatively infrequently compared to most other tissues and when they did, it was usually with a great handicap because often the first step in preparation of the tissue after fixation was to remove its principle component, the mineral. It was difficult to identify accurately the microanatomy and nearly impossible to appreciate that dynamic processes were ongoing in vivo let alone to measure them.

With the introduction of plastic embedding material and heavy duty microtomes and other methods of specimen preparation, it became possible to overcome the difficulty in having to examine bone tissue without its mineral. Bone workers began to become more fully aware of the dynamic processes responsible for producing the pictures they were seeing in the microscope.

In the late 1950s and early 1960s the key that unlocked the door to these various bone dynamic processes came in the form of tetracycline.[2-4] This rather humble antibiotic could be found trapped in the bone tissue and the reason it could be identified was because of its excitation by fluorescent light at 360-nm wavelength.

Extensive investigation of this phenomenon, led by Frost in this country, resulted in some remarkable discoveries about the biology of bone. A few pertinent examples are illustrative. Growth (elongation), modeling (sculpting), and remodeling (renewal) are three fundamentally different dynamic systems[5] all performed by bone cells (osteoclasts and osteoblasts), but under different control and for different biologic reasons. Defects in each produce unique clinical consequences and a disease of one system cannot be used as a model to study a disease of another. Thus a rat model (growth and modeling) should not be used to study postmenopausal osteoporosis (a disease of remodeling).

The remodeling process in higher vertebrates became understood as a surface phenomenon and following a biologically programmed sequence. Activation of osteoclast precursors to form osteoclasts is followed by resorption of mineralized bone matrix and in turn is followed by activation of osteoblast precursors to form osteoblasts which replace the resorbed bone. The rate of appearance of new remodeling sites, the rate of resorption and formation at the individual site, and the length of time required for completion of the remodeling cycle (called sigma by Frost) are dynamic rate measurements that could be made using tetracycline as a tissue-time marker. Previously confusing findings could be explained. For example, an agent which suppressed the rate of appearance of new remodeling sites as its primary effect (estrogen) can be shown to inhibit resorption if the bone is examined soon after its administration during the resorption phase of remodeling. Or it can be shown to inhibit both formation and resorption if the bone is examined later when enough time has elapsed to allow the skeleton to pass through the formation phase of its remodeling cycles. The resultant steady state many months later may show no imbalance between resorption and formation. The point illustrated by this example is that the steady-state effect of an agent or perturbation and not the acute transient effect causes lasting changes in bone (disease). It was information gathered from the tetracycline data that clarified this and allowed a method of measuring the length of transients in bone.

The earlier work by Frost was done using the rib biopsy. The rib specimen gave a very good sample of cortical bone and since the remodeling packets were all oriented in the long axis of the rib, they were all sampled in cross section. There was no need to account for malorientation of the section plane with the remodeling surface or tetracycline labels.

The transilial biopsy site, however, offers several advantages over the rib and has largely supplanted it. There is less patient discomfort, no body cavity is threatened, it is less dangerous, and a trained surgeon is not required. The ilium near the crest contains largely trabecular bone and it behaves more like the vertebrae where metabolic bone disease is prominently expressed.

However, there are a number of different problems to be confronted with the iliac specimen. For example, the orientation of the section plane would deviate variably and unpredictably from perpendicular to the remodeling surface. A correction factor had to be derived.

Another problem was the extrapolation of area and perimeter measurements from the two-dimensional section plane to the real world of three dimensions. This generated much controversy.

Other difficulties emerged such as how to deal with singly labeled surfaces, how to select an optimum labeling schedule, how to distinguish ''old'' from ''new'' label, and many more.

The solutions to many of these technical problems and the application of bone biopsy and bone histomorphometry to investigation and treatment of metabolic bone disease have advanced to the point where many workers in the field agreed that a monograph summarizing the state-of-the-art would be useful. Accordingly, the contributors set about this task and the results are recorded in the following pages.

REFERENCES

1. Arnold, J. S. and Jee, W. S. S., Embedding and sectioning undecalcified bone and its application to radioautography, *Stain Technol.*, 29, 225, 1954.
2. Milch, R. A., Rall, D. P., and Tobie, J. E., Bone localization of the tetracyclines, *J. Natl. Cancer Inst.*, 19, 87, 1957.
3. Frost, H. M., Villanueva, A. R., and Roth, H., Tetracycline staining of newly forming bone and mineralizing cartilage in vivo, *Stain Technol.*, 35, 135, 1960.
4. LeBlond, C. P., Lacroix, P., Ponlot, P., and Dhem, D., Les stades initiaux de l'osteogenese: nouvelles donnees histochimiques et autoradiographiques, *Bull. Acad. R. Med. Belg.*, 24, 421, 1959.
5. Frost, H. M., *Bone Remodeling and Its Relationship to Metabolic Bone Diseases*, Charles C Thomas, Springfield, Ill., 1973.

Chapter 2

PRACTICAL APPROACH TO BONE BIOPSY

D. Sudhaker Rao

TABLE OF CONTENTS

I. INTRODUCTION

Histologic examination of an abnormal tissue is often essential for precise diagnosis and indeed the biopsy has become an integral part of the practice of medicine. With improved instrumentation and techniques, for most sites the biopsy can now be performed in outpatients and does not require elaborate physical facilities. While bone biopsy has been a routine procedure for identification of localized bone lesions for decades, it is less often used in evaluating metabolic bone disease (MBD). As a research tool, however, bone histology has contributed immensely to the understanding of pathogenesis and to the study of the effects of various forms of therapy in MBD.

In this chapter, the instruments, the techniques, complications, and acceptability of various types of bone biopsy procedures will be discussed. Other sections will deal with specific aspects of bone histology as it pertains to diagnosis and treatment.

II. INSTRUMENTATION

For biopsy of the ilium it is essential that a suitable instrument is used. The standard orthopedic biopsy using a hammer and chisel or equivalent instruments causes considerable postoperative pain and is useless for quantitative histology. The Jamshidi needle is satisfactory for diagnosis, but the sample is too small to be representative for measurement. There are several types of bone biopsy trephines that are now commercially available and they are essentially based on the instrument used for a bone marrow biopsy. The trephines are of two general categories: iliac crest and transiliac. No special instrument is required to obtain a rib biopsy, which consists of a segmental surgical resection of the eleventh rib.

Iliac crest trephines are obviously limited in size by the thickness of the ilium[1-5] (Table 1); they have an internal diameter of 3 to 5 mm. Specimens obtained via the crest consist mainly of trabecular bone; the small amount of cortical bone at the upper end is too close to ligaments and muscle attachments to be useful. There is usually some distortion of trabecular architecture because of the rotary to and fro motion of the instrument. Preservation of trabecular architecture is usually better with an electrically driven trephine such as that of Burkhardt.[6] There may be less sample variation than with transiliac biopsy because the biopsy site can be located more accurately. Transiliac trephines are of larger internal diameter (6 to 10 mm) and the biopsy core contains both cortical (external and internal) and trabecular bone of adequate quantity and of good quality[7-10] (Table 1). There is essentially no distortion of bone architecture since it is a cylindrical plug of bone drilled out of the ilium *in toto* (Figure 1). However, since the biopsy site has to be accurately located vertically as well as horizontally, there is greater sample variation than for crest biopsy. The various types of instruments and their major advantages and disadvantages are summarized in Table 1.

III. PROCEDURE FOR OBTAINING BONE BIOPSY

Irrespective of the approach used, iliac bone biopsy can safely be performed in outpatients and neither hospitalization nor operating room time is required. It is essential to demonstrate a normal coagulation profile (normal prothrobmin and partial thromboplastin time and platelet count) prior to the procedure. This is particularly important in hemodialized patients or in those with hematogenous malignancies. In patients on maintenance hemodialysis, biopsy should be performed on an interdialysis day and the postbiopsy dialysis should be delayed by 1 day.

After an overnight fast, the patient is premedicated with intramuscular (i.m.) injec-

Table 1
BONE BIOPSY TREPHINES

Internal diameter (mm)	Comments	Ref./ manufacturer
	Iliac Crest Trephines	
5.0	Advantages: less invasive; quicker to perform; fewer complications and less pain after procedure	1
5.0		2[a]
3.0		3[b]
3.0	Disadvantages: mainly trabecular bone; limited quantity of specimen; architecture may be distorted	4[c]
4.0		5[d]
	Transiliac Trephines	
6.8	Advantages: both cortical and trabecular bone good quality and adequate quantity; less distortion of bone architecture	7[e]
7.2		8[f]
7.5—10.0		9[g]
5.0	Disadvantages: more invasive; slightly more painful and more complications; takes slightly more time to perform	10

[a] Edwards Surgical Supplies Ltd., 289 City Road, London EC1 Engl.
[b] Institute of Biophysics of the Ministry of Health, Moscow, U.S.S.R.
[c] Accurate Surgical Instruments, 588-590 Richmond St. W., Toronto 2B Canada.
[d] "Myelotomiegerat" F Strauman Institute, Waldenburg BL Switzerland.
[e] Lépine à Lyon, Instruments de Chirurge BP #8, 69394 Lyon-Cedek 3 France.
[f] Downs Surgical Inc., 2500 Park Central Blvd., Decatur, Ga.
[g] Zimer Company, Warsaw, Ind.

tions of meperidine hydrochloride (Demerol®) 75 to 100 mg (or 1 mg/kg body weight) and diazepam (Valium®) 10 to 15 mg (or 0.15 mg/kg body weight) about 15 to 30 min prior to starting the procedure. This will assure adequate sedation without causing respiratory depression. The skin over the appropriate biopsy site is then prepared with Phisohex® and Betadine® taking the usual antiseptic precautions.

A. Iliac Crest Biopsy

With the superior (or vertical) approach the patient lies in the supine position and the anterior iliac crest area is prepared; 10 to 15 mℓ of 1% lidocaine hydrochloride (Xylocaine®) is used to anesthetize the skin and subcutaneous (s.c.) tissue. A 1- to 2-cm long incision is made directly over the iliac crest border, about 2 to 3 cm posterior to the anterior superior iliac spine. The periosteum is exposed, anesthetized, and in-

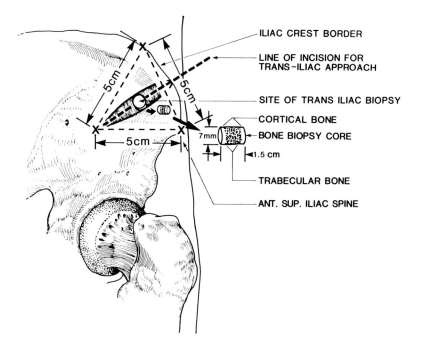

FIGURE 1.   Anatomic orientation for transiliac bone biopsy procedure.

cised in a similar fashion. The outer guide of the instrument is firmly applied (vertically) over the exposed bone. Holding the device in the plane of the body of ilium, the inner trephine is slowly advanced with a rotary motion applying a moderate pressure. The depth is determined by the instrument size and is usually 1 to 2 cm long (Figure 2). The specimen is freed by fracturing its base with a gentle rocking motion and the instrument is withdrawn with specimen held in the trephine. Technically, this is the easiest of the three types of bone biopsy procedures, but the specimen quantity and quality are not always satisfactory. More than one attempt may have to be made to obtain an adequate biopsy sample.

B. Transiliac Bone Biopsy

This approach is slightly more invasive than through the crest, but with proper anatomic orientation, the procedure can safely be done in 30 to 60 min. During this lateral approach the patient assumes a supine lateral position with semiflexion at both hip and knee. A triangular area* is outlined with the anterior superior iliac spine as a guide. Roughly, it is a 5-cm isolateral inverted triangle (Figure 1), the base of which is formed by the iliac crest border. This geometric outlining is important to minimize the site variation between sequential biopsies. More local anesthetic is needed (approximately 15 to 30 ml) because of the bulk of subcutaneous and muscle tissue in this location. A 2- to 3-cm long incision is made along the line indicated in Figure 1, and the glistening fascia lata is exposed. The fascia itself is then incised along its fiber direction, and the dissection is continued until the scalpel touches the periosteum and

---

*   In view of the pioneering efforts in the area of bone histomorphometry and in particular the current technique of transiliac bone biopsy by the late Dr. Phillipe Bordier, it would seem appropriate to coin the term, ''Bordier's Triangle'' for the anatomic site from which the biopsy is taken.

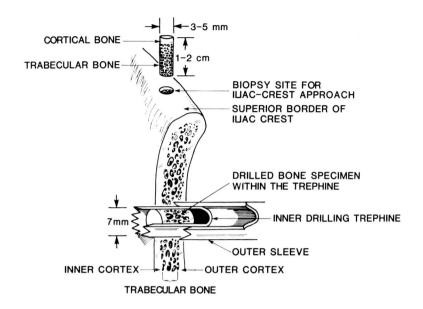

FIGURE 2.   Details of two types of iliac bone biopsies.

bone underneath. The patient will feel slight pain at this point. Scraping of the periosteum after further local anesthesia is helpful in minimizing the discomfort. With the
index finger, the surface of the bone is freed of its muscle attachments, and the plane
of the body of the ilium is ascertained (an important step). The instrument is introduced and the outer sleeve is fixed on the bone perpendicular to its plane (Figure 3).
The procedure is then completed with the inner drilling trephine by rotary (90 to 180°
clockwise and anticlockwise) motion while applying steady but gentle pressure. Although there is a guard to prevent excessive advance of the trephine, it is necessary to
feel and distinguish the differences in resistance of the outer cortex, the intervening
trabecular bone, and the inner cortex to avoid injury to the internal organs. After the
trephine has pierced the inner cortex it is rotated 360° 2 or 3 times to achieve separation
of the specimen from the inner muscle attachments. The trephine is withdrawn with
circular, clockwise motions and the outer sleeve is removed. Incised fascia lata must
be sutured with interrupted absorbable chromic material to avoid herniation of the
underlying muscle. The wound is then closed in two layers using chromic material for
the s.c. tissue and nylon or silk for the skin. A pressure dressing is then applied with
elastic tape and the patient is asked to turn over and lie on the incision site for 15 to
20 min. The patient is asked to lie on that side as much as possible for the next 24 hr,
but this is not an absolute necessity. Minor analgesics are usually needed for 1 or 2
days. Although the transiliac approach appears to be more elaborate and time consuming, it can safely be performed in only 30 to 60 min.

C. Repeat Biopsies

Whenever a follow-up biopsy is needed, it should be performed on the opposite side
to avoid the previously biopsied area. For the third biopsy, the first side should be
used again and for the fourth biopsy the second side again. Repeat biopsies on the
same side should be at least 2 cm from the site of the first biopsy on that side to avoid
callus and should be delayed for at least 1 year to avoid local acceleration of bone
turnover produced by the earlier biopsy. Up to five serial transiliac bone biopsies have
been performed by the author in the same subject without complications.

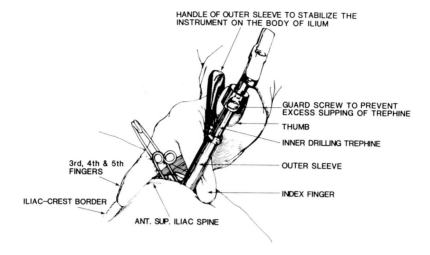

FIGURE 3.   Illustrating the positioning of transiliac trephine and its components.

D. Rib Biopsy

This is best performed by a surgeon in the operating room. The right eleventh rib is preferred; as one of the "floating ribs" of the thoracic cage, it is almost completely below the pleural reflection and is relatively easy to resect under local anesthesia.[11,12] The patient is positioned in supine lateral decubitus. Local anesthesia is achieved by infiltration of 10 to 15 mℓ of 1.0% Xylocaine® just above and below the rib at the posterior axillary line. A 4- to 5-cm long incision parallel to and directly over the rib is made. The underlying muscle layers (latissimus dorsi and quadratus lumborum) are split open in the direction of their fibers and the periosteum is exposed. The periosteum is then stripped off for a distance of 4 to 5 cm with a small periosteal elevator. A thin grooved director is passed under the rib so that the underlying pleura is protected. With a Stryker® oscillating bone saw, a segment of the rib is cut. The periosteal bed from which the bone segment has been removed is not closed. The wound is then closed in two layers using absorbable chromic for muscle and s.c. tissues and silk or nylon for the skin. During the entire procedure extreme care must be exercised not to injure the pleura in order to avoid pneumothorax. The procedure takes about 30 min. Although the rib biopsy was extensively used in the early 1960s, its use is now limited to experimental studies in animals in view of the accompanying discomfort, morbidity, and formidable nature of the procedure in human subjects. It has the advantage, however, of providing a more satisfactory sample of cortical bone. This site should be used only in patients with rare forms of MBD for which only rib histomorphometric data are available, or when iliac biopsy has failed to provide a satisfactory specimen and bone biopsy is judged to be essential for beginning treatment.

IV. COMPLICATIONS

Since all three types of bone biopsy procedures are invasive, local hematoma and wound infection are the two most frequent complications and tend to occur more commonly in patients on maintenance hemodialysis. Hematoma is more common with the transiliac approach and less with either vertical or iliac crest approach. Although Bordier's triangle is essentially free of any important arteries, veins, or nerves, one may occasionally encounter anomalous venous drainage in this location beneath the fascia lata which could result in severe bleeding. In the author's experience of over 500 transiliac bone biopsies, this complication arose only once and the procedure had to be aborted. This particular patient was on maintenance hemodialysis and had had a successful transiliac bone biopsy on the opposite side the previous year. Wound infection, like hematoma, is also more likely to occur in the hemodialyzed patient, and occasionally in the nondialyzed patient. Proper antiseptic precautions and prompt recognition and treatment will prevent this complication. Transient neuropathy due to severing or entrapment of one of the cutaneous branches of the femoral nerve may cause hyperesthesia at the biopsy site. This usually lasts about 2 to 3 weeks, but occasionally the patient may feel local paresthesia for up to 4 to 6 months. The symptom disappears spontaneously and no specific therapy is needed. One case of femoral nerve palsy has been reported[13] and is probably due to compression of the main femoral nerve trunk by an internal hematoma. Fracture of the pelvis has been reported in two patients with osteomalacia.[14] In one patient it was directly related to the excessive pressure applied during the procedure, and in the other the fracture occurred through the biopsy site following trauma 2 weeks after the procedure. In neither case were there any serious long-term sequelae. Osteomyelitis has been reported in one patient[14] and one patient developed serious respiratory depression following the use of intravenous (i.v.) diazepam.

There are two complications that are specific to transiliac bone biopsy. First, biopsy specimens may dislodge from the trephine during withdrawal.[15] A total of ten such specimens were dislodged; two of these were lost internally (i.e., between inner aspect of ilium and iliacus muscle). These were left in vivo and neither patient developed any long-term complications as a result. Eight specimens were lost externally (i.e., external to the bone in s.c. tissue or under the fascia lata) and were retrieved easily by digital exploration without additional discomfort.[15] It is therefore important to explore the operating area for dislodged specimens before attempting a second biopsy. Second, the cutting end of the trephine (a 2 cm-long piece) may separate from the shaft after penetrating the outer cortex of ilium.[15] Since the newer trephines are now made of a single cylinder of good quality stainless steel, it is no longer a problem. In a few patients the trephine teeth were bent or distorted as a result of undue pressure applied during the procedure. For this reason the trephine teeth need sharpening from time to time (usually once in 20 to 30 attempts). This is easily accomplished by hand file, but if the teeth are badly damaged it is best to replace the instrument.

Pneumothorax, obviously a complication specific to rib biopsy, has been reported in 5 of 136 attempts and fortunately none were serious.[11] This potentially hazardous complication has limited the use of rib biopsy procedure to experimental animals or to the most unusual cases of MBD.

Pain during and after the bone biopsy procedure is an expected complication. The pain is slightly more with the transiliac than with the iliac crest approach. It is usually mild in the majority of patients and does not last for more than 2 to 3 days. In only a minority of patients (less than 1%) does the pain last more than 7 days (*vide-infra*).

V. ACCEPTABILITY

To be practical, an ideal bone biopsy procedure should have the following qualities: the biopsy site should be easily accessible, the procedure should be least invasive and easily performed, it should not require elaborate physical facilities or operating room time, can be performed in outpatients, should have the least number of complications, and the specimen should be adequate in quantity and should be of good quality. Most of all, the procedure should be well tolerated with the least amount of pain and discomfort. Obviously it is impossible to incorporate all the above qualities in any biopsy procedure.

Table 2

COMPLICATIONS OF THREE TYPES OF
BONE BIOPSY PROCEDURES

	Type of bone biopsy		
	Transiliac	Iliac crest	Rib
Total cases	9131	5780	136
Complication:			
Hematoma	22	14	—
Pain (>7 days)	17	—	—
Neuropathy (transient)	11	2	—
Wound infection	6	4	—
Fracture	2	—	—
Osteomyelitis	1	—	—
Specimens lost[a]	2	—	—
Instrument problem	3	—	—
Pneumothorax	—	—	5[b]
Drug reaction	—	1	—
Total complications	64 (0.7%)	21 (0.36%)	5 (4%)

[a]   See text for explanation.
[b]   Transient.

Data compiled from the literature and from the multicenter survey of 18 national and international bone mineral centers.[15]

Table 3

PAIN SCORE (TRANSILIAC
BIOPSY ONLY)[15]

	Number of patients in each category			
Pain score	First survey (1979)	Second survey (1980)	Total	Total (%)
0	12	15	27	23
1	26	44	70	58
2	11	10	21	17
3	1	1	2	2
4	0	0	—	—

Since the transiliac approach is more widely used, (Table 2) and since it is more invasive, its acceptability was determined by assessing the discomfort experienced by patients. A pain score was developed on a scale of 0 to 4,* as previously described.[15] In two different surveys (Table 3) 80% of patients considered their discomfort as 0 or 1 and 17% scored 2. Only 2% scored the discomfort as 3, and none scored as 4. The results in these two surveys were similar to those reported by other investigators.[9]

Despite its disadvantages, the transiliac approach in experienced hands offers the best method of obtaining a full-thickness bone biopsy with minimal morbidity. Rib biopsy is technically difficult to perform and potentially hazardous complications may ensue.

*   0—No discomfort; 1—mild; 2—moderate; 3—severe; 4—excruciating or unbearable.

REFERENCES

1. Sacker, L. S. and Nordin, B. E. C., A simple bone biopsy needle, *Lancet,* 1, 347, 1954.
2. Williams, J. A. and Nicholson, G. I., A modified bone biopsy drill for outpatient use, *Lancet,* 1, 1408, 1963.
3. Smirnov, A. N. and Baranov, A. E., Trephine for iliac crest biopsy, *Lancet,* 1, 1353, 1971.
4. Fornasier, V. L. and Vilaghy, M. I., Laboratory suggestions: the results of bone biopsy with a new instrument, *Am. J. Clin. Pathol.,* 60, 570, 1973.
5. Sherrard, D. J., Baylink, D. J., Wergedal, J. E., and Maloney, N. A., Quantitative histological studies on the pathogenesis of uremic bone disease, *J. Clin. Endocrinol. Metab.,* 39, 119, 1974.
6. Ritz, E., Malluche, H., Krempien, B., and Mehls, O., Bone histology in renal insufficiency, in *Calcium Metabolism in Renal Failure and Nephrolithiasis,* David, D. S., Ed., John Wiley & Sons, New York, 1977, 197.
7. Bordier, P., Matrajt, H., Miravet, B., and Hioco, D., Mesure histologigue de la masse et de la resorption des travees osseuse, *Pathol. Biol. (Paris),* 12, 1238, 1964.
8. Byers, P. D. and Smith, R., Trephine for full-thickness iliac-crest biopsy, *Br. Med. J.,* 1, 682, 1967.
9. Johnson, K. A., Kelly, P. J., and Jowsey, J., Percutaneous biopsy of the iliac-crest, *Clin. Orthop.,* 123, 34, 1977.
10. Teitelbaum, S. L., Rosenberg, E. M., Bates, M., and Avioli, L. V., The effects of phosphate and vitamin D. therapy on osteopenic, hypophosphatemic, osteomalacia of childhood, *Clin. Orthop.,* 116, 38, 1976.
11. Jett, S., Wu, K., and Frost, H. M., Tetracycline-based histological measurement of cortical-endosteal bone formation in normal and osteoporotic rib, *Henry Ford Hosp. Med. J.,* 15, 325, 1967.
12. Sedlin, E. D., Frost, H. M., and Villanueva, A. R., The eleventh rib biopsy in the study of metabolic bone disease, *Henry Ford Hosp. Med. Bull.,* 11, 217, 1963.
13. Walton, R. J., Femoral palsy complicating iliac bone biopsy, *Lancet,* 2, 497, 1975.
14. Duncan, H., Rao, D. S., and Parfitt, A. M., Complications of bone biopsy, *Metab. Bone Dis. Relat. Res.,* Suppl. 2, 475, 1980.
15. Rao, D. S., Matkovic, V., and Duncan, H., Transiliac bone biopsy: complications and diagnostic value, *Henry Ford Hosp. Med. J.,* 28, 112, 1980.

Chapter 3

PROCESSING OF UNDECALCIFIED BONE SPECIMENS FOR BONE HISTOMORPHOMETRY

Roland Baron, Agnès Vignery, Lynn Neff, Ann Silverglate, and Angelica Santa Maria

TABLE OF CONTENTS

I. INTRODUCTION

The performance of bone histomorphometry requires obtaining undecalcified sections of bone of excellent quality. If the first and important step is to get a good biopsy specimen, each of the following steps leading to the histological sections is of major importance. This is the reason that this chapter has been written, and we believe that it should help other investigators to obtain good enough undecalcified sections of bone to start their histomorphometric work. The reasons for avoiding decalcification are obvious and numerous; most important is the conservation of the differences between calcified and uncalcified bone (osteoid) in order to be able to diagnose and measure abnormalities in the bone formation and/or bone mineralization processes. However, other reasons also have to be kept in mind; the ability to make dynamic measurements of the mineralization rate through fluorescent labels and optimal conservation of the cell structure and minimal shrinkage of the bone marrow to avoid artifacts at the bone-bone marrow interface are among the most important. Undoubtedly, bone histomorphometry cannot be performed satisfactorily on decalcified specimens.

Numerous methods for embedding, processing, and staining undecalcified bone specimens have been described in the past 25 years.[1-8] Our purpose in this chapter is not to review and compare all of these techniques but instead to give a detailed description of the methods we are currently using in our laboratory at Yale which are the results of 10 years of trials, both in Paris and Yale, and of numerous visits and discussions with other people in the field.

The central idea that guided us during this search and that we want to convey here is the following: the end result of bone histomorphometry is a set of numbers and standard deviations that always look nice in table form when published. However, the *only* way to get reliable counting is to have excellent morphology with a highly reproducible technique, therefore enhancing the chances of having highly reliable and reproducible quantitative results. Too often one is surprised by the poor quality of the morphology, usually only seen at meetings, from groups which have published numerous and impressive sets of ''numbers''.

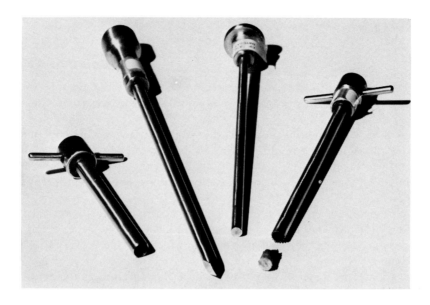

FIGURE 1. View of an 8-mm diameter trephine for iliac crest biopsies. The biopsy core is seen at the right bottom.

The methods described in this chapter work, consistently and well. All the little "gimmicks" that we are aware of using are given, so that this should be reproducible by anyone who wishes to use the same procedures. When appropriate, brand names of the products we use are given; this is because our experience has clearly shown that they often vary from one manufacturer to another.

Before entering in the detailed description of the procedures we recommend, we would like to give to the reader an overview of the principles that led to the selection of these methods.

II. GENERAL COMMENTS ON THE METHODS DESCRIBED

A. Taking the Biopsy

Over the years, we have found that there are a few important details to keep in mind when taking the biopsy. The *trephine* (Figure 1) should be a Bordier's needle with a minimum internal diameter of 6 mm or, for better results, 8 mm. It is critical that the teeth always be perfectly sharp. If not, it will be impossible to go through the bone without pushing. This results in compression and fracture of the specimen, markedly compromising the histomorphometry. When taking the biopsy, the physician has to be strictly instructed not to push through the bone, but, instead, to turn very gently back and forth. The same precautions apply when pushing the biopsy out of the needle.

B. Fixation

Fixation has to be immediate and the fixative solution should be cold (4°C). Of all the fixatives we have used, 40% ethanol has proven to be the best for bone. We believe that the main advantage of this procedure is that dehydration is carried out at the same time as fixation and therefore avoids the overnight washing in water that has to be performed with formalin. This usually results in swelling of tissues, followed by shrinking during dehydration, and often leads to more retraction at the bone marrow-bone interface.

C. Embedding

Embedding of the specimen after fixation has to be performed according to the following goals:

1.   The bone specimen must be infiltrated and not only surrounded by the embedding material; the infiltration procedure must therefore be carried out slowly for best results.
2.   The embedding material should ideally be as hard as the calcified bone itself to avoid vibration and fractures during sectioning. It should, however, be supple enough to facilitate sectioning and handling of the sections. This is the reason why a "softener" is added to the methyl methacrylate in the procedure we describe here.
3.   The temperature reached during the polymerization should not exceed 45 to 50°C; it therefore has to be moderated by adding a minimum amount of catalyzer and by leaving the specimen at room temperature at the beginning of the polymerization process.
4.   The embedding material should be easily dissolved after sectioning, thereby allowing the use of any desired staining technique.

These are the reasons why the procedures we describe in this chapter have been selected. For instance, Epon embedding is good but staining of the sections is very difficult because of the relative insolubility of the plastic; Bioplastic (Ward) embedding material is impossible to dissolve after polymerization, therefore preventing good staining of the cells; and methyl methacrylate without softener is brittle, therefore not permitting good sections to be obtained. One may, however, find out that methyl methacrylate with a softener does not prevent shattering of calcified bone as much as Bioplastic. If the priority is put on avoiding shattering, for instance to study osteocytes, the latter is a better embedding medium. If the priority is put on cellular activity at the bone-bone marrow interface, what is described here gives better results.

D. Sectioning

In addition to the critical importance of the embedding medium for obtaining good sections, some other factors are of great importance. The microtome has to have enough power for the size of the specimen to be sectioned. The Jung® K sliding microtome is best but the Jung Autocut is very sufficient for small pieces (up to 1 cm). The knives have to be of the HK-2 profile; the angle of the HK-1 is too small, resulting in frequent damage to the edge during sectioning and the angle of the HK-3 is too wide, resulting in increased shattering due to the marked change in direction that the section has to take to go over the knife. The thickness of the section which is best for good cellular detail is 4 to 5 μm; 8 μm is a bit too thick. For fluorescence, on the other hand, 8 μm is too thin for a good contrast and better results are obtained between 10 and 15 μm. Most important for quality of the morphology is the flatness of the sections at the end of the procedure. For this reason, coating of the slides with chromium-alum gelatin, flattening with 95% ethanol, the use of photoroller, and, finally, drying under the pressure of C-clamps are essential for getting a perfectly flat section before staining.

E. Staining

The goal one wants to achieve when staining sections of undecalcified bone is to make clear and reliable distinction between calcified bone and osteoid tissue and, at the same time, to get as much cellular detail as possible, especially for those cells which are at the bone-bone marrow interface. Obviously, as discussed above, high quality of the section itself (including fixation and embedding) is required for good staining. Most important then is the ability to completely dissolve the plastic embedding medium. Finally, one has to select a stain or a set of stains. Undisputedly the most reliable stain for the distinction between osteoid and calcified bone is the Von Kossa silver impregnation (Figures 7 and 8C); we use it routinely. However, it has the drawback of being time consuming and is not ideal for cellular detail. The most commonly used stain in bone laboratories is the Goldner modification of the Masson trichrome (Figure 8A); we also use it routinely but it is very time consuming, not best for cellular detail, and not very reliable for staining of osteoid, especially when this latter is thin or normal in thickness. Other stains have been recommended such as the Solochrome Cyanin R,[9] which is good for osteoid but which does not stain the cells, or the Villanueva osteochrome and tetrachrome,[6] excellent but very time consuming procedures. In our laboratory, we have selected toluidine blue at acid pH as the stain giving us the most satisfaction; it is very rapid, excellent for cellular detail, and reliable for osteoid staining (Figures 8B and 9). However, and mostly for increased security, a set of sections is always stained with the Von Kossa if there would be a need for double checking the osteoid in a specimen. The Goldner trichrome is also routinely performed on a set of sections for each specimen, but we never use it for counting; instead, these sections are used for color slide presentations.

Finally, because none of these stains gives clear and reliable enough staining of cement lines for mean wall thickness (MWT) measurements, a set of sections is processed differently and stained under different conditions (see below) for this purpose. Other laboratories perform this measurement under polarized light with an unstained section.

Having explained our goals and the reasoning underlying the procedures used in our laboratory, we can now proceed to describe in more detail all the methods we have selected.

III. FIXING, EMBEDDING, AND SECTIONING UNDECALCIFIED BONE SPECIMENS

A. Material

Supplies
   24-mm glass scintillation vials
   Buehler Ecomet II Polisher-Grinder
   120 grit Carbimet wet/dry discs
   Jung® K sliding microtome
   Jung® Autocut microtome
   Jung® HK-2 profile, tungsten-carbide tipped knife (minimum 3)
   Imperial® II Radiant Heat Oven (Labline, Inc.)
Chemicals
   Methyl methacrylate — Merck® #12244*
   Dibutyl phthalate — Merck® #12487
   Benzoyl peroxide — Merck® #12435
   Gelatin, 275 bloom — Fisher
   Chromium-alum (chromium potassium sulfide) — Fisher

B. Procedures
1. Fixation-Dehydration
   Place biopsy immediately in 40% ethanol, at 4°C, and follow the schedule:

---

* Merck products are available in the U.S. through MC/B, Cincinnati, Ohio.

40% Ethanol	1—2 Days
70% Ethanol	1—2 Days
95% Ethanol	2 Days
100% Ethanol	1 Day
100% Ethanol	1 Day
Xylene	1 Day

If specimen was formalin-fixed, wash overnight with tap water and proceed with 40% ethanol; however, better results are obtained with ethanol if all steps are maintained at 4°C. If it is necessary to store material or prepare it for transit it is best brought to 70 or 95% ethanol first.

2. Infiltration and Embedding

Methacrylate Preparation

Solution I:

Methyl methacrylate	75 cc
Dibutyl phthalate	25 cc

Stir continuously for several hours

Solution II:

Methyl methacrylate	75 cc
Dibutyl phthalate	25 cc
Benzoyl peroxide	1.0 g

Stir continuously for several hours

Solution III:

Methyl methacrylate	75 cc
Dibutyl phthalate	25 cc
Benzoyl peroxide	2.5 g

Stir continuously for 4 to 6 hr

Solutions I, II, and III may be stored in the refrigerator for about 1 week when used for infiltration; Solution III for embedding should be prepared on day of use.

Infiltration Schedule

Solution I	2 days
Solution II	2 days
Solution III	1—2 days

Infiltration is also carried out at 4°C.

The embedding procedure (Figure 2) is as follows:

1. Prepare a polymerized layer in the bottom of a 25-m*l* glass vial by pouring 5 m*l* of fresh Solution III. Cap and leave at room temperature overnight, then place in a 42°C oven in the morning. Polymerization will occur in 1 to 3 days. (A soft layer may remain at the plastic-air interface but this is not a big inconvenience.)
2. Place the infiltrated specimen on the polymerized layer and fill the bottle almost completely with fresh Solution III. Cap and leave at room temperature overnight. Place in oven at 42°C the following morning. Polymerization should be complete in 2 to 3 days but may take a little longer. Again, a sticky layer may persist on top of the polymerized plastic.

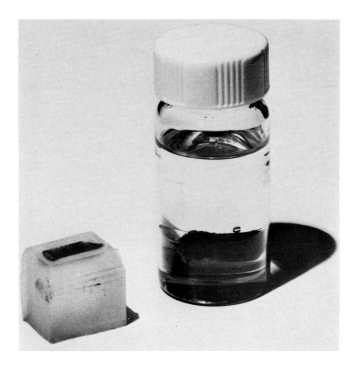

FIGURE 2. View of the biopsy core during the embedding procedure in a glass vial (right) and after trimming of the block for sectioning (left).

The embedded specimen can be stored at room temperature for a very long time. Avoid direct sunlight if the bone has been labeled with tetracyclines.

3. Sectioning
a. Preparation for Sectioning
Vials with polymerized plastic are refrigerated briefly to facilitate removal of glass vial. Cold vials are wrapped in a paper towel and gently tapped with a hammer to break and remove the glass. Blocks are trimmed with a bandsaw to an appropriate size ($\sim$ 1 by ½ in.) and shaped for the microtome holder; several millimeters of plastic are left around each side of the biopsy during trimming to allow handling of the sections (Figure 2). A grinder with 120 grit carborundum wet/dry paper is used to sand into the bone and to prepare an even block face to shorten trimming time with the microtome.

b. Sectioning
We presently use one of two microtome models; routinely, the Jung® K sliding microtome is used for biopsies and large research specimens. The Jung® Autocut may be used for small specimens, especially mouse tissue.

In both cases a tungsten-carbide tipped Jung® HK-2 profile knife is employed. These knives must undergo frequent sharpening* as cutting bone causes the rapid loss of a good edge. The block is oriented in the holder so that the cortices are perpendicular to the knife edge. Sections are cut at 4 to 5 μm for routine light microscopy and at 10 to 15 μm for fluorescence microscopy.

* Dorn and Hart, Microedge Inc., 131 Home Avenue, Villa Park, Ill. 60181.

FIGURE 3.   View of the HK2 tungsten carbide knife on the Jung® k sliding microtome during sectioning. Notice the method used to collect the section on the glass slide with a brush.

Method — The knife edge and slide are moistened with 40% ethanol while cutting and 95% ethanol is used on the sections during straightening and flattening. A single section is allowed to slide onto the knife as it is cut, the microtome is stopped, and the section collected with the aid of additional 40% ethanol and a fine paintbrush (Figure 3). The brush and a dissecting needle are used to remove folds and wrinkles from the wet section (Figure 4). A strip of clear ($\sim$ 0.1-mm thick) plastic film the size and shape of a slide is placed on the sections and excess alcohol is drained on a paper towel. While exerting gentle pressure, a photoroller is rolled over the plastic-coated slide (Figure 5). A group of slides is placed between two pieces of wood cut to a slide's dimensions and clamped with slight pressure with 2 C-clamps (Figure 6). Slides are then placed in a 40 to 50°C oven overnight. The following morning the plastic strips are gently removed and the slides may be stored at room temperature. Slides (as well as embedded specimens) from biopsies labeled with fluorescent material should be stored in the dark. (Note: To facilitate the adherence of sections, the glass slides are coated with the following mixture: Solution A: 4.5-g gelatin dissolved in 1000 mℓ distilled water at 75°C and Solution B: 4% chromium-alum solution prepared in distilled water.)

Working Solution — 3.85 mℓ Solution B to 100 mℓ Solution A. Stir well. Heat mixture to about 50°C and dip baskets of slides for 2 min; drain and dry in low oven or at room temperature.

c. Sequences

Sectioning Sequence — A set of 15 slides is prepared from each biopsy as follows: slides 1 through 9: 4 μm thick, taking every 4th to 10th section, 2 sections per slide; 10: 4 μm thick, 2 consecutive sections; 11 and 12: 10 to 15 μm thick, consecutive sections, 2 per slide; 13, 14, and 15: 4 μm thick, every 4th to 10th section, 2 per slide.

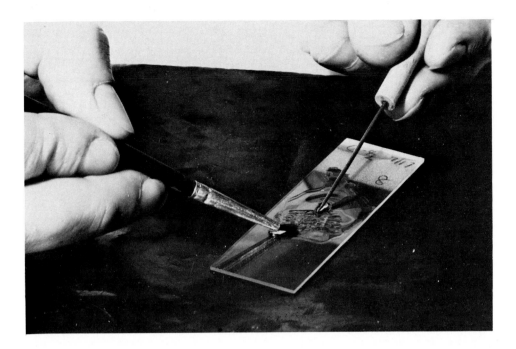

FIGURE 4. After putting the section on the glass slide, further flattening is performed in ethanol 95%
with a probe and a brush.

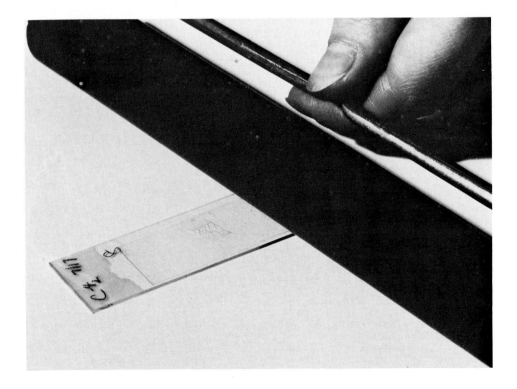

FIGURE 5. The flattened section is covered with a plastic strip and further flattened by rolling a rubber
photoroll over it.

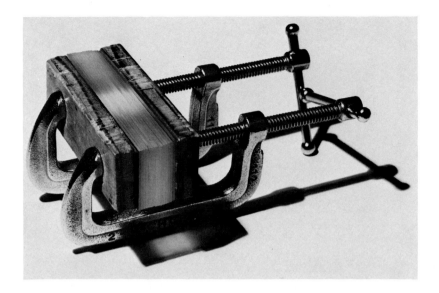

FIGURE 6.   Final flattening of the section is performed during drying under the pressure of two C-clamps.

Staining Sequence — slides 1, 3, 5, 7, and 9: undecalcified toluidine blue; slides 2 and 4: Goldner's trichrome; slide 6: Von Kossa; slide 8: decalcified toluidine blue; slides 11 and 12: unstained, for fluorescent labels; slides 10, 13, 14, and 15: extra sections, stored undeplastified.

d. Deplastification of Sections Before Staining

Cellosolve (ethylene glycol monomethyl ether acetate, Fisher)	25 min
Cellosolve	25 min
70% ethanol	5 min
40% ethanol	5 min
dH$_2$O	5 min

Thick sections for fluorescence microscopy are not deplastified, but directly mounted in UVinert medium (Gurr).

IV. STAINING PROCEDURES

A. Toluidine Blue for Undecalcified Sections (Figures 8B and 9)
1. Buffer

Citric acid	0.63 g
di-sodium phosphate	0.30 g
dH$_2$O	400 cc
pH	3.7

2. Toluidine Blue Stain

Buffer	100 cc
Toluidine blue (Fisher)	2 g

Filter and adjust pH to 3.7 with 1 M NaOH.

1. Deplastify sections and bring to dH$_2$O.
2. Stain in toluidine blue for 10 min (human tissue; animal tissue shorter).
3. Rinse in two changes of buffer — 1 min each.
4. Blot dry.
5. Dehydrate in butyl alcohol — two changes, 1 min each. ½ butyl alcohol and ½ toluene — 1 min.
6. Clear in two changes toluene, 1 min each and mount in Permount.

Total time for procedure: 20 min.

Result — Calcified bone: dark blue to dark purple; osteoid: light blue; calcification front: dark purple grains; osteoblasts: dark blue with Golgi complex area light blue; osteoclasts: light blue; mast cells: metachromatic purple.

Goal — Rapid but reliable staining of osteoid with excellent cellular detail.

B. Von Kossa Stain (Figures 7 and 8C)
1. Solutions

Silver nitrate

Silver nitrate	5 g	
dH$_2$O	100 ml	

Filter and keep refrigerated.

Sodium carbonate and formaldehyde

Sodium carbonate	5 g
Formaldehyde	25 ml
dH$_2$O	75 ml

Methyl Green Pyronin (PolyScientific)

2. Stain

1. Deplastify section and bring to dH$_2$O.
2. Stain in 5% silver nitrate — 30 min (in the dark).
3. Rinse in three changes dH$_2$O.
4. Reduce with sodium carbonate — formaldehyde solution — 2 min.
5. Wash gently in running tap water — 10 min.
6. Stain in methyl green pyronin — 20 min.
7. Wash twice with cooled freshly boiled distilled water — 1 min.
8. Dehydrate once with 95% alcohol — 1 min.
9. Dehydrate in two changes absolute alcohol.
10. Clear in two changes xylene — mount.

Total time for procedure: 1 hr 15 min.

Result — Calcified bone: dark black; osteoid: light pink; calcification front: black granules; osteoblasts: pink/red with Golgi complex area light pink; osteoclasts: pink/red.

Goal — Control of osteoid.

C. Masson-Goldner Trichrome (Figure 8A)
1. Solutions

Weigert hematoxylin
   Solution A

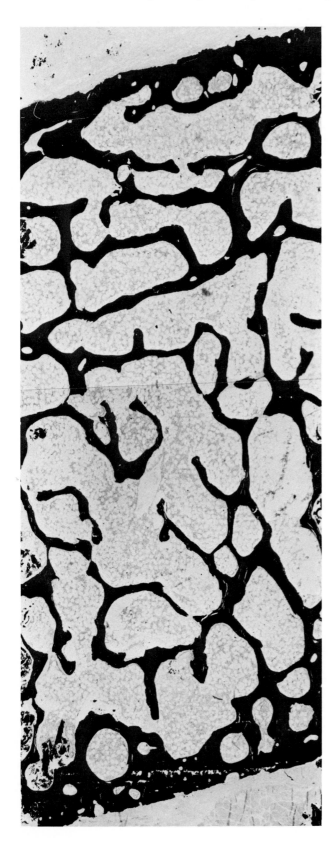

FIGURE 7. General view of a 4-μm thick section of an iliac crest biopsy. Notice the minimal extent of shattering despite a cortex to cortex sectioning area (Von Kossa stain).

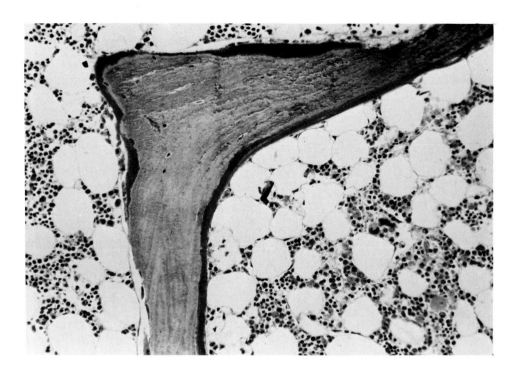

A

FIGURE 8.   (A) Higher magnification of trabecular bone on semiserial 4-μm thick sections showing the Goldner stain; (B) the toluidine blue stain; (C) the Von Kossa stain. Notice that best cellular details are obtained with toluidine blue with also an excellent contrast between osteoid and calcified bone. Notice also the absence of marrow retraction allowing characterization of all the cells at the bone surface.

Hematoxylin (Gurr)	1 g
95% Alcohol	100 mℓ
Filter	

Solution B
29% Ferric chloride	4 mℓ
Diluted HCl[a]	1 mℓ
Bring to 100 mℓ with dH₂O	

Working solution — equal parts A and B.

[a]   Use 40% HCl, dilute 1:4 with distilled water.

Ponceau Fuchsin stock
   Solution A
Ponceau	1 g
dH₂O	100 mℓ

   Solution B
Acid Fuchsin	1 g
dH₂O	100 mℓ

3 Parts Solution A and 1 part Solution B.
Working solution — dilute stock 1:5 with 0.2% acetic acid.

Phosphotungstic acid — Orange G
Orange G (Gurr)	2 g
Phosphotungstic acid (Fisher)	4 g
dH₂O	100 mℓ

Filter before use.

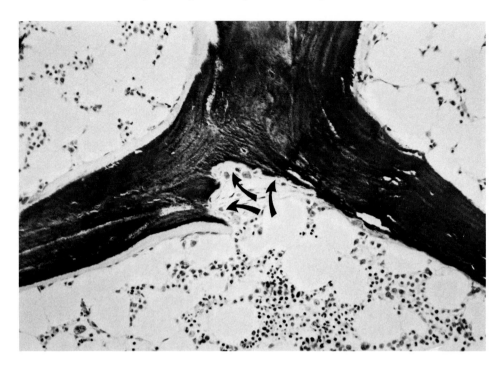

FIGURE 9C.   *Reversal surface.* Howship's lacunae not lined with osteoclasts but containing seemingly active mononuclear cells.

FIGURE 9D.   *Active formation surface* covered with osteoid lined with active osteoblasts (straight closed arrows) which show very typical polarized nuclei and lightly stained Golgi complex areas. The curved open arrows point at an *inactive osteoid surface* lined with inactive cells. On the right (closed curved arrow), a reversal surface.

Light green (SF, yellowish)

Light green	0.2 g
Acetic acid	0.2 ml
dH$_2$O	100 ml

Filter before use.

2. Stain

1. Deplastify section and bring to dH$_2$O.
2. Stain in working solution of Weigerts hematoxylin — 25 min.
3. Rinse with distilled water.
4. Wash gently in running tap water — 10 min.
5. Rinse with distilled water.
6. Stain in Ponceau-Fuchsin working solution — 17 min.
7. Rinse in two changes 1% acetic acid — 1 min each.
8. Stain in freshly filtered Orange G — 7 min.
9. Rinse in two changes 1% acetic acid — 1 min each.
10. Stain in freshly filtered light green — 20 min.
11. Rinse in two changes 1% acetic acid — 1 min each.
12. Dehydrate in one change 95% alcohol — 1 min.
13. Dehydrate in two changes absolute alcohol — 1 min each.
14. Clear in two changes xylene — 2 min each. Mount in Permount.

Total time for procedure: 1 hr 40 min.

Result — Calcified bone: green; osteoid: red and/or green; osteoblasts: red with Golgi complex area pink; osteoclasts: red with ruffle border light and sealing zones dark red.

Goal — Best for color slides.

D. Special Staining Procedures for Demonstration of Cement Lines

Both techniques described here have to be performed on sections that have been decalcified after sectioning. This gives the best results for demonstration of cement lines for measurement of the MWT.

1. Toluidine Blue for Decalcified Sections
a. Reagents

Decalcifying Solution

Sodium formate (Mallinckrodt)	34 g
Formic acid (Baker)	170 ml

Bring to 1l with distilled water.

b. Buffer

The buffer is the same as the one used for toluidine blue stain for undecalcified sections except that the pH is adjusted to 6 with 1 M NaOH.

Toluidine blue — The stain is the same as the one used for undecalcified sections except that the final pH is adjusted to 6 with 1 M NaOH.

1. Deplastify sections and bring them to distilled water.
2. Place the sections in the decalcifying solution overnight.
3. Wash sections in gently running tap water for 3 hr.

4.   Stain in toluidine blue (pH 6) for 8 min (this is for human tissue, animal tissue — slightly less time).
5.   Rinse in two changes of buffer, 1 min each.
6.   Blot dry.
7.   Dehydrate in butyl alcohol two changes, 1 min each; ½ butyl alcohol and ½ toluene, one change for 1 min.
8.   Clear in two changes toluene 2 min each and mount in Permount.

Result — Cement lines appear as dark blue to purple on a light blue background.

2. Cajal-Gallego Trichrome
a. Solutions

1.   Basic Fuchsin (Ziehl): (1) Stock solution — grind with mortar and pestle 1 g of basic Fuchsin (Gurr) and 5 g of phenol in 10 mℓ of ethanol 95%, add progressively, 90 mℓ of dH$_2$O, and then filter; (2) Staining solution — dilute 18 mℓ of the stock solution in 100 mℓ of dH$_2$O and add 4 to 6 drops of glacial acetic acid.
2.   Acetified formalin: Dilute 5 mℓ of formalin in 95 mℓ of dH$_2$O containing 10 drops of glacial acetic acid.
3.   Picric acid solution: in dH$_2$O until saturation.
4.   Indigo carmine: 0.4 g of indigo carmine; 100 mℓ dH$_2$O.
5.   Mix 1:1 solutions 3 and 4 just before use.

b. Staining

1.   Proceed through decalcification as before.
2.   Stain in diluted basic Fuchsin — 1 min.
3.   Fix in acetified formalin — 1.5 min.
4.   Stain in Picric acid-Indigo carmine — 1 min.
5.   Dehydrate in absolute ethanol — 3 × 1 min.
6.   Clear in toluene — 2 × 2 min.
7.   Mount.

Results — Cement lines appear as dark red on a light green or light blue background.

V. RAPID EMBEDDING TECHNIQUE FOR DIAGNOSIS AND/OR ENZYME HISTOCHEMISTRY

Recently glycol methacrylate embedding at low temperatures (4°C) has been found to be useful for rapid processing of specimens and is particularly suited for enzyme histochemistry.[10] Enzyme activity is not diminished to the extent that it is with methyl methacrylate because of glycol methacrylate's water soluble nature and the ability to polymerize at 4°C.

The following procedure using the JB4 Embedding Kit (Polysciences), is a modification of the manufacturer's suggested technique to make it more suitable for mineralized tissue.

A. Regular Processing
1. Fixation

Formal calcium—1 day
(0.11 *M* Calcium chloride in 4% formaldehyde)
Rinse in phosphate buffer 7.4 for 1 hr at 4°C

2. Dehydration and Infiltration

50:50 catalyzed Solution A: distilled water	2 days
75:25 catalyzed Solution A: distilled water	2 days
100% catalyzed Solution A	2 days

These steps are carried out at 4°C.

3. Embedding

The tissue is embedded in catalyzed Solution A and Solution B at a concentration of 40:1. Peel-A-Way disposable plastic molds (Peel-A-Way Scientific) are used, filled to the top with the embedding mixture, and covered with a thin plastic film. The embeddings will polymerize overnight at 4°C.

4. Sectioning

The polymerized blocks are sectioned on the Jung® K microtome and the sections are lifted from the dry knife with a forceps. After floating on distilled water on a glass slide, the water is drained and the section allowed to air dry. (Slides may then be stored at 4°C.)

5. Staining

Routine histologic stains and enzyme histochemistry can be done on the sections without deplastifying.

B. Rapid Processing

Rapid processing of small pieces of human bone biopsies can be accomplished using glycol methacrylate.[8] The biopsy pieces (3 mm by 6 mm) are dehydrated in 40, 70, 95%, and absolute alcohol at approximately 6-hr changes and then infiltrated in catalyzed Solution A overnight. All of these steps are carried out at 4°C under vacuum. The tissue is embedded in catalyzed Solution A and Solution B at 40:1 and may be sectioned and stained as previously described.

C. Enzyme Histochemical Techniques

Although not routinely used in our laboratory, these techniques are given here for the reader's convenience. These staining procedures use undecalcified sections prepared after embedding in glycol-methacrylate.

1. Acid Phosphatase[11-13]

Reagents

Pararosanaline stock solution

Pararosanaline (Eastman Kodak®)	1 g
dH$_2$O	20 ml
HCl acid (concentrated)	5 ml

Dissolve the pararosanaline in the dH$_2$O. Add the HCl while warming the solution. Filter and store at 4°C in a dark bottle.

Sodium nitrite

Sodium nitrite	25 mg
dH$_2$O	1 ml

Dissolve sodium nitrite in dH$_2$O.

Naphthol ASTR phosphate

Naphthol ASTR phosphate (Sigma®)	16 mg
Dimethylformamide (MC/B)	1 ml

Dissolve naphthol ASTR phosphate in dimethylformamide.

Acetate Buffer 0.1 M at pH 5
   Solution A
      Sodium acetate   0.82 g
      dH$_2$O   100 mℓ
   Solution B
      Acetic acid   0.6 mℓ
      dH$_2$O   100 mℓ
Combine 70 mℓ Solution A and 30 mℓ of Solution B.

Manganese Sulfate
   Manganese Sulfate   1 g
   dH$_2$O   10 mℓ
Dissolve manganese sulfate in dH$_2$O.

The procedure is as follows:

1.   Place glycol methacrylate sections (on slides) in dH$_2$O.
2.   Prepare hexazotized pararosanaline by slowly mixing 0.5 mℓ of stock solution with the sodium nitrite under a fume hood.
3.   Add 10 mℓ 0.1 M acetate buffer pH 5.0 and adjust pH to 5 with 1 M NaOH.
4.   Add the naphthol ASTR phosphate.
5.   Add 3 drops of the manganese sulfate.
6.   Place the wet slides on a staining tray. Cover the sections with 2 to 3 drops of the staining solution and cover the staining tray to keep sections from drying. Incubate at 37°C for 30 min to 1 hr. Sections may be rinsed in dH$_2$O to check staining.
7.   Rinse slides gently in dH$_2$O for 2 min.
8.   Place slides in 70% alcohol at room temperature for 30 min.
9.   Rinse slides gently in 2 changes dH$_2$O for 2 min each.
10.   Counterstain section in toluidine blue pH 3.7, as previously described (or sections may be counterstained in Harris hematoxylin).
11.   Sections may be dehydrated in butyl alcohol and mounted in Permount.

Result — Acid phosphate-rich structures are stained dark red on a light blue background.

2. Alkaline Phosphatase[10,14]

Reagents

Naphthol AS-MX Phosphate Concentrated Solution — pH 8.6 (Sigma®)
Fast Blue RR Salt (Sigma®)
Nuclear Fast Red
   Nuclear Fast Red   0.1 g
   Aluminum sulfate 5% aqueous   100 mℓ

Dissolve the Nuclear Fast Red in the aluminum sulfate solution with the aid of heat. Cool and filter the solution and add a crystal of thymol; may be stored at room temperature and reused.
   The procedure is as follows:

1.   Pour 2 mℓ of Naphthol AS-MX phosphate concentrate, pH 8.6, into a Coplin jar.
2.   Add 48 mℓ distilled water.
3.   Add 25 mg Fast Blue RR salt to the Coplin jar and stir to dissolve.

4. Place slides in Coplin jar and incubate at room temperature for 15 min to 1 hr. Slide may be rinsed in distilled water and checked for staining.
5. Rinse slides in distilled water for 1 min.
6. Wash slides gently in tap water for 5 min.
7. Counterstain in Nuclear Fast Red for 8 min.
8. Wash gently in distilled water for 1 min.
9. Mount in buffered glycerol.

Result — Alkaline phosphatase-rich structures are stained dark blue-purple on a light red background.

3. Nonspecific Esterase[15]

Reagents

α Naphthyl acetate (Sigma®)
   α Naphthyl acetate   10 mg
   Acetone   0.25 ml
Phosphate buffer 0.1 M at pH 7.4
   Solution A
      Na_2HPO_4 1.42 g
      dH_2O 100 ml
   Solution B
      KH_2PO_4 1.36 g
      dH_2O 100 ml

Combine 82 ml Solution A with 18 ml Solution B.
   Fast Blue B Salt (Sigma®, O-dianisidine, tetrazotized)
   Nuclear Fast Red (the same as for alkaline phosphatase)

The procedure is as follows:

1. Place glycol methacrylate sections (on slides) in dH_2O.
2. Add 20 ml of 0.1 M phosphate buffer to the α Naphthyl acetate in acetone.
3. Shake thoroughly until the initial cloudiness disappears.
4. Add 50 to 100 mg Fast Blue B Salt; shake.
5. Place wet slides on staining tray. Filter the staining solution directly into the slides and cover staining tray to keep sections from drying. Incubate at room temperature for 15 to 30 min. Section may be rinsed in dH_2O and checked for staining.
6. Wash gently in tap water for 2 min.
7. Counterstain in Nuclear Fast Red for 8 min (sections may also be counterstained in Mayer's hemalum).
8. Wash gently in distilled water for 1 min.
9. Mount in buffered glycerol.

Result — Nonspecific esterase-rich structures are stained dark blue on a light red background.

VI. CONCLUSION

The accumulation of years of progressive improvements in the methods used in different laboratories and aimed at obtaining thin and well preserved sections of undecalcified bone have led to the definite possibility of performing such sections routinely and in a very reliable fashion. In parallel, the extraordinary development of stereological techniques and of histological and pathological concepts regarding bone remodeling have led to the definite possibility of measuring the various components of this process and assessing them dynamically in various bone diseases. It is, however, clear that good histology and good stereology have to be associated for bone histomorphometric studies.

In this chapter, we have described in detail the ways to routinely obtain a good enough morphology to proceed to its quantification, using the information contained in the following chapters of this publication.

It is our deep belief that one should not proceed to bone histomorphometry without having first attained a very good morphological quality of undecalcified sections of the tissue to be studied (Figures 8 and 9). This chapter should help in realizing that goal.

ACKNOWLEDGMENTS

The authors wish to acknowledge here the fact that this chapter is the result not only of their own experience and creativity but of numerous contacts and discussions with other researchers in the field both at meetings and during visits in their laboratories. Although it seems impossible here to thank and acknowledge all of them, we are more particularly indebted to Dr. M. Juster (CNRS, Paris) who helped us start in this field; the late Dr. P. Bordier (INSERM, Paris) whose friendship and experience were of invaluable help in our laboratory; Dr. R. Schenk (Anatomische Institut, Bern, Switzerland) who is the undisputed master of morphological perfection in undecalcified bone and who so generously told us during a visit to his laboratory in 1974 all the "tricks" he uses; Dr. P. Meunier (INSERM, Lyon, France) for his support, advice and experience in processing undecalcified bone, and, finally, to Drs. M. Mailland, J. L. Saffar, and F. Guyomard, our colleagues in the Paris 5 Dental School, with whom most of these techniques have been tried and developed.

REFERENCES

1. Arnold, J. S. and Jee, W. S. S., Embedding and sectioning undecalcified bone and its application to radioautography, *Stain Technol.,* 29, 225, 1954.
2. Frost, H. M., Preparation of thin undecalcified bone sections by rapid method, *Stain Technol.,* 33, 273, 1958.
3. Burkhardt, V. R., Praparative Vorausset zungen zur Klinischen Histologie des Menschlichen Knochenmarkes, *Blut,* 14, 30, 1966.
4. Meunier, P., Vignon, G., and Vauzelle, J. L., Methodes histologiques quantitatives en pathologie osseuse, *Lyon Med.,* 18, 133, 1969.
5. Bordier, P. J. and Tun Chot, S., Quantitative histology of metabolic bone disease, *Clin. Endocrinol. Metab.,* 1, 197, 1972.
6. Villanueva, A. R., Methods for preparing and interpreting mineralized sections of bone, in *Bone Morphometry,* Jaworski, Z. F. G., Ed., University of Ottawa Press, Ottawa, Canada, 1976, 341.
7. TeVelde, J., Burkhardt, R., Kleiverda, K., Leenheers-Binnendijk, L., and Sommerfeld, W., Methylmethacrylate as an embedding medium in histopathology, *Histopathology,* 1, 319, 1977.
8. Fallon, M. D. and Teitelbaum, S. L., A simple procedure for the rapid histologic diagnosis of metabolic bone disease, *Calcif. Tissue Int.,* 33, 281, 1981.
9. Matrajt, M. and Hioco, D., Solochrome cyanin R as an indicator dye of bone morphology, *Stain Technol.,* 41, 97, 1966.
10. Brinn, N. T. and Pickett, J. P., Glycol methacrylate for routine special stains, histochemistry, enzyme histochemistry and immunohistochemistry, *J. Histotechnol.,* 2, 125, 1979.

11. **Burstone, M. S.**, Histochemical demonstration of acid phosphatase activity in osteoclasts, *J. Histochem. Cytochem.*, 7, 39, 1959.

12. **Hayashi, M.**, Comments on the cytochemical techniques using Naphthol AS-BI substrates and hexazonium pararosanaline for the demonstration of acid phosphatase, β-glucuronidase and N-acetyl-β-glucosaminidase activities, *J. Histochem. Cytochem.*, 25, 1021, 1977.

13. **Evans, R. A., Dunstan, C. R., and Baylink, D. J.**, Histochemical identification of osteoclasts in undecalcified sections of human bone, *Mineral. Electr. Metab.*, 2, 179, 1979.

14. **Burstone, M. S.**, *Histochemistry and Its Applications in the Study of Neoplasms*, Academic Press, New York, 1962, 276.

15. **Pearse, A. G. E.**, *Histochemistry, Theoretical and Applied*, Vol. 2, 3rd ed., Williams & Wilkins, Baltimore, 1972, 1303.

Chapter 4

BONE HISTOMORPHOMETRY: CHOICE OF MARKING AGENT AND LABELING SCHEDULE

H. M. Frost

TABLE OF CONTENTS

I. PURPOSE

This chapter describes how to label a living skeleton with tissue-time markers in order to make usefully accurate measurements of bone tissue turnover in a subsequent bone biopsy. This information is necessary to understand the dynamics of bone remodeling and to interpret static morphological, biochemical, and physical data.

II. INTRODUCTION

Tetracycline tissue-time markers, introduced in the late 1950s, allow measurement of the rate of new bone formation at bone-forming surfaces. The method was first applied to Haversion bone formation in ribs where the formation surfaces are oriented perpendicular to the plane of section (provided cross sections of ribs are viewed under the microscope). More recently, the system has been used for measurement of bone formation in iliac trabecular bone. In this case, the bone-forming surface is more randomly oriented to the plane of section causing a "geometric projection error" which is described in Chapter 5 and in the cited Reference 16. This chapter will describe the optimum choice of marking agent and labeling schedule in order to minimize this geometric problem as well as a number of other difficulties which can arise in attempting to measure bone formation rates in iliac trabecular bone.

In addition, a group of potential errors in experimental design will be identified which may be eliminated or minimized by the proper choice of labeling agent, schedule, and other factors.

III. DEFINITIONS

Before beginning this discussion, two fundamental definitions must be made: "bone label" and "bone tissue-time marker".

A. Bone Label

This is defined as any subsequently identifiable substance which deposits in bone tissue. To be useful, it must identify some chemical or anatomical component of bone, and that component must then be distinguishable from all other components.

B. Bone Tissue-Time Marker

This is any identifiable feature, naturally occurring or artificially induced, which permits one to locate a given bone *surface* in anatomical *space* at a known moment in *time.*

There are numerous bone labels and bone tissue-time markers which can service diverse purposes. Not all bone labels are good bone tissue-time markers, and conversely not all bone tissue-time markers are good bone labels. For instance, serial measurements of the medullary diameter of a metacarpal will permit location of a surface in anatomical space at specified moments in time, but that surface is not a bone label. Radiocalcium may label the mineral phase of bone and may be useful in many experiments involving bone mineral. However, it is a poor tissue-time marker in part because of its diffuse distribution and relatively indistinct identification of any particular bone surface.

Morphometric measurement of bone remodeling dynamics (rates) calls for a bone label which will operate as a bone tissue-time marker at a very specific bone surface location, namely, the bone formation surface at each bone remodeling site.

C. The Optimal Bone Label

The following attributes are required of an optimal bone label for this purpose: (1) nontoxic, (2) inexpensive, (3) simple, (4) widely available, (5) stable, (6) detectable, (7) labels new bone, and (8) useable as a tissue-time marker.

In further detail, the agent must be relatively nontoxic to the organism and cause few unpleasant side effects. It must not seriously threaten a patient's health or well being. It must not interfere with normal osteoblast function, nor the function of any other cells which may affect the remodeling process under study.

The label must not be burdensomely expensive. The acquisition and preparation of bone specimens from patients are already quite expensive.

Simplicity is necessary for widespread, standardized use. The agent must be widely available and its use free from complicated, time-consuming regulation of patient safety matters.

The label must be stable, meaning that once deposited it must remain *in situ* both in the living organism and afterwards during preparation of a bone sample for analysis.

It must be detectable by some means. Obviously an agent which cannot be seen, photographed, touched, or quantified by some convenient method will not work. Further, the detection process must not alter the label or marker characteristics of the agent. The label to be used must localize in new bone, preferably exclusively in new bone.

At the present time only the tetracycline antibiotics meet all of the above criteria. Indeed, most of the knowledge concerning, bone tissue remodeling dynamics was discovered because tetracyclines met these requirements before anyone knew what an appropriate bone tissue-time marker might be, or that one might be useful in studying bone tissue.

Tetracycline incorporates itself into new bone at the time it is formed, and it remains *in situ* as long as the crystal remains intact both in vivo and ex vivo. It is easily detected by fluorescence under excitation by blue light at 360-mm wavelength; it is cheap, nontoxic, simple, and widely available. It is an excellent bone tissue label and if properly used, a nearly ideal tissue-time marker. There are several ways in which it may be used as a tissue-time marker and several options will be described.

The drug accumulates at the plane of initial calcification (the calcification front, zone of demarcation, ligne frontière) during the time it is present in the circulation, and has a reasonably consistent spatial relationship to this plane. It accumulates in the form of a band at this plane and if the circulating levels are maintained for a period of the time, the width of the band can function as a tissue-time marker. That is, the two edges of the band can locate the bone-forming surface at two points in time.

However, the edges are relatively indistinct and hard to locate accurately. In fact, the center of the band formed by drug when administered for 2 to 6 days can be localized as much as 20 times more accurately than its edges. One key then to using tetracycline successfully as a tissue-time marker is to administer it for a short period, then allow a drug-free interval, followed by another short period of administration. Then the center of each band as a measuring point can mark the location of the bone-forming surface accurately at two points in time, allowing measurement of the appositional rate. The ability to measure the appositional rate is the fundamental property that allows access to important kinetic information about bone turnover and related cellular activities.

Other potential labeling agents are listed in Table 1. Special mention might be made of radiocalcium. It is quite a useful label of the mineral phase of bone and the first direct measurement of a bone appositional rate was performed with it in rabbits at the Argonne National Laboratory in 1953 by Dr. J. S. Arnold and W. S. S. Jee.[1] In order to locate the bone-forming surface with this (or any other) beta emitter, autoradi-

Table 1

PROPERTIES OF SELECTED BONE LABELING AGENTS

Labeling Agent	Labels osteoid	Labels the calcification front	Stable in sections	Provides good markers	Stable in the bone in vivo	Stable in the bone ex vivo	Toxic to subject	Toxic to osteoblasts	Inexpensive	Reliable
Tetracyclines	0	+	+	+	+	+	0	0	+	+
DCF (Suzuki)	0	+	+	+	+	+	+	+	+	+
Procyon dyes	+	0	0	+	?	+	+	+	+	+
Alizarin	0	+	+	+	+	+	+	+	+	0
Chlorazol dyes	+	0	0	?	?	0	+	+	+	0
Lead	0	+	+	?	+	0	+	+	+	+
Fluorescein	0	+	+	0	0	0	0	?	+	+
Hematoporphyrin	0	+	+	0	0	0	+	+	0	0
Uroporphyrin	0	+	+	0	0	0	+	+	0	0
Suzuki's agent	0	+	+	+	+	+	+	+	0	+
Cobalt	0	+	+	0	0	0	+	+	+	+
Quinine	0	+	+	0	?	0	+	?	+	0
Tritiated[a] proline	+	0	+	+	+	+	+	0	0	+
Radiosulfur[a]	+	0	+	+	+	+	+	0	0	+
Radiocalcium[a]	0	+	0	+	+	+	+	0	0	+
Toluidine blue	0	0	0	0	0	0	0	0	+	0

Note: This table compares some properties of various agents used in vivo bone labels. At present the best available agents are the tetracyclines.

[a]   Requires use of developed autoradiographs.

ographs must be prepared from thin, undecalcified sections. This application of 45-Calcium for determination of bone formation rates must be distinguished from its use as a plasma calcium tracer where measurements of mineral accretion can be made over the entire skeleton.[2]

IV. LABELING SCHEDULES

The above discussion indicated that the center of a tetracycline band can be more accurately located in a bone section than its edges. Therefore, an effective way to use tetracycline as a tissue-time marker is to administer it for a short time, allow a drug-free interval, then give it again for a short time. Sometime later, a biopsy may be performed and the appropriate measurements made using a fluorescence microscope. In newly formed bone tissue, this schedule deposits two bands or "labels" which appear much like growth rings seen on the cross section of a tree stump (see Figure 1).

A. Labeling Schedule Code

The unit of time used for scheduling is the day. For example, a scheduling code may be written, 1-5-2:5, which means that a continuous label was given for 1 day followed by a 5-day period without label, another 2 days of label, followed by 5 days of no label prior to biopsy. The order of events in the labeling schedule is termed its staging and the permissible durations are termed staging limits. The duration of label admin-

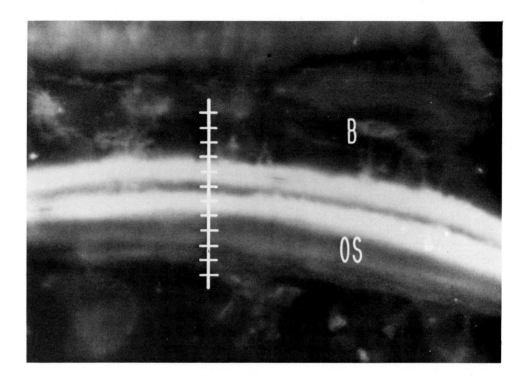

FIGURE 1. Photomicrograph of BMU which took tetracycline during the labeling prior to biopsy near the iliac crest. The section is 10 μm thick, undecalcified, unstained, and viewed at magnification × 320. Mineralized bone is identified as "B" and the osteoid seam is identified as "os". An eyepiece micrometer is superimposed over the labels in the manner used to measure the marker separation. Epifluorescence microscopy is used with a Xenon light source.

istration is the labeling period and the time interval between the tissue-time markers, the marker interval.

B. Optimal Staging

The optimal staging limits for human use depend on certain characteristics of bone formation at remodeling sites (BMU, or Basic Multicellular Units named by Frost, also called BRU by Drs. Parfitt and Jaworski in other chapters) in iliac trabecular bone, and on certain chemical, physical, and physiological characteristics of tetracycline. These act in combination such that an optimum schedule for human use ranges between 1-5-1:5 to 3-14-6:14 or any variation within these staging limits.

Two characteristics of bone formation at the BMU govern the choice of an optimum schedule: (1) the rate of new bone apposition and (2) the length of time taken to complete the formation phase of a typical new BMU. Marker intervals that are too short will not allow enough separation between markers for accurate measurement of double-label separation, while those that are too long will allow too many BMU to escape one of the pairs of markers.

Other characteristics of bone formation may complicate this picture, such as the fact that bone formation may temporarily pause sometimes or even halt permanently prior to completion of a BMU. Further, as the appositional rate decreases the two labels come closer together. These and other factors affecting the choice of an optimum labeling schedule are listed in Table 2 and described in the paragraphs that follow.

Table 2
ATTRIBUTES OF SOME TECHNIQUES FOR STUDYING BONE TISSUE TURNOVER

Technique allows one to measure an accurate value of:	Microradiography with morphometric analysis	Light microscopy, no time markers	Tetrcycline and 3 or more markers	Tetracycline, 2 markers	Tetracycline, 1 marker	Alizarin labeling	Procyon dyes	Lead, cobalt	Toluidine blue	Gamma ray densitometry	Tracer kinetics	Neutron activation
Amount of resorption surface	0	+	+	+	+	+	+	+	+	0	0	0
Amount of osteoid surface	0	+	+	+	+	+	+	+	+	0	0	0
Amount of active formation surface	0	0	+	+	0	0	0	0	0	0	0	0
Appositional rate	0	0	0	+	0	0	0	0	0	0	0	0
Resorption rate (cellular level)	0	0	0	+	0	0	0	0	0	0	0	0
MWT	+	+	NA	NA	NA	NA	NA	NA	NA	0	0	0
Sigma	0	0	0	+	0	0	0	0	0	0	0	0
Bone balance (tissue)	+	+	NA	NA	NA	NA	NA	NA	NA	+	0	0
Bone formation rate (tissue level)	0	0	0	+	+	0	0	0	0	0	0	0
Bone resorption rate (tissue level)	0	0	0	+	0	0	0	0	0	0	0	0
Inactive formation surface	0	0	+	+	0	0	0	0	0	0	0	0
Inactive resorption surface	0	+	0	0	0	0	0	0	0	0	0	0
Perimeters	+	+	0	0	0	0	0	0	0	0	0	0
Areas	+	+	0	0	0	0	0	0	0	0	0	0
Amount of osteoid tissue	0	+	0	0	0	0	0	0	0	0	0	0
Location of the calcification front, point in time	0	0	0	+	+	0	0	0	0	0	0	0

Note: A comparison of attributes of various methods of studying bone tissue turnover. Some of the techniques mentioned have the potential implied by the plus signs even though many who have used the techniques did not choose to make use of that potential.

C. Factors Governing Choice of Schedule
1. The Labeling Escape Error

In order to measure an appositional rate and its dependent rate parameters, two tissue-time markers must be deposited within each bone-forming BMU (and each marker must have been deposited at a time known to the investigator).

Not all BMU will take both markers even if the optimal staging limits described above are used. For example, if a given marker interval is 14 days, and it is assumed that all BMU form in the same time period (60 to 75 days average in adult humans), a certain portion will not take the initial marker because they began to mineralize after it was given. They will usually take the second marker at the end of the labeling period because formation remains active then. Yet other BMU finish forming after the first marker is given and so fail to display the second marker. In either case, such BMU contain only one tissue-time marker and will not permit measurement of an appositional rate.

One feature of this "marker escape" is that the longer the marker interval, the greater the fraction of singly labeled BMU. At marker intervals of 40 days or longer,

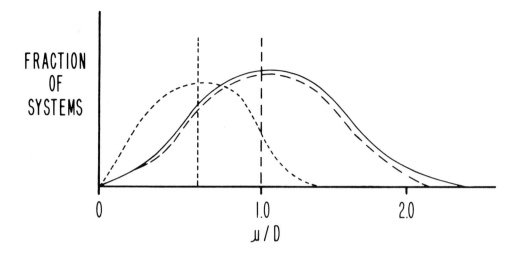

FRACTION
OF
SYSTEMS

0   1.0   2.0

μ / D

FIGURE 2. The skewed sampling error. This graph plots the distribution of appositional rates in the many different bone forming centers lying within a bone biopsy. Microns per day distribute along the horizontal axis, and relative frequencies along the vertical. The solid line represents the true distribution or "mix" within the bulk tissue. The vertical dashed lines represent the means of these curves. The dashed curve to the left represents the observed mix when marker intervals are too long; the mean is smaller than the true mean shown by the bold vertical dashed line on the right.

The closely parallel dashed-line curve signifies the "sample" of the above true situation which one sees if he uses a marker interval of 6 to 8 days in a double labeling schedule. The sample already shows skewing so that the mean of this curve will lie somewhat to the left of the true mean. But the difference between them remains on the order of the accuracy of measurement. The dotted curve represents the distribution seen by the microscopist in the same material if the marker interval equals 30 days. He sees a grossly nonrepresentative sample simply because about half of the bone-forming systems finish depositing new bone too quickly to "catch" both tissue-time markers. Worse, more than 90% of the slower forming systems accept them. Thus, the mean appositional rate obtained in such a situation proves spuriously low by a factor of about two because of the effect illustrated here. The errors caused by this effect would add to others described in the text and create serious errors in all the derived parameters which include bone formation and resorption rates, σ, σ-f, σ-r, the appositional rate, the fraction of bone surface actively making bone, etc.

the phenomenon may result in single labels appearing in over 90% of all bone-forming sites that were "active" some time during the labeling interval. Higher appositional rates will also tend to result in more single labels than lower rates.

Marker intervals of days or less usually will reduce this error to from one to three times the error in measurement of bone formation itself. This is by no means a trivial error even if the staging limits recommended above are used. Chapter 6 is devoted to minimizing or correcting this labeling escape error, a task made difficult because "marker escape" is probably not the only mechanism responsible for the appearance of singly labeled surfaces (see below).

2. The Skewed Sampling Error

Two properties of BMU formation tend to skew the process of sampling their formation rates by the double-label method: (1) the individual BMU form at different appositional rates and (2) as each BMU nears completion, its appositional rate tends to slow.[3] Figure 2 illustrates hypothetical frequency distributions of appositional rates in BMU in the entire skeleton of a human. The range is great enough that the BMU completion times range from around 35 to 250 days. In Figure 2 the solid curved line represents the true frequency distribution and the dashed curve the skewed distribution found when the marker interval is too long (i.e., greater than 14 days).

What has happened here is that when a long marker interval is applied to this mix of different appositional rates, the slower centers were more likely to take both markers than the speedy centers. Thus mean appositional rates determined from measurements of many individual double-label sites will come from BMU clustering about the slower end of the distribution curve as in the dotted curve. The speedy systems are more likely to escape taking one of the markers (label escape error) thereby eliminating themselves from measurement. Consequently, the observed mean rates will be erroneously low.

To minimize this problem one must use the shortest marker interval which still allows sufficient label separation for acurate measurement. At present there is no known reliable method of correcting for the skewed sampling error.

3. Intermittent Apposition

The rate of new bone apposition at individual BMU can vary throughout the formation period.[3] It may speed up, slow down, stop briefly, or remain rather steady. Further, matrix formation may at times proceed out of step with mineralization.[4] During its inactive stage, the calcification front of some such systems can still take a permanent label.

4. Resting Seam Label

A related phenomenon occurs sometime when a BMU stops formation activity before the final portion of matrix becomes mineralized. The resultant "resting" osteoid seam may accept a tetracycline label, yet there is no new bone being formed. This phenomenon was suspected by LaCroix as early as 1955 and was demonstrated in human bone in 1959 by the writer and co-workers.[5]

A single label appearing in an inactive BMU cannot be distinguished from a single label in an active one which escaped the second. What is most important, however, given the above problems and the accompanying uncertainties, is that a single label appearing in a patient given a proper double labeling schedule cannot indicate reliably the presence or absence of local bone-forming activity, nor can it allow any calculation of any parameter which depends on knowledge of the appositional rate.

5. The Band Width Error

This has been called the thickness of the instantaneously staining zone error (TISZ).

When viewed in the fluorescence microscope, the two borders of a tetracycline band in a thin section do not define sharply. Instead they fade off from maximal to zero over a linear and variable distance of about 3 μm. It is difficult to determine the exact spatial plane that corresponds to the exact beginning (zero time) or end of the interval of tetracycline administration. This border "fuzziness" varies from place to place in the section, probably partly because of the different geometric projections in the section. However, a number of other factors may also affect it: (1) the intensity of source illumination, (2) the numerical aperture and/or the transmission of the objective and/or substage condensor, (3) the magnification, (4) the quality and appropriateness of the filters used in the microscope, (5) ambient illumination (or darkness) in the lab, (6) thickness of the section, (7) daily dose of tetracycline, (8) the kind of tetracycline used, and (9) the degree of dark adaptation by the microscopist.

6. Geometric Projection Overlap

Geometric projection overlap is briefly mentioned here to emphasize the fact that "double" labels may appear "single" to the microscopist at some BMU because of it, and thus prohibit measurement of double-label width at such sites (see Chapter 5). These sites may be indistinguishable from others containing only one label.

Bone-forming centers on trabecular surfaces are planar, and the orientation of the planes in three dimensions is random relative to the plane of section seen in the microscope.* When the plane of section is an estimated 75° or more away from perpendicular to the plane of the bone-forming center, two short-duration tetracycline labels may not be resolved and so may appear as a single label to the microscopist. There is no method currently available to correct for this. However, with labeling schedules as described above, the number of unresolvable doubly labeled surfaces is estimated to be less than 10% of the total.

7. Label Recognition Error

About half of all people in past years have tetracycline trapped in their bone as result of previous medical treatment for infections.[7] These previously deposited labels cannot yield measurements of appositional rates unless one can know precisely when the labels were administered. These "old" labels can sometimes be confused with a recent double labeling performed with optimal staging limits; however, the presence of overlying osteoid identifies a given pair of labels as recent, and also as part of a forming surface at the time the labels were given. The advantage of recognizing overlying osteoid is lost if too much time elapses between the second label and the biopsy. This allows more of the centers to finish mineralizing any osteoid overlying recent labels.

This problem is usually trivial when a patient is doubly labeled the first time because the odds are remote that two successive courses of tetracycline would have been previously given during a 20- to 60-day interval, to cause some old doubly labeled surfaces to mix with new ones. It is not trivial when one wishes to label and biopsy the skeleton on two occasions before and after 1 year of treatment with a test medication. In the latter case, old and new double labels may both occupy trabecular surface. However, there are clues which help distinguish them: (1) old labels may terminate abruptly when an interim BMU overlapped while recent double labels generally blend into a single fluorescence band at either end (see Figure 3), (2) resting trabecular surface many times shows a very thin single band of fluorescence (when this overlies a double label, one may assume it is old — see Figure 3), and (3) the presence of overlying osteoid is strong evidence that an underlying double label is of recent origin (see Figure 1).

The label recognition problem may be partially resolved by using different tetracyclines. For instance, demethylchlortetracycline fluoresces a slightly different color than tetracycline itself in some fluorescence systems. The distinction between them depends somewhat on the color acuity of the microscopist which varies from person to person, and in the same person from time to time.

The problem will be eliminated when suitable labels can be found which fluoresce in obviously different colors.

8. Labeling Agent Toxicity

All known labeling agents including tetracycline suppress osteoblast work efficiency to some extent, thus interfering with the cell activity under measurement. A double-label schedule with a short labeling period such as 2-10-2:7 minimizes this problem when compared to a longer labeling period such as 6-10-6:7 because labeling agent circulated in the blood for only 16% of the time covered by the measurements in the first case compared to 38% in the second. The usual dose of tetracycline used for labeling is about 15 mg/kg/day. This will reversibly suppress osteoblastic activity, typ-

---

* Strictly speaking, the orientation is not quite random because iliac trabecular bone is slightly anisotropic. It is near enough to isotropy that the assumption of randomness can be made without causing more than trivial error.[6]

rates during the time they are present in the circulation. In continuous labeling systems tetracycline circulates during the entire period of labeling, thus suppressing the rate under measurement. At the doses usually given for bone labeling (15 mg/kg/day), appositional rates will be suppressed typically 10 to 30%.[7] As noted earlier this problem is minimized with double label schedules because the labeling agent circulates for only a small fraction of the time during which appositional rates are actually measured.

11. Multiple Labeling Schedules

Multiple labels (more than two) have been used sequentially during a series of interventions in a given experimental animal in hopes of storing the information in the skeleton for later retrieval at a single biopsy. For example, the effects of a drug would be studied by giving a series of labels, each 10 days apart. The first two would be given before treatment, the next two during, and the final two after. A biopsy would then follow.

This method is subject to all of the errors mentioned above but particularly the recognition error and so cannot be reliable until labeling agents are developed that fluoresce different colors so that the timing of each label can be accurately identified.

V. EXPERIMENTAL DESIGN

The focus of this chapter has been on optimal labeling techniques for the use of tetracyclines as tissue-time markers in bone. The following discussion will concentrate on strategy and tactics for using the method in human and animal experiments. A brief discussion of skeletal physiology will be offered first. Much of the information was discovered by the use of tetracycline tissue-time markers and has been covered in detail in this volume and elsewhere.

A. Growth, Modeling, and Remodeling

These three skeletal activities must be understood. They differ fundamentally from each other, operate under different controls, in different locations, and at different ages in the human even though the same types of bone cells are involved in all three instances. Derangement in each results in different clinical syndromes. The term growth applies to the activity beneath an epiphyseal plate prior to its closure and results in elongation of the bone involved. It is characterized by deposition of cartilage followed by its mineralization and then its division into strands by eroding osteoclasts which keep up with the advancing epiphysis though trailing it by several millimeters. The mineralized cartilage is removed by the chondroclasts and ultimately is replaced by lamellar bone to form the permanent or secondary spongiosa.

Modeling signifies the sculpting processes that follow the elongation of growth. It shapes a growing bone as it elongates. Thus, for instance, a tibia continues to resemble a tibia at any point during its growth. The same types of bone cells are involved as in remodeling, but they seem to "know" how to resorb and replace bone tissue according to a preprogrammed architectural mold.

Remodeling renews the skeleton and repairs microdamage before it accumulates to the point of loss of skeletal integrity. Most adult human metabolic bone disease results from its derangement. Remodeling occurs in packets on surfaces of bone and follows a preprogrammed Activation→Resorption→Formation (ARF) sequence. The remodeling that occurs in different areas of bone behaves somewhat differently. For instance, periosteal remodeling typically results in a net gain of bone so that the external volume of the skeleton continuously enlarges until death.[10] Endosteal, Haversian, and corticoendosteal remodeling all behave differently, probably have separate control mechanisms, and can malfunction individually or in combination to produce disease. These different remodeling surfaces have been called "envelopes",[9,10] the individual remodeling packets are known as Basic Multicellular Units (BMU), the individual life span of the BMU from activation through completed formation is known as sigma, and the frequency (or rate) of activation of new BMU is known as μ (Greek lower case mu).

With this brief review as background, a number of recommendations can be made to aid in designing and analyzing experiments in skeletal pathophysiology and pharmacology.

B. Factors Influencing Strategy

1. Choice of Model

Animal models should be chosen with care before beginning an experiment. A common mistake is to try to study growth, modeling, or remodeling in a system in which the process under study isn't happening. If so, no valid conclusion can be drawn (it should be noted that none of these highly integrated processes occurs in tissue culture systems). The most frequent example of this mistake is to attempt the study of remodeling in animals which possess almost no remodeling activity. Rats, mice, hamsters, rabbits, chickens, etc. are inexpensive but lack usable amounts of BMU-based remodeling and so cannot provide valid models of the physiology, pharmacology, or pathology of human remodeling.

2. Sigma

The concept of sigma must be understood prior to beginning any experiment. It is the length of time from activation of a BMU until its completion. The components are sigma-r (σ-r), the resorption period, and sigma-f (σ-f), the formation period. Sigma can be thought of as the "life span" of the BMU. A disease or pharmacologic perturbation may change the length of sigma or change the frequency of activation of new BMU. In either case, a *transient* state will be induced which remains in effect until one new sigma has passed, usually from 3 to 6 months in the human, or much longer in patients with diseases such as osteoporosis when sigma may sometimes be prolonged to 1 or more years.

One can be misled by measurements made during a transient. For example, a perturbation which increases μ, without changing a preexisting negative bone tissue balance at the individual BMU, will lead to an increase in resorption surface over formation surface along with an increase in negative external calcium balance if examined during σ-r, yet just the opposite if examined during σ-f. A new *steady state* of overall negative balance and increased remodeling cannot be found until the system has passed one new sigma.

Transient states usually cannot cause disease *nor can they cure a diseased skeleton*. Therefore, one must examine the skeleton during steady states in order to predict the effect on any disease of a therapy; this requires knowing what the particular sigma value in question is.

The following recommendations, then, are suggested for studying the effects of an agent on BMU-based bone remodeling.

1. Use a test animal possessing such activity.
2. The animal must be observed during one sigma at baseline, one sigma of treatment, and one sigma of recovery which typically adds up to about 9 months.
3. One set of double labels must be given during each time period, each followed by a biopsy for a total of three biopsies.

3. Skeletal Envelopes

Since the four bone surface envelopes (endosteal, corticoendosteal, Haversian, and periosteal) can each exhibit different behavior, one must use methods which resolve envelope-specific activity when studying BMU-based remodeling. For example, in postmenopausal osteoporosis, the abnormality in remodeling is most severe in the corticoendosteal envelope, less so on the endosteal, minimal on the Haversian, and absent on the periosteal envelope. When studying this disease, or a potential treatment for it, the most important information will come from corticoendosteal remodeling so one should be able to focus on it to the exclusion of the other three. Of all the methods currently used for studying bone remodeling kinetics, only the double tetracycline labeling system permits isolation of envelope-specific activity.

4. Levels of Organization

Bone remodeling can be viewed at several different levels of activity: (1) cell-level, (2) BMU-level, (3) tissue-level, and (4) whole organ-level. When studying BMU-based remodeling one must realize before beginning which level of organization contains the information desired, and then choose a method which will resolve the level of interest from the others. For example, a skeleton may remain in mineral balance, but do so with vastly increased numbers of cells working very inefficiently. In this case, looking at the whole organ or tissue levels will not detect the abnormality. Perhaps the BMU level will also reveal nothing and the real remodeling defect will only be seen when bone cell work efficiency is resolved using the double tetracycline labeling system coupled with light microscopy and histomorphometric examination. A common error committed here is to look at static histologic features of bone, find increased numbers of remodeling cells or surfaces, and conclude that there are increased rates of remodeling. The opposite is often true but can be discerned only by coupling static and dynamic measurements.

5. Regional Acceleratory Phenomenon (RAP)

RAP is an increase in all metabolic activities in a part of the body (including remodeling activity in the skeleton) which has been attacked by any noxious stimulus[12] such as denervation, acute paralysis, arterial puncture, biopsy, fracture, crush injury, burn, local surgery, bone grafting, etc. Note that many of these stimuli are actually part of the procedure used in many skeletal experiments. Denervation of an extremity, for example, is a common model for osteoporosis. Results of denervation experiments are almost hopelessly confounded by the RAP which typically lasts for months following the procedure.

6. Static vs. Dynamic Measurements

Static parameters are such things as the amount of trabecular surface covered by osteoblasts, osteoclasts, osteoid, Howship's lacunae, and the like. Dynamic measurements are *rate* measurements. The principal rate measurement from which all others are calculated in bone remodeling is the appositional rate. The fundamental difference between a static and a dynamic (rate) measurement must be reemphasized. Static measurements have proven very unreliable as indicators of cell, BMU, whole-tissue, or whole-organ remodeling activity unless they are coupled with appropriate tissue-time marking measurements. Static measurements of bone histology alone are of marginal value in making diagnostic or therapeutic decisions. In fact, excepting for purposes of oncologic diagnosis, the discomfort and expense of a bone biopsy are probably not warranted without prelabeling with tetracycline and performing dynamic measurements. Combining measurements made with several methods can greatly augment the usefulness of each. For instance, the combination of tetracycline dynamics, static measurements, and microradiography on a biopsy coupled with radiogrammetry, photon absorptiometry, and radiocalcium kinetics can yield information at all levels of skeletal organization that is much more comprehensive than any measurement alone.

7. Tetracycline Administration

There are several recommendations regarding the administration of tetracycline that should be mentioned. It should be given to humans on an empty stomach, meaning nothing to eat for 1 hr before or 30 min after each of four divided doses, usually 250 mg in adults. Aluminum-containing antacids especially interfere with absorption and should be avoided. Outdated drug should be avoided because it will not accumulate and fluoresce normally (besides being toxic).

Storage of biopsy specimens in conventional fixatives will cause the labels to leach out. Ordinary fixatives such as formalin, osmic acid, Bouin's, Zenker's, picrate, gluteraldehyde, etc., or an unusual pH will destroy the labels. Absolute acetone or 70% alcohol definitely, or phosphate-buffered formalin possibly will preserve the labels if 5 or more days elapse between the last label and the biopsy.

Parenteral tetracycline at a dose of 15 mg/kg/day given by slow (5-min) i.v. infusion seems to yield more distinct labels than the oral route in dogs for obscure reasons. The infusion must be very slow in a volume of at least 100 mℓ in order to avoid severe, transient, and sometimes fatal lowering of ionized calcium by forming complexes with the tetracycline.

VI. SUMMARY

This chapter describes the choice of an optimum tissue-time marking and labeling schedule for measurement of bone remodeling rates. The definitions of ''bone label'' and ''bone tissue-time markers'' are clarified along with desirable properties of such agents. Several errors and pitfalls of measurement of bone remodeling rates are described along with recommendations for experimental design based on an accumulated experience with this technology. Other chapters will elaborate on two of the problems in the use of tetracycline tissue-time markers, the ''geometric projection'' phenomenon and the ''labeling escape effect''.

ACKNOWLEDGMENTS

The writer is grateful to Dr. R. R. Recker for obtaining professionally executed illustrations for this chapter and for his editorial assistance in making a presentable opus of the original manuscript.

REFERENCES

1. Arnold, J. S. and Jee, W. S. S., personal communication.
2. Heaney, R. P., Evaluation and interpretation of calcium kinetic data in man, *Clin. Orthoped.,* 31, 153, 1963.
3. Manson, J. D. and Waters, N. E., Observation on the rate of maturation of the osteon, *J. Anat.,* 99, 539, 1965.
4. Parfitt, A. M., Villanueva, A. R., Crouch, M. M., Mathews, C. H. E., and Duncan, H., Classification of osteoid seams by combined use of cell morphology and tetracycline labeling. Evidence for intermittency of mineralization, in *Proc. 2nd Int. Workshop, Bone Histomorphometry,* Meunier, P. J., Ed., Armour-Montagu, Paris, 1976.
5. Frost, H. M., Observations on osteoid seams: the existence of a resting state, *Henry Ford Hosp. Med. Bull.,* 8, 220, 1960.
6. Schwartz, M. P. and Recker, R. R., Comparison of surface density and volume of human iliac trabecular bone measured directly and by applied stereology, *Calcif. Tissue Int.,* in press, 1980.
7. Frost, H. M., Villanueva, A. R., Roth, H., and Stanisavljevic, S., Tetracycline bone labeling, *J. New Drugs,* p. 206, Oct. 1961.
8. Garn, S. M., *The Earlier Gain and Later Loss of Cortical Bone in Nutritional Perspective,* Charles C Thomas, Springfield, Ill., 1970.
9. Frost, H. M., *Bone Remodeling and Its Relationship to Metabolic Bone Diseases,* Charles C Thomas, Springfield, Ill., 1973.
10. Frost, H. M., *Bone Modeling and Skeletal Modeling Errors,* Charles C Thomas, Springfield, Ill., 1973.
11. Frost, H. M., *The Physiology of Bone, Fibrous and Cartilaginous Tissues,* Charles C Thomas, Springfield, Ill., 1972.
12. Frost, H. M., *Orthopaedic Biomechanics,* Charles C Thomas, Springfield, Ill., 1973.
13. Frost, H. M., Tetracycline based analysis of bone dynamics, *Calcif. Tissue Res.,* 3, 211, 1969.
14. Frost, H. M., *Clinical Orthopaedics and Related Research,* Vol. 49, 1969.
15. Frost, H. M., *Mathematical Elements of Lamellar Bone Remodeling,* Charles C Thomas, Springfield, Ill., 1964.
16. Frost, H. M., Bone histomorphometry: theoretical correction of appositional rate measurements in trabecular bone, in *Proc. Lyon Bone Morphometry Conference,* Meunier, P., Ed., University of Lyon Press, Lyon, 1977, 361.
17. Frost, H. M., Bone histomorphometry: a method of analysis of trabecular bone dynamics, in *Proc. Lyon Bone Morphometry Conference,* Meunier, P., Ed., University of Lyon Press, Lyon, 1977, 445.
18. Jaworski, A. F. G., Ed., *Bone Morphometry,* University of Ottawa Press, Ottawa, Canada, 1976, 1.
19. Meunier, P. Ed., *Proc. 2nd Bone Morphometry Conference,* University of Lyon Press, Lyon, 1977, 1.

Chapter 5

STEREOLOGIC BASIS OF BONE HISTOMORPHOMETRY; THEORY OF QUANTITATIVE MICROSCOPY AND RECONSTRUCTION OF THE THIRD DIMENSION

A. M. Parfitt

TABLE OF CONTENTS

I. INTRODUCTION

A slice of tissue on a slide is a three-dimensional object, but if cut thinly enough it approximates a true geometrical section, and when looked at through the microscope it produces a two-dimensional image. This image consists of profiles which are the projections of structures in three dimensions on to a plane. Stereology is the study of how these profiles in the two-dimensional image are related to the three-dimensional structure of the organ which was sampled.[1] Misunderstanding of this relationship abounds in the medical literature; an example is the commonly held but erroneous notion that the liver is made up of cords or cylinders rather than of plates or walls.[2] The three-dimensional structure of cortical bone has been deduced by studying serial sections[3] and by comparing sections of different orientation.[4] This approach is less successful for the liver because its structural elements are evenly dispersed and randomly oriented in space, properties which are summarized by the term "isotropic". Cortical bone, by contrast, is strongly anisotropic because of the longitudinal orientation of the osteons and Haversian canals. The three-dimensional structure of trabecular bone can be examined directly after removal of the marrow, either by stereoscopic light microscopy[5] or by scanning electron microscopy.[6] It is mainly a continuous system of curved plates of varying thickness enclosing communicating chambers of varying size and shape in which lies the marrow.[7] The plates are connected in places by bars or pillars (trabeculae or cancelli), but in normal bone there are none of the spicules (needles) wrongly deduced from histologic sections.[2] It seems inappropriate to describe a noncompressible structure as "spongy", so that the descriptive term "trabecular" will be retained despite its inaccuracy. Trabecular bone is also highly anisotropic, but in a more complex way than cortical bone, and the resulting stereologic problems are more difficult. There are important differences in three-dimensional structure between trabecular bone in different locations, such as rib, sternum, vertebral bodies, and ilium,[7] but only the latter is commonly used for biopsy.

Many treatments of stereology derive formulae which move directly from primary data obtained at the microscope to three-dimensional quantities, but it makes for clearer exposition if two separate operations are distinguished. The calculation of two-dimensional quantities such as area and perimeter from the primary data is the province of quantitative microscopy, whereas the extrapolation of these two-dimensional quantities to three-dimensional quantities such as volume and surface area is the province of stereology. The first operation is unaffected by anisotropy provided that randomization is adequate. The formulae derived can be applied without modification to any type of tissue, are easy to verify experimentally, and are subject to general agreement. By contrast, the second operation may be greatly affected by anisotropy. The formulae derived may require modification for particular cases, are often difficult to verify experimentally, and in some cases are subject to continuing dispute.

II. QUANTITATIVE MICROSCOPY

Before the problems of stereology proper can arise, measurements must be made on the profiles which exist in the two-dimensional world of the microscopic image. The profile of a tissue volume is an area in the section, and the profile of a tissue surface or more generally, of the interface between one type of structure and another, is a boundary or perimeter. Measurements of these quantities are most readily accomplished by randomly superimposing on the image a set of lines and a set of points known as a grid or graticule. This is either built into the eyepiece of the microscope, or drawn to a much larger scale on a flat surface on to which the microscopic image can be projected; the former method is more commonly used, although the latter has a number of advantages.[1] For quantitative scanning electron microscopy an electronically generated grid can be superimposed on the visual display.[8] Areas and perimeters can also be determined on a suitably enlarged photomicrograph using map measuring equipment,[9] but this is a cumbersome and inaccurate procedure.[1] The use of digitizing tablets and of automatic scanning methods is described in Chapter 12.

A. Grid Geometry

The original Zeiss[10] integrating eyepieces 1 and 2 had separate grids for lines and for points, but for the present purpose a grid is a two-dimensional system of lines and points with specifiable geometric properties symbolized as follows:

Total number of test points	P_T
Total length of test lines (mm)	L_T
Area associated with each point (mm²)	a
Total test area (mm²)	$A_T = P_T a$
Distance between points in object space (mm)	d

The grid must be calibrated for the magnification used by measuring the distance between points (d) with a stage micrometer.

Four representative types of grid will be illustrated (Figure 1 A to D), and the relationships among these quantities will be given for each type. The grid may be circumscribed by a circle, which facilitates multiple trials in the same field by random rotation of the eyepiece, or by a square, which facilitates examination of the entire section.

1. Rectangular Grid — Horizontal Lines Only

In Figure 1A the points are located at the crossings of the short vertical lines with the long horizontal lines. Grids of this type are exemplified by the Zeiss Integrationsplatte Series I to IV for which P_T is respectively 25 (illustrated) 100, 400, and 900. No. II of the series is most suitable for bone.

2. Rectangular Grid — Horizontal and Vertical Lines

In Figure 1B the points are located at the crossings of the lines. Alternatively, the Zeiss Integrationsplatte Series I to IV can be used in this way by mentally extrapolating between the short vertical lines.

3. Hexagonal Grid — Interrupted Horizontal Lines

In Figure 1C the points are located at the ends of the lines. This grid was designed by Weibel[11] for use mainly with lung and is supplied for use with Wild microscopes. Note that the distance between lines ($\sqrt{3}/2 \cdot$ d) is not the same as the distance between points (d). In the example shown $P_T = 42$.

4. Rectangular Grid — Alternating Hemispherical Lines

In Figure 1D the points are located at the crossings of the short horizontal lines with the inflections between hemispheres. This grid was designed by Merz[12] for use with anisotropic tissues like trabecular bone; it is also supplied for use with Wild microscopes but can be made by special order for any microscope. It has the advantage that perpendiculars to the test lines are distributed evenly in all directions. In the example shown $P_T = 36$.

B. Relationship Between Primary Data and Two-Dimensional Quantities

Four types of primary numerical data can be obtained: (1) the numbers of test points

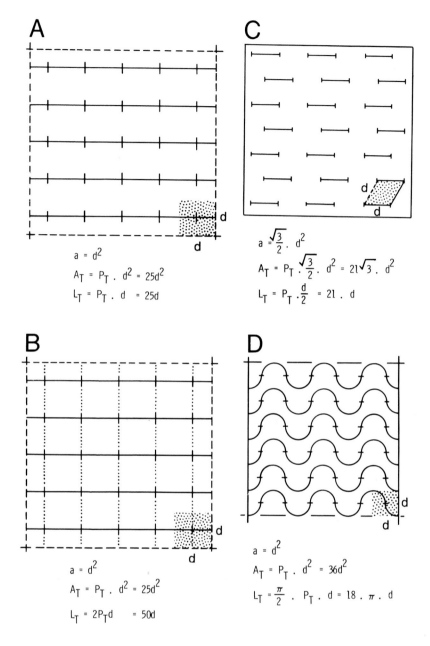

FIGURE 1.   Diagrammatic representation of four types of grid used for quantitative microscopy, showing relationships between test area and test line length and numbers and spacing of points for each type; (A) Zeiss Integrationsplatte II, (B) same, with interpolated vertical as well as horizontal lines, (C) Weibel grid, (D) Merz grid.

overlying different structures for determination of profile areas, (2) the numbers of intersections between test lines and different interfaces for determination of profile boundary or perimeter lengths, (3) directly measured distances between points or between or along lines, and (4) the number of profiles of discrete objects. Note that "intersection" is the approved stereologic term for the crossing of a test line with an interface profile. An "intercept" is a segment of a test line overlying a profile area in the section; this distinction is illustrated in Figure 2.

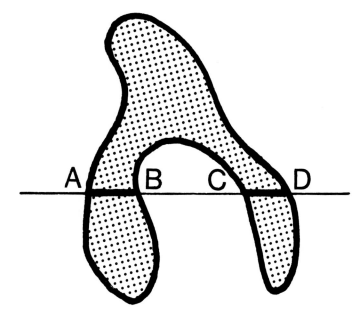

FIGURE 2. Diagram to show difference between intersection and intercept. The shaded area represents the structure to be examined with superimposed test line. Points A, B, C, and D are intersections with perimeter of the structure; line segments AB and CD are intercepts (or chords) overlying the structure; line segment BC is intercept overlying other material or tissue.

The relationships between the primary data and derived quantities which apply within a single field will be presented, and then extended for use with multiple fields. Working formulae will then be given for each of the four types of grid, for both relative and absolute measurements. The symbol $\stackrel{e}{=}$ devised by Martin[13] will be used in the sense "can be expected to equal exactly with a sufficient number of trials".

1. Profile Areas

Profile areas in the section are determined by numbers of superimposed test points, referred to as "hits" (Figure 3). If h_x = number of hits in a single field overlying structure x (or the mean of multiple trials in the same field) then the area occupied by structure x per unit test area ($A_{A(x)}$) is given by:

$$A_{A(x)} \ (mm^2/mm^2) \stackrel{e}{=} \frac{h_x}{P_T} \tag{1}$$

Theorem 1 is intuitively obvious. It is usually attributed to Glagoleff but was derived independently by several earlier investigators. Rigorous proofs are cited by Weibel[14] and Underwood[15] together with the history of its derivation and use. It is based on an earlier theorem by Rosiwal relating area to the length of linear intercepts (or chords) overlying the profile. This is little used because it is more time-consuming to measure lengths than to count points, but the method is more accurate for very small areas, and can be used for individual structures.[16] Various automated and semiautomated scanning methods for measuring area also make use of the Rosiwal theorem[17] (Chapter 12). Calibration of the grid is not needed for determination of relative areas, but is

The actual perimeter length (APL) of x in a single field is given by:

$$APL_x \text{ (mm)} \underset{=}{e} B_A \cdot A_T$$

With appropriate magnification, this formula can be used for measuring specific structures such as individual osteoid seam perimeters in cortical bone. The ratio of perimeter to area in the section (P/A) is given by:

$$P/A \text{ (mm/mm}^2\text{)} = B_A/A_A$$

This is an indication of the shape of the profiles in the section; for the same cross-sectional area, P/A is least for circles and most for thin structures or those with extremely convoluted perimeters.

For the derivation of working formulae, L_T must be expressed as a function of P_T and d for the particular grid (Figure 1). If n fields are examined, the total intersections in all fields on the perimeter of x ($i_{x(t)}$) is given by:

$$i_{x(t)} = i_{x(1)} + i_{x(2)} + \cdots + i_{x(n)}$$

The mean value of $i_x = i_{x(t)}/n$ so that:

$$\frac{i_x}{P_T} = \frac{i_{x(t)}}{nP_T} = \frac{i_{x(t)}}{\Sigma_h}$$

The working formulae for each of the grids are given in Table 1 in terms of the total number of intersections counted in all fields; the total for all subdivisions of the perimeter = Σ_i.

The derivations from Theorem 2 in Table 1 enable it to be verified experimentally. This was recently done once more by A. R. Villanueva using Grid B. He prepared five small pieces of fine thread of known length between 1.6 and 10.0 cm and mounted them on slides in random orientation. Counting intersections along only one edge of the magnified image of the thread, the lengths were determined using the formula $APL_{(T)} \underset{=}{e} \Sigma_i \cdot \pi/4 \cdot d$, and the measured lengths were within 0.5% of the true length in each case.

In contrast to area measurements, boundary profile measurements in bone are strongly influenced by the magnification. Olah found that B_A for trabecular bone increased nearly 30% when the magnification was increased from 25 × to 400 × because of increasing resolution of small irregularities of the surface.[18] This would not necessarily affect fractional perimeter measurements but standardization of magnification is essential for any absolute perimeter measurements.

3. Distances

The distances of most importance in bone are the width of osteoid seams, the distance between fluorescent labels, and the wall thickness of completed cortical or trabecular osteons, referred to as bone structural units (BSU).[19] These distances are all usually measured perpendicular to the bone surface, which can be done either directly or indirectly.

Table 1
WORKING FORMULAE FOR PERIMETERS[a]

Grid	A	B	C	D
$B_{A(x)}$	$\dfrac{i_{x(t)}}{\Sigma_h} \cdot \dfrac{\pi}{2d}$	$\dfrac{i_{x(t)}}{\Sigma_h} \cdot \dfrac{\pi}{4d}$	$\dfrac{i_{x(t)}}{\Sigma_h} \cdot \dfrac{\pi}{d}$	$\dfrac{i_{x(t)}}{\Sigma_h} \cdot \dfrac{1}{d}$
$B_{A(T)}$	$\dfrac{\Sigma_i}{\Sigma_h} \cdot \dfrac{\pi}{2d}$	$\dfrac{\Sigma_i}{\Sigma_h} \cdot \dfrac{\pi}{4d}$	$\dfrac{\Sigma_i}{\Sigma_h} \cdot \dfrac{\pi}{d}$	$\dfrac{\Sigma_i}{\Sigma_h} \cdot \dfrac{1}{d}$
$APL_{(x)}$	$i_{x(t)} \cdot \pi/2 \cdot d$	$i_{x(t)} \cdot \pi/4 \cdot d$	$\sqrt{3}/2 \cdot \pi \cdot d$ or $\sqrt{3} \cdot \pi/2{:}d$	$i_{x(t)} \cdot d$
$APL_{(T)}$	$\Sigma_i \cdot \pi/2 \cdot d$	$\Sigma_i \cdot \pi/4 \cdot d$	$\sqrt{3}/2 \cdot \pi \cdot d$ or $\sqrt{3} \cdot \pi/2{:}d$	$\Sigma_i \cdot d$
P/A	$\dfrac{\Sigma_i}{h_{x(t)}} \cdot \dfrac{\pi}{2d}$	$\dfrac{\Sigma_i}{h_{x(t)}} \cdot \dfrac{\pi}{4d}$	$\dfrac{\Sigma_i}{h_{x(t)}} \cdot \dfrac{\pi}{d}$	$\dfrac{\Sigma_i}{h_{x(t)}} \cdot \dfrac{1}{d}$

[a] Working formulae for perimeter length per unit test area (B_A), actual perimeter length (APL), and perimeter/area ratio (P/A) for Grids A through D shown in Figure 1. Subscript notation: x = one subdivision of the perimeter; T = total perimeter; Σ_h = total hits on all structures in all fields; Σ_i = total number of intersections on all surfaces in all fields; $h_{x(t)}$ = total hits on structure for which P/A is determined.

a. Direct Measurement

The distance between points or the mean distance between parallel lines can be measured directly using an eyepiece micrometer. The usual kind has a built-in scale of 100 divisions, but for more accurate work a filar micrometer can be used; this has 10 divisions and a movable hairline, the position of which is determined by a rotating Vernier scale. Both types of eyepiece micrometers must have previously been calibrated using a stage micrometer. Alternatively, distances can be determined directly with a two-dimensional scanning stage which enables the relative movement of the section with respect to an eyepiece crossline to be measured.[20] A similar principle is used by various automated scanning devices.

The main difficulty with the direct methods is in the choice of sites for measurement. In cortical bone both osteoid seam width and interlabel distance are usually measured at four equally spaced intervals in each focus of bone formation. During radial closure the perimeters of both the seam and of the label get progressively smaller, the osteoid seam gets thinner, and the labels get closer together. Consequently, if each focus is given the same weight, small seams are overrepresented and the true mean seam thickness and mean interlabel distance are systematically underestimated. This problem does not arise in the measurement of mean wall thickness of BSU. To eliminate this bias, measurement sites must be selected at random, which is most easily done by making all measurements at grid line intersections. In trabecular bone the individual bone-forming sites are less clearly defined and the need for randomizing the sites of measurement is more evident.

b. Indirect Measurement

For profiles whose width is small in relation to length, the approximate mean width

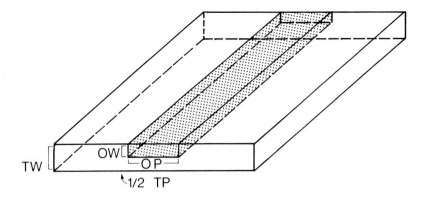

FIGURE 4.   Relationship of profile width to perimeter and area measurements. Diagram of a thin plate representing a single element of trabecular bone with an inlay on the surface representing an osteoid seam, and their profiles in two dimensions. If the width of the trabecula (TW) or of the osteoid seam (OW) is small in relation to the perimeter length of the trabecula (TP) or osteoid seam (OP), then the mean trabecular width will be approximately equal to profile area divided by half perimeter, and the mean osteoid seam width approximately equal to profile area divided by perimeter. Similar relationships will exist in three dimensions between mean thickness and volume and surface area values.

can be determined indirectly from perimeter and area measurements, which automatically eliminates the problem of randomizing the sites of measurement. When the perimeter is measured on only one side, as in osteoid seams, area = perimeter · width, so that width = area/perimeter. When the perimeter is measured on both sides, as in whole trabeculae, area = 1/2 perimeter · width, so that width = 2 · area/perimeter (Figure 4). Mean profile width (MPW) is given by:

$$\text{MPW (mm)} = \frac{2A_A(\text{mm}^2/\text{mm}^2)}{B_A(\text{mm}/\text{mm}^2)}$$

Calculated distance measurements decrease with increasing magnification in inverse proportion to the increase in perimeter measurement. Working formulae for MPW for each of the four types of grid are shown in Table 2. If $h_{x(t)}$ = total hits on both mineralized bone and osteoid in trabecular bone (h_{TB}; Table 2), these working formulae give mean trabecular profile width (MTW), usually expressed in μm. MTW estimated in this way correlated well with and did not differ significantly from MTW measured directly on scanning electron micrographs.[21] In addition, values for MTW measured directly in iliac bone samples were 140 μm using the Leiz scanning stage[20] and 139 μm using automatic computerized scanning of microradiographs.[22] These values agree reasonably well with a mean of 148 μm determined indirectly from area and perimeter measurements.[22] If necessary, the direct (D) and indirect (I) measurements can be identified by appropriate subscripts.

The formula for indirect determination of mean profile width is a particular instance of the more general problem of determining mean distances by linear scanning, and three principal cases must be distinguished. When scanning is performed in any one direction, the total lengths of the intercepts on different structures are proportional to their areas — the theorem of Rosiwal already mentioned; the mean intercept length (L) is equal to the area divided by the projected length perpendicular to the direction

Table 2
WORKING FORMULAE FOR INDIRECT DISTANCE DETERMINATIONS[a]

Grid	A	B	C	D
MPW	$\dfrac{h_{x(t)}}{i_{x(t)}} \cdot \dfrac{4d}{\pi}$	$\dfrac{h_{x(t)}}{i_{x(t)}} \cdot \dfrac{8d}{\pi}$	$\dfrac{h_{x(t)}}{i_{x(t)}} \cdot \dfrac{2d}{\pi}$	$\dfrac{h_{x(t)}}{i_{x(t)}} \cdot 2d$
MTW	$\dfrac{h_{TB}}{\Sigma_i} \cdot \dfrac{4d}{\pi}$	$\dfrac{h_{TB}}{\Sigma_i} \cdot \dfrac{8d}{\pi}$	$\dfrac{h_{TB}}{\Sigma_i} \cdot \dfrac{2d}{\pi}$	$\dfrac{h_{TB}}{\Sigma_i} \cdot 2d$
MMCL	$\dfrac{h_v}{\Sigma_i} \cdot 2d$	$\dfrac{h_v}{\Sigma_i} \cdot 4d$	$\dfrac{h_v}{\Sigma_i} \cdot d$	$\dfrac{h_v}{\Sigma_i} \cdot \pi d$
MOSW	$\dfrac{h_o}{i_o} \cdot \dfrac{2d}{\pi}$	$\dfrac{h_o}{i_o} \cdot \dfrac{4d}{\pi}$	$\dfrac{h_o}{i_o} \cdot \dfrac{d}{\pi}$	$\dfrac{h_o}{i_o} \cdot d$
MILW	$\dfrac{h_{IL}}{i_{DL}} \cdot \dfrac{2d}{\pi}$	$\dfrac{h_{IL}}{i_{DL}} \cdot \dfrac{4d}{\pi}$	$\dfrac{h_{IL}}{i_{DL}} \cdot \dfrac{d}{\pi}$	$\dfrac{h_{IL}}{i_{DL}} \cdot d$

[a] Working formulae for indirect determinations of mean profile width (MPW), mean trabecular profile width (MTW), mean marrow chord length (MMCL) or mean intertrabecular distance, mean osteoid seam profile width (MOSW), and mean interlabel width (MILW). Subscript notation: $h_{x(t)}$ = total hits and $i_{x(t)}$ = total intersections on structure for which MPW is determined; h_{TB} = total hits on both mineralized bone and osteoid; h_v = total hits on marrow (void); h_o = total hits on osteoid; i_o = total intersections on osteoid perimeter; h_{IL} = total hits on bone between midpoints of double labels; i_{DL} = total intersections on double labeled perimeter.

of scanning (Figure 5). When scanning is carried out in a direction perpendicular to the principal axis of orientation, the projected length is at a maximum and the mean intercept length (L) is equivalent to the mean length of all possible chords without regard to the direction to the measurement. When applied to bone (Figure 6), MTW is equivalent to the mean intercept length over bone (L_b) when scanning of each segment is carried out in a direction perpendicular to the principal axis of orientation.[21] If L_b is averaged for all directions of scanning ($\overline{L}_b$), the following relationship holds:[21]

$$\overline{L}_b \quad \pi \cdot \frac{A_{A(TB)}}{B_{A(TB)}} = \frac{h_{TB}}{\Sigma_i} \cdot 2d$$

from which it follows that:

$$MTW = \overline{L}_b \cdot 2/\pi$$

Similarly, the mean intercept length over marrow averaged for all directions of scanning ($\overline{L}_m$) gives the mean marrow chord length (MMCL). This is equivalent to the mean intertrabecular distance and is given by:

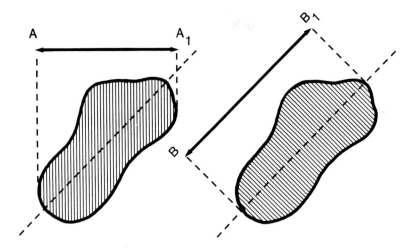

FIGURE 5.   Linear scanning; relationships between area, mean intercept length and projected length. On left, the closed figure is scanned at an acute angle to the major axis (bold interrupted line), with projected length AA$_1$ perpendicular to scanning direction. On right, figure is scanned perpendicular to major axis with projected length BB$_1$. Area of figure is proportional to the total intercept length (Rosiwal Theorem; see Figure 2 for definition of intercept); the proportionality constant is the interval between the scan lines. For a given direction of scanning, mean intercept length (L) = area/projected length. For perpendicular scanning (as on right) L = mean profile width (MPW). L averaged for all directions of scanning (L) = mean chord length. It can be shown that MPW = $2/\pi \cdot \bar{L}$.

$$\text{MMCL} = \bar{L}_m = \frac{\pi\left(1 - A_{A(TB)}\right)}{B_{A(TB)}} = \frac{\pi}{2} \cdot \frac{\text{MTW}}{A_{A(TB)}} \left(1 - A_{A(TB)}\right)$$

With representative values for $A_{A(TB)}$ of 0.2 and for MTW of 150 μm, the mean intertrabecular distance calculated from this formula is 940 μm. Working formulae for MMCL, for each of the four grids (Figure 1) are given in Table 2.

Mean osteoid seam width (MOSW) can be determined indirectly by the same principle as for MTW.[24] Since the osteoid perimeter is determined from intersections with the osteoid:marrow interface but not the osteoid:bone interface, the appropriate expression is A_A/B_A, not $2A_A/B_A$. Corresponding working formulae are given in Table 2. This method has not been compared with direct measurements on the same subjects, but mean values[24] ranged between 11 μm at age 20 to 40 and 6.7 μm at age 70 to 80, which are comparable to those reported for direct measurements.[25] Meunier[26] calculates the osteoid thickness index (OTI) as the ratio of fractional osteoid area (h$_o$/h$_{TB}$) to fractional osteoid perimeter (i$_o$/Σ_i) multiplied by 100. The working formulae for OTI for all four grids is (h$_o$/i$_o$)·(Σ_i/h$_{TB}$)·100. This is equivalent to 2MOSW/MTW·100 and so expresses mean osteoid seam width as a percentage of mean trabecular half width. When patients with osteomalacia are included, OTI correlates well (r > 0.95) with directly measured seam width in the same subjects, but within the normal range the correlation between direct and indirect measurements of MOSW is poor.[59] This indicates that the indirect method is insensitive and can only detect large deviations from normal.

Although possible in principle, the indirect method is less suitable for determining the mean distance between tetracycline labels. It is normal practice to count intersections with the double-labeled perimeter, but to count hits overlying the area between

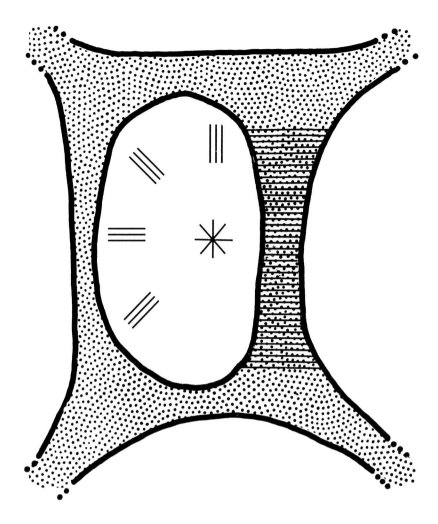

FIGURE 6. Application of linear scanning to trabecular bone. Interconnected trabeculae (shaded) enclose a marrow space (unshaded). Mean trabecular width is given by mean bone profile width, scanning perpendicular to major axis. Mean intertrabecular distance is given by mean marrow profile width averaged for all directions of scanning.

the midpoints of the labels would be very time-consuming since the boundaries are not visually demarcated. Nevertheless, for some purposes it may be useful to compare the direct and indirect methods and working formulae given in Table 2.

4. Numbers of Profiles

Numbers of profiles can be counted directly and referred either to the total area of a grid or the area of some particular tissue (for example the number of osteoid seams or number of osteocyte lacunae per unit profile area of cortical bone or the number of osteoclasts per unit profile area of marrow space). The number of profiles may also be related to a perimeter length (for example the number of osteoblasts per unit of length of osteoid seam perimeter, or the number of osteoclasts per unit length of trabecular perimeter). If the entities being counted are large in relation to the area of the grid, there may be a significant edge effect.[27] The usual convention, which is similar to that used in cell counting in hematology, is to ignore profiles intersected by the lower and left-hand border and to include profiles intersected by the upper and right-

For determining the precision of area and perimeter measurements the number of hits (h_x) or intersections (i_x) on one type of structure can be regarded as the number of successes (n) in a series of N binomial trials constituting a sample.[34] The proportion of successes (p = n/N) in such a series expressed as a decimal fraction is a random variable for which are obtained particular values $\hat{p}$ ($= \hat{n}/\hat{N} = h_x/\Sigma_h$ or i_x/Σ_i). If $\hat{N}$ (Σ_h or Σ_i) is large, the theoretical distribution of p is approximately Gaussian; this has been confirmed experimentally for grid point counting methods.[35,36] The different points of a regular array cannot provide a completely random sample[36] but the method effectively divides the area into squares within which the tissue components can be randomly distributed.[37] Some investigators have used random arrays of points[35] but this precludes the measurement of absolute area or of measuring perimeters at the same time. With one exception[38] regular arrays have been found to give as good precision as random arrays or better[1,15] provided that the grid is not in register with any regularity in the structure examined.[37] Another disadvantage of random arrays is that it is much more difficult to be sure that all points have been scanned.[39] A regular grid also makes it easier to determine tissue components whose proportion is small since the field can be scanned more quickly for hits.

The standard deviation of the number of successes in a series of binomial trials[34] is given by: $SD_n = \sqrt{Np(1-p)}$. Since n = Np and N is a constant for a given series, the standard deviation of the proportion of successes is given by:

$$SD_p = \sqrt{\frac{p(1-p)}{N}} \qquad \text{(a)}$$

The relative standard deviation (SD as a percentage of p) is given by:

$$RSD_p = \sqrt{\frac{(1-p)}{p \cdot N}} \cdot 100 \qquad \text{(b)}$$

An estimate of the SD can be obtained from a particular sample (series of trials). This may be termed the standard error (SE) of the proportion, by analogy with the standard error of the mean of a sample, which is an estimate of the standard deviation of the means of successive samples of a particular size. Taking the sample proportion (p) as an estimate of the true proportion, this is given by:

$$SE_{\hat{p}} = \sqrt{\frac{\hat{p}(1-\hat{p})}{\hat{N}}} \qquad \text{(c)}$$

The estimate of relative standard deviation is given by:

$$RSE_{\hat{p}} = \sqrt{\frac{(1-\hat{p})}{\hat{p} \cdot \hat{N}}} \cdot 100 \qquad \text{(d)}$$

A graphical representation of Equation c is given in Figure 8A; graphical representation of Equation d is easiest if $p \cdot N$ is replaced by n[39] (Figure 8B). Equations c and d can be rewritten in terms of numbers of trials and successes,[37] replacing $\hat{p}$ by $\hat{n}/\hat{N}$. If $\hat{n}$ is replaced by h_x or i_x and $\hat{N}$ is replaced by Σ_h or Σ_i we have:

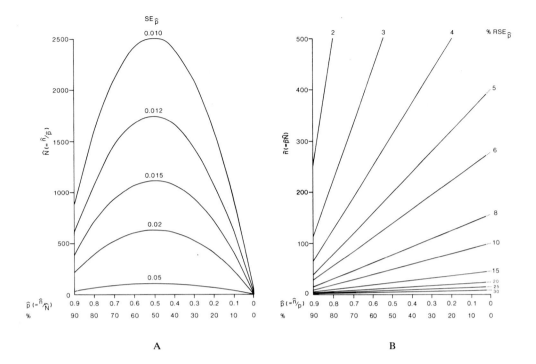

FIGURE 8A. Minimum values for $\hat{N}$ ($= \hat{n}/\hat{p}$) required to obtain particular values of $SE_{\hat{p}}$ at different values of $\hat{p}$ ($= \hat{n}/\hat{N}$). Although shown as continuous, both $\hat{N}$ and $\hat{n}$ are discrete variables, with limiting rather than actual values at $\hat{p} = 1.0$ or 0. $\hat{N} =$ total number of hits or intersections; $\hat{n} =$ number of "successes"; $\hat{p} =$ proportion of "successes". (B) Minimum values for $\hat{n}$ ($= \hat{p}\hat{N}$) required to attain particular values of RSE_p at different values of $\hat{p}$ ($= \hat{n}/\hat{N}$), based on Gaussian approximately to a binomial distribution.

$$SE_{\hat{p}} = \sqrt{\frac{h_x/\Sigma_h(1-h_x/\Sigma_h)}{\Sigma_h}} \quad \text{or} \quad \sqrt{\frac{i_x/\Sigma_i(1-i_x/\Sigma_i)}{\Sigma_i}}$$

$$RSE_{\hat{p}} = \sqrt{\frac{(1-h_x/\Sigma_h)}{h_x} \cdot 100} \quad \text{or} \quad \sqrt{\frac{(1-i_x/\Sigma_i)}{i_x} \cdot 100}$$

$$= \sqrt{1/h_x - 1/\Sigma_h \cdot 100} \quad \text{or} \quad \sqrt{1/i_x - 1/\Sigma_i \cdot 100}$$

When h_x is small in relation to Σ_h, or i_x small in relation to Σ_i these can be approximated by $\sqrt{1/h_x} \cdot 100$ or $\sqrt{1/i_x} \cdot 100$.

The equations for SE or RSE can be used to calculate confidence intervals for P. The 95% confidence limits are given by $\hat{P} \pm t_{0.025}SE_p$ and the 99% confidence limits by $\hat{p} \pm t_{0.005}SE_p$; if $\hat{n}(h_x$ or $i_x)$ is >100, the limiting values for t of 1.96 and 2.58 can be used. For example if $\Sigma_h = 5000$, h_x for total mineralized bone (h_{TB}) $= 1000$, and h_x for osteoid (h_o) $= 30$, then the estimate of mineralized bone as a proportion of total marrow area is 0.2. The RSE for this estimate $= 100 \sqrt{1/1000 - 1/5000} = 2.83\%$, so that the 95% confidence limits for the proportion are 0.189 to 0.211 and the 99% confidence limits are 0.185 to 0.215. The estimate of osteoid as a proportion of miner-alized bone area is 0.03; the RSE of this estimate $= 100 \sqrt{1/30 - 1/1000} = 18\%$, so that the 95% confidence limits for the proportion are 0.019 to 0.041 and the 99% confidence limits are 0.0162 to 0.0438.

Equations c or d can be inverted so that the minimum numbers of trials and successes required to attain a specified degree of precision can be calculated. This is given by:

$$N = \frac{10,000(1-\hat{p})}{(RSE_{\hat{p}})^2 \cdot \hat{p}} \quad \text{and} \quad n = \frac{10,000}{RSE_{\hat{p}}^2} \cdot (1-\hat{p})$$

For example, to estimate osteoid area as a proportion of mineralized bone area with an $RSE_{\hat{p}}$ of 10%, if the estimated proportion is 0.03 then minimum values for total hits (Σ_h) and hits on osteoid (h_o) are

$$\Sigma_h = \frac{10,000(0.97)}{100(0.03)} = 3233$$

$$h_o = \frac{10,000(0.97)}{100} = 97$$

Values for h_x and i_x corresponding to different values for $\hat{p}$ and $RSE_{\hat{p}}$ are also shown graphically in Figure 8B. The limiting values of n for p = 0 are equal to $1/RSE_{\hat{p}}^2$.

For distance measurements, precision depends both on the interval between microm-eter-scale divisions and on the magnification. With a 100-division scale the value of one division is approximately 2.5 to 5 μm at total magnifications in the range of 250 × to 400 ×. The precision of a single measurement will normally be equal to a single division, but visual interpolation to the nearest half division may be possible. For the mean of several measurements at different locations, if the range of values is large in relation to one division, the number of divisions can be treated as a continuous variable and the standard error of the mean calculated in the usual way.[34] If the range of values includes less than five divisions, the standard error can be estimated by fitting the observed values to a binomial distribution, but it is more satisfactory to use a filar micrometer, particularly if there are many values of less than three divisions as with thin osteoid seams and low appositional rates.

If the number of profiles per unit cross-sectional area is randomly distributed it will conform to a Poisson distribution, for which the variance is equal to the mean. If N profiles are counted in a standard unit area, the standard deviation for repeated counts in the same area will be $\sqrt{N}$, and the RSE will be $1/\sqrt{N}$. This assumes that the error in measuring the area in which the counts are made can be disregarded. The formula $1/\sqrt{N}$ is also the simplist estimate of RSE for nonrandomly distributed profiles such as osteocyte lacunae.

III. RECONSTRUCTION OF THE THIRD DIMENSION

The extrapolation of two-dimensional quantities measured in a section to three-di-mensional quantities existing in the sample is inevitably subject to error and uncer-tainty so that the need for it may legitimately be questioned. For determining whether diagnostic criteria are satisfied in individual patients or for characterizing particular disease states by studying groups of patients, the measured two-dimensional quantities would serve just as well. It makes no difference whether the extrapolation is carried out or not, or even if the correct method is used, so long as all data are treated alike. However, for proper understanding of bone remodeling in health and disease, solution of the stereologic problem is essential. If bone cell function at different sites is to be compared, or if bone formation rate determined by double tetracycline labeling is to be compared with measurement of bone turnover by radiocalcium kinetics, histomorphometric measurements must be expressed in three-dimensional terms. Furthermore, the interactions between biomechanics and bone turnover are critically dependent on the relationship between porosity and surface area in cortical and trabecular bone in different locations, at different ages, and in different diseases.

A. Volume of a Structure Determined From an Area in the Section

If $V_{V(x)}$ is the volume density or fractional volume of structure x (mm³/mm³) then:

$$V_{V(x)} = A_{A(x)} \tag{3}$$

Theorem 3 was first proposed by DeLesse for studying the composition of rocks. Many have expressed doubts about its validity but several rigorous proofs have been presented.[40,41] It is unaffected by the three-dimensional geometry or orientation of the structure examined. The only restriction on the general validity of the DeLesse theorem occurs when sampling must be biased by the need for unequivocal identification, as (for example) when the nuclear volume fraction of particular cell types is being determined.[42] Combining Theorems 1 and 3 we have:

$$V_{V(x)} \text{ (mm}^3\text{/mm}^3) \; \underset{=}{e} \; \frac{h_{x(t)}}{\Sigma_h}$$

The validity of Theorem 3 and the accuracy of the point count method for trabecular bone have been confirmed experimentally by comparing fractional volume measured directly using Archimedes' principle with the results of quantitative microscopy on the same bone samples.[43] The methods gave virtually identical results with a correlation coefficient of 0.980 and a mean difference of less than 6%.

B. Surface Area of a Structure Determined From a Perimeter in the Section

The area of a surface per unit volume of tissue is known as specific surface or surface density, symbolized by S_V. The relationship of this to perimeter is given by:

$$S_V = B_A \cdot k_{PS}$$

where k is a constant of dimensional addition (perimeter to surface), which depends on the three-dimensional orientation. For a structure which is oriented perpendicular to the plane of section, such as osteons in cortical bone, the value of k_{PS} is 1, so that $S_V = B_A$.[12,40,44] For a perfectly isotropic structure (one in which a perpendicular erected on the plane tangential to any element of area of the surface has an equal probability of lying in any equal element of solid angle) the value of k_{PS} is $4/\pi$[12,14,15,37] so that $S_V = B_A \cdot 4/\pi$. Combining this with Theorem 2 we have:

$$S_V \; \underset{=}{e} \; 2 \cdot \frac{i_x}{L_T} \tag{4}$$

Note that π in the expression for k_{PS} has cancelled out the π in the expression for B_A. Rigorous proofs of Theorem 4 are offered by several authors[1,14,15,45] and many other derivations are cited. Some stereologists have placed restrictions on the shape of the structure examined but a completely general proof has been given by Martin.[40] All proofs relating S_v directly to the primary data rest on the assumption of isotropy, and the more general proofs require more than elementary mathematics. A proof based on simple geometry for the specific example of the sphere is given in Appendix 1.

Surface area can be expressed per unit volume of structure rather than per unit volume of tissue; this is known as the surface:volume ratio (S/V), which is the three-dimensional counterpart of the perimeter:area ratio. It is given by:

$$S/V = \frac{S_v}{V_v} = k_{PS} \cdot P/A$$

For isotropic structures (for which Theorem 4 applies) it is given by:

$$S/V \underset{=}{e} \frac{i_x}{h_x} \cdot \frac{P_T}{L_T} \cdot 2$$

Because extension in the third dimension cannot be defined, there is no three-dimensional counterpart to absolute perimeter length. The working formulae for S_v and S/V for each grid based on the total number of hits and intersections in all fields are given in Table 3.

These derivations of Theorem 4 enable its truth to be verified by experiment. Three-dimensional bodies of known shapes and dimensions were carved from balsa wood and embedded randomly in paraffin.[46] The block was cut into parallel sections from which counts of intersections and hits were accumulated using a single needle of known length dropped 2000 times. The formula used for calculating the results was shown by Weibel[14] to be equivalent to Theorem 4, and the S/V ratio so calculated from the experimental data agreed closely with those calculated from the known dimensions of the three-dimensional objects.[46]

For trabecular bone in general the degree of anisotropy can be described by approximating the curvature of the surface to a prolate or an oblate spheroid.[44] These are produced by rotation of an ellipse about its long and short axes respectively, and the value of k_{PS} is a function of the ratio of the hemiaxes, of the ellipse generating the spheroid.[44] This approach has been applied to the lumbar vertebra and to the sternum. In general k_{PS} is larger rather than smaller than $4/\pi$, but will only rarely exceed $4/\pi$ by more than 25 to 30%.[21,44] The three-dimensional structure of the trabecular bone in the region of the anterior superior iliac spine (including the area where transilial biopsies are normally taken) is much more complex than in the sternum,[47] and trabecular orientation could not be inferred with any accuracy from two-dimensional sections.[48] Consequently, extrapolation from perimeter to surface area is fraught with even more uncertainty than in other bones.

The methods of measuring anisotropy devised by Whitehouse[21,44] cannot be routinely applied to biopsy specimens. The bone structure of the ilium shows preferred orientation both at the gross macroscopic level[49] and at the microscopic level.[48] If the section plane through the biopsy core is randomized with respect to rotation about the long axis, the effect of this orientation and consequent anisotropy may be removed in a large group of subjects, but the effect in a single subject will remain. Furthermore, most biopsy cores are taken somewhat oblique to the true plane of the ilium so that in vivo orientation is still discernible. Consequently, complete randomness of the section plane may be more difficult to achieve than has been assumed.

Fortunately, the problem may not be as serious in practice as these theoretical uncertainties suggest, since experimental determination of k_{PS} by Schwartz and Recker[43]

Table 3
WORKING FORMULAE FOR DERIVED THREE-DIMENSIONAL QUANTITIES[a]

Grid	A	B	C	D
$S_{v(x)}$	$\dfrac{i_{x(t)}}{\Sigma_h} \cdot \dfrac{2}{d}$	$\dfrac{i_{x(t)}}{\Sigma_h} \cdot \dfrac{1}{d}$	$\dfrac{i_{x(t)}}{\Sigma_h} \cdot \dfrac{4}{d}$	$\dfrac{i_{x(t)}}{\Sigma_h} \cdot \dfrac{4}{\pi d}$
$S_{v(T)}$	$\dfrac{\Sigma_i}{\Sigma_h} \cdot \dfrac{2}{d}$	$\dfrac{\Sigma_i}{\Sigma_h} \cdot \dfrac{1}{d}$	$\dfrac{\Sigma_i}{\Sigma_h} \cdot \dfrac{4}{d}$	$\dfrac{\Sigma_i}{\Sigma_h} \cdot \dfrac{4}{\pi d}$
S/V	$\dfrac{\Sigma_i}{h_{x(t)}} \cdot \dfrac{2}{d}$	$\dfrac{\Sigma_i}{h_{x(t)}} \cdot \dfrac{1}{d}$	$\dfrac{\Sigma_i}{h_{x(t)}} \cdot \dfrac{4}{d}$	$\dfrac{\Sigma_i}{h_{x(t)}} \cdot \dfrac{4}{\pi d}$
MST	$\dfrac{h_{x(t)}}{i_{x(t)}} \cdot d$	$\dfrac{h_{x(t)}}{i_{x(t)}} \cdot 2d$	$\dfrac{h_{x(t)}}{i_{x(t)}} \cdot \dfrac{d}{2}$	$\dfrac{h_{x(t)}}{i_{x(t)}} \cdot \dfrac{\pi d}{2}$
MTT	$\dfrac{h_{TB}}{\Sigma_i} \cdot d$	$\dfrac{h_{TB}}{\Sigma_i} \cdot 2d$	$\dfrac{h_{TB}}{\Sigma_i} \cdot d$	$\dfrac{h_{TB}}{\Sigma_i} \cdot \dfrac{\pi}{2} \cdot d$
MOST	$\dfrac{h_o}{i_o} \cdot \dfrac{d}{2}$	$\dfrac{h_o}{i_o} \cdot d$	$\dfrac{h_o}{i_o} \cdot \dfrac{d}{4}$	$\dfrac{h_o}{i_o} \cdot \dfrac{\pi}{4} \cdot d$
MILT	$\dfrac{h_{IL}}{i_{DL}} \cdot \dfrac{d}{2}$	$\dfrac{h_{IL}}{i_{DL}} \cdot d$	$\dfrac{h_{IL}}{i_{DL}} \cdot \dfrac{d}{4}$	$\dfrac{h_{IL}}{i_{DL}} \cdot \dfrac{\pi}{4} \cdot d$

[a] Working formulae for surface density (S_v), surface:volume ratio (S/V), and indirect determination of mean structural thickness (MST), mean trabecular thickness (MTT), mean osteoid seam thickness (MOST), and mean interlabel thickness (MILT) for Grids A through D shown in Figure 1. Mean marrow cavity thickness or mean intertrabecular distance is the same as mean marrow cavity width. Subscript notations: X = one subdivision of surface; T = total surface. Other symbols as in Tables 1 and 2.

for trabecular bone in the ilium gave a value very close to $4/\pi$. Specimens were taken at autopsy, and thoroughly cleaned and weighed. Each specimen was sealed and dipped in fast green stain, dried, and reweighed. The difference in weight was converted to a surface area using a relationship determined with stain applied to paper and corrected for differences in stain thickness between bone and paper. The values for surface density and surface:volume ratio for trabecular bone measured in this way were almost identical with those obtained from histomorphometric analysis using Grid D, based on Theorem 4. The correction factor determined empirically was 1.199 compared to the theoretical value of 1.273. This ingenious experiment suggests that the value of $4/\pi$ should continue to be used for the dimensional addition factor k_{PS} for human trabecular bone.

C. Distances Between Surfaces or Planes Determined From Distances Between Perimeters or Lines in the Section

In accordance with the principle of dimensional reduction[1,2] perimeters or lines in

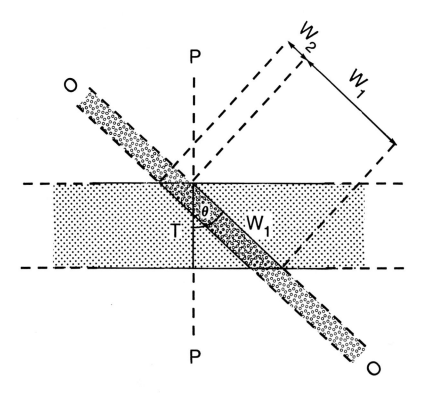

FIGURE 9.   Effect of section obliquity on profile width or apparent thickness — flat surface; the horizontal solid lines represent parallel tissue planes, either surfaces or interfaces, and the oblique solid lines with enclosed shading represent a histologic section of finite thickness. Varying obliquity of the section plane is equivalent to rotation about an axis perpendicular to the figure; θ = angle between ideal perpendicular section plane (P − P) and real oblique section plane (O − O). Such planes produce linear profiles in the sections separated by varying distances. For an infinitely thin histologic section, the relationship between true thickness (T) and apparent thickness or profile width (W_1) is given by: $T = W_1 \cos\theta$ and $W_1 = T \sec\theta$). For a real histologic section of finite thickness the profile width is further increased by an amount W_2 which varies with section thickness. The derivation of various values for k_{WT} from this relationship is given in Appendix 2.

the section are the profiles of surfaces or interfaces which are planes in three dimensions. Consequently, the mean width (W) or apparent thickness of osteoid seam profiles is the two-dimensional counterpart of the true mean thickness (T) of osteoid seams, which is the mean distance between the osteoid-marrow interface and the osteoid-mineralized bone interface. Similar interpretations apply to the apparent mean distance between fluorescent labels, the apparent mean wall thickness, and the apparent mean thickness of trabeculae. It is evident that an oblique section will increase the profile width or apparent distance between two perimeters (Figure 9) and that the value of the dimensional addition factor needed to convert width to thickness (k_{WT}) will be less than 1. This occurs even with an infinitely thin section, but there will be a further increase in profile width due to the Holmes' projection effect, as with profile area.

A simple and effective way to obtain an estimate of osteoid seam width which is independent of section obliquity is to count the number of alternating bright and dark lines within the osteoid seam using polarizing microscopy;[50] seams of normal width do not contain more than three such pairs of lines. It would be useful to calibrate this method more accurately by determining the mean interlamellar separation in trabecular bone, taking only the lowest 20% of the values to exclude oblique sections. It could be determined how much this quantity varies in different locations in the same biopsy specimen, between different sites and between different persons. If the separation between lines of birefringence was reasonably constant, it might be possible to use the same method to obtain correct values for mean wall thickness and interlabel distance, but for the present, less straightforward methods must be used for this purpose.

1. Distances Determined Directly

The first derivation of k_{wT} for direct measurements was by Frost.[51] The problem had been alluded to briefly several years earlier,[28] but only reently has been given explicit treatment in the stereologic literature.[52,53] A variety of different solutions have been proposed (Table 4). The simplest approach is to consider rotation of the section plane about a single axis which is parallel to the interfaces whose true separation is to be determined, and perpendicular to the plane of Figure 9. Rotation about this axis is defined by the angle θ between the actual section plane and the ideal perpendicular section plane. The relationship between T and W is given by:

$$T = W\cos\theta$$

$$\text{and } W = T\sec\theta$$

If rotation is random, any value of θ between 0 and $\pi/2$ is equally likely, but in practice there must be an upper limit to the obliquity which is possible in trabecular bone because of deviation of the trabecular plates from perfect flatness, and the likelihood of shattering if a trabecular plate is split down the middle. Frost took this upper limit to be $75°$ or $5\pi/12$ radians. However, in the ilium there is a preferred orientation of trabeculae with most plates lying between 30 and $75°$ from the horizontal.[20] Taking a core of bone approximately perpendicular to the major plane of the body of the ilium and randomizing the section plane about the axis of rotation of this core, it would seem that the angle of obliquity is unlikely to exceed $60°$ or $\pi/3$.

The value of k_{wT} required to obtain T from W in a particular case is evidently $\cos\theta$, and Frost's approach was to compute the mean value $(\overline{\cos\theta})$ for different ranges of θ. As is shown in Appendix 2, $\overline{\cos\theta}$ is $2/\pi$ for all angles between 0 and $\pi/2$. The values for k_{wT} calculated in this way for different upper limits of section obliquity are given in Table 4. This simple approach was a useful beginning but embodies two errors. First, if the mean of the individual correction factors is applied to the mean of the individual profile widths, the true value for thickness is not obtained. The problem is analogous to that of combining data for fractional area or fractional perimeter from different fields, where the quotient of the means of the numerator and denominator must be used, not the mean of the quotients.[54] The apparent thicknesses or profile widths in the section will take a range of values such that the mean profile width $(\overline{W})$ is related to the mean value of $\sec\theta$, $(\overline{\sec\theta})$ by:

$$\overline{W} = T \cdot \overline{\sec\theta}$$

Consequently, the correction factor needed to obtain T from $\overline{W}$ is $1/\overline{\sec\theta}$. The values of k_{wT} calculated in this way are smaller and approach a limit of zero rather than $2/\pi$ (Appendix 2, Table 4). Second, the three-dimensional relationship between the section plane and a flat surface involves rotation about two axes not just one, so that a line perpendicular to the section plane may point in any direction in three-dimensional space. The mean value of $\sec\theta$ must be calculated in a different way which takes this

Table 4[a]
VALUES FOR k_{WT}

Type of surface	Formula and assumptions		Upper limits of θ		
			60°	75°	90°
Flat	$\overline{COS\theta}$	One axis of rotation	0.827	0.738	0.637 (2/π)
Flat	$1/\overline{Sec\theta}$	One axis of rotation	0.795	0.646	0
Flat	$1/Sec\theta$	Two axes of rotation	0.827	0.738	0.637 (2/π)
Flat		$\dfrac{2\theta + Sin\,2\theta^b}{4\,Sin\theta}$	0.855	0.807	0.785 (π/4)

		(R − r) radius of curvature		
		0.1	0.02	0
Spherical	$\dfrac{(R - R) \text{ Parallel sections}}{(R_1 - r_1)}$	0.812	0.792	0.785 (π/4)
Isotropic	Reciprocal of k_{ps} · Indirect measurement		Invariant	0.785 (π/4)

[a] Values for k_{WT} obtained by different methods described in text. For explanations of symbols see text and Appendix 2.

[b] See Reference 52.

into account (Appendix 2). This leads to the same values for k_{WT} as the Frost method (Appendix 2, Table 4), so that the two errors in that method cancel one another out!

Most surfaces and interfaces in bone are curved rather than flat as depicted in Figure 9. Consequently, the methods just given for calculating k_{WT} all rest on an erroneous assumption. Unfortunately, more realistic assumptions are more difficult to analyze. The profile of a cylindrical object in an oblique section is an ellipse, and the apparent separation between interfaces depends not only on the angle of sectioning but on the position of the measurement site in relation to the axes of the ellipse; a correction factor has not yet been derived for this case. For structures known to be cylindrical, such as blood vessels[55] or osteons in the cortex of long bones, measurements can be made only along the short axis of the ellipse, but the osteons are of much more variable shape in the ilium. The method is also not applicable to trabecular bone because the structures to be measured do not form closed systems and their three-dimensional geometry is unknown.

If trabecular bone is perfectly isotropic, all elements of the bone surface can be conceptually reassembled on the surface of a sphere.[44] If the two surfaces or interfaces whose separation is to be measured are similarly reassembled on the surfaces of two spheres, all possible section planes in any direction are defined by their perpendicular distance from the center of the sphere (Figure 10). The vertical distance between the spheres will increase until the distance of the section plane from the center equals the radius of the smaller sphere; beyond this point only one of the surfaces or interfaces will be seen, as in the smallest and outermost slice of an orange. With such extreme obliquity, osteoid would be seen as a broad sheet viewed from above with no free edge, surrounded by a single fluorescent label seen as a broad and indistinct smear, surrounded in turn by mineralized bone. Such appearances are only rarely seen in bone because of the relative coarseness of the sectioning process and because of the persistence of in vivo trabecular orientation through the usual biopsy and sectioning procedure, as mentioned earlier.

The value for k_{WT} calculated for this model depends on the radius of curvature of

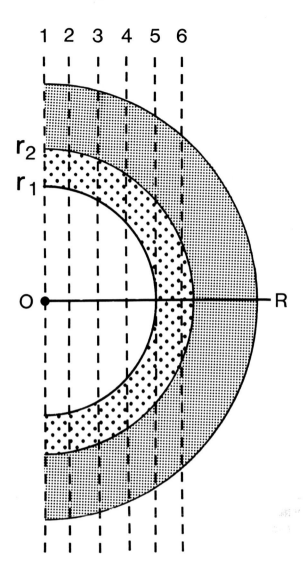

FIGURE 10. The effect of section obliquity on profile width or apparent thickness — curved (spherical) surface. Semicircular solid lines represent a cross-section through a curved trabecular plate (darker shading) with an osteoid seam on the surface (lighter shading); r_1 = radius of osteoid-marrow interface; r_2 = radius of osteoid-bone interface. Vertical dotted lines represent different planes of section defined by their distances along the radius OR. Plane 1 is an ideal perpendicular section, plane 5 is the furthest plane from the center in which both faces of the osteoid seam will be visible.

the surface and approaches a limit of $\pi/4$ with the infinite radius of a flat surface. Two other approaches, based on three-dimensional geometric probability, also lead to values for k_{wT} of $4/\pi$ if the upper limit of θ (angle of section) is not restricted.[52,53] For the spherical model, the apparent separation between the interfaces for a particular section plane and the mean apparent separation for all section planes are shown in

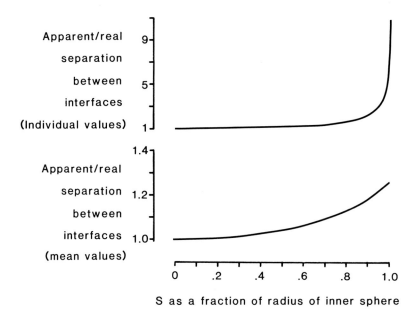

FIGURE 11.   Values for profile width or apparent thickness predicted by spherical model. Upper panel shows relative increase in profile width in an individual section plane as a function of the distance (s) of section plane from center of sphere, expressed as ratio of s to r (Figure 10). Lower panel shows mean relative increase in profile width integrated for all section planes as a function of the upper limit of section plane distance from the center; the reciprocal of these numbers gives corresponding values for k_{wT}, computed on the assumption that $r_1 = 500$ μm and $r_2 = 510$ μm. Individual values for profile width are very large when s/r_1 exceeds 0.95, but this has much less effect on the mean profile width because the prevalence of particular sections falls with increasing distance of the section plane from the center. The method of computing the values of this figure is explained in Appendix 2.

Figure 11 as functions of the distance of the section plane from the center of the sphere, assuming a mean radius of curvature of the trabecular surface of 500 μm. The upper curve supports the subjective impression that only modest increases in separation occur in most locations and very large increases in a few locations. The correction factor cannot be less than 0.792 for this model (Table 4).

Despite the different theoretical approaches, the values proposed for k_{wT} are all in the range of 0.65 to 0.8 (Table 4). In practice the higher values are more likely, further reducing the magnitude of the correction, for two reasons. First, as already mentioned, the preservation of some in vivo trabecular orientation restricts the degree of section obliquity which needs to be considered. Second, the microscopist can exclude certain sites for measurement. As explained earlier, ideally the choice of measurement sites should be determined by the random superimposition of grid line intersections. However, unless the number of measurements is very large, the chance inclusion of several sites of extreme obliquity could seriously bias the mean value. Consciously or unconsciously, most microscopists will avoid making measurements at such sites. With reasonable judgment it should be possible to exclude at least the worst 5% of cases, which would reduce the magnitude of correction necessary (Figure 8). This heretical departure from stereologic rigor applies ONLY to direct measurements of distance and not to any other procedure!

2. Distances Determined Indirectly

By an extension of the argument in Section III. C, for sheets or plates whose thickness is small in relation to the other dimensions, volume = 1/2 total surface area · thickness, so that mean structural thickness (MST) = 2 · volume/surface area (Figure 4). This is given by:

$$MST = \frac{2V}{S} = \frac{2V_v}{S_v}$$

This formula can be applied directly for mean trabecular thickness (MTT), but for mean osteoid seam thickness (MOST) and mean interlabel thickness (MILT) the appropriate formula is V_v/S_v because intersections are only counted on one interface. For mean marrow cavity diameter or mean intertrabecular distance the value of k_{WT} is unity, because the formula for MMCL is based on the marrow intercept length averaged for all directions of scanning. Working formulae for MST, MTT, MOST, and MILT are given in Table 3. Note that in each case the estimated thickness in three dimensions is obtained from the mean profile width in two dimensions (Table 2) by multiplying by $\pi/4$. Consequently, the value of k_{WT} for indirectly determined distances $(\pi/4)$ is the reciprocal of k_{PS}, the dimensional addition factor required to convert perimeter in two dimensions to surface area in three dimensions for isotropic structures $(4/\pi)$. This relationship has recently been given formal mathematical proof.[60] Consequently, since all the estimates of k_{WT} are uncertain and give values which approximate $\pi/4$ or 0.785 (Table 4), the simplest approach to the whole problem is to use the reciprocal of k_{PS} as the value for k_{WT} for direct a well as indirect distance measurements; the product of k_{WT} and k_{PS} would then be unity. This makes the calculation of three-dimensional volume-based bone formation rate (F_V) independent of both factors:

$$F_V \text{ (mm}^3\text{/mm}^3\text{/yr)} = \frac{\text{Double labeled surface (mm}^2) \cdot \text{Mean interlabel thickness (mm)}}{\text{Total bone volume (mm}^3) \cdot \text{Interlabel time (ILT) (yr)}}$$

or in symbols:

$$F_V = \frac{^iDL}{\Sigma_i} \cdot \frac{S_{V(TB)}}{V_{V(TB)}} \cdot \frac{MILT_D}{ILT}$$

$$= \frac{^iDL}{\Sigma_i} \cdot \frac{B_{A(TB)}}{V_{V(TB)}} \cdot k_{PS} \cdot \frac{MILW_D}{ILT} \cdot k_{WT}$$

Clearly, if $k_{PS} \cdot k_{WT} = 1$, F_V can be calculated directly from the two-dimensional values for B_A and MILW without any correction factors for the addition of the third dimension.

The value of $k_{PS} \cdot k_{WT}$ (but not of either alone) can be estimated by comparing direct and indirect distance measurements; the following analysis is modified from that of Frost.[61] If the expression for bone formation rate is expanded in terms of the working formula for B_A (Grid A, Table 1):

$$F_V = \frac{^iDL}{\Sigma_i} \cdot \frac{\Sigma_i}{\Sigma_h} \cdot \frac{\pi}{2d} \frac{\Sigma_h}{h_{TB}} \cdot \frac{MILW_D}{ILT} \cdot k_{PS} \cdot k_{WT}$$

$$= \frac{^iDL}{h_{TB}} \cdot \frac{\pi}{2d} \cdot \frac{MILW_D}{ILT} \cdot k_{PS} \cdot k_{WT}$$

The volume of bone formed between the two labels may also be determined directly as a fraction of the total bone volume:

$$F_V = \frac{h_{IL}}{h_{TB}} \cdot \frac{1}{ILT}$$

Combining the two equations for F_V:

$$\frac{i_{DL}}{h_{TB}} \cdot \frac{\pi}{2d} \cdot \frac{MILW_D}{ILT} \cdot k_{PS} \cdot k_{WT} = \frac{h_{IL}}{h_{TB}} \cdot \frac{1}{ILT}$$

which on rearrangement gives:

$$MILW_D \cdot k_{PS} \cdot k_{WT} = \frac{h_{IL}}{i_{DL}} \cdot \frac{2d}{\pi}$$

But the expression on the right-hand side is the working formula (Grid A, Table 2) for indirectly determining interlabel width ($MILW_I$), so that:

$$k_{PS} \cdot k_{WT} = \frac{MILW_I}{MILW_D}$$

Stated in words, the product of the perimeter-surface and width-thickness dimensional addition factors is equal to the ratio of the indirect and direct methods for determining the mean width between labels, and by an analogous argument, the same ratio for osteoid seam width. By comparing direct and indirect measurements made with sufficient precision, individual values for $k_{PS} \cdot k_{WT}$ could be obtained and the assumption that this expression is close to unity subjected to experimental verification. Furthermore, if either k_{PS} or k_{WT} could be measured independently in the real world of anisotropic trabecular bone, the other could then be calculated.

D. Number of Objects Per Unit Volume Determined From Number of Profiles Per Unit Area in the Section

This is the least studied of the four stereologic problems presented. Weibel[14,56] derived a formula applicable to objects which are (1) of well-defined and known shape, (2) of known size variation, (3) randomly distributed, and (4) large in relation to the thickness of the section. However, none of these conditions is met by the discrete objects of interest in bone. Aherne[57] derived a formula for objects of similar magnitude to the thickness of the section:

$$N = \frac{2n}{(fid + 2t)}$$

where N is the number of objects per mm^3 of tissue, n is the number of profiles per mm^2 of section area, t is the section thickness in mm, d is the grid constant in mm, i is the mean number of grid line intersections per object perimeter, and f is the ratio of maximum object perimeter length:mean perimeter length in the section, which depends on the three-dimensional shape of the object. The formula given applies to Grid A (Figure 1). The section must be examined at high enough magnification to count grid line intersections with the individual object perimeter. With this proviso, the formula could be applied to bone, although no such data have been reported. It is universal practice not to attempt three-dimensional extrapolation, but simply to report the data in the appropriate two-dimensional format and units.

APPENDIX 1
Calculation of k_{PS}, the dimensional addition factor for converting perimeter to surface area for the specific example of a sphere

In Figure 12, the circle of unit radius represents a midline section through a sphere. Parallel chords at distance F (between 0 and 1) from the center represent the diameters of circles of radius r in planes perpendicular to the plane of the diagram, produced by sections through the sphere at varying distances from the center. As a function of f, r is given by: $r = 1 - f^2$.

1. If the sections of the sphere are randomly distributed any value of f is equally likely, so that the mean value of r is equal to the area of the quadrant of the circle divided by its radius or $\pi/4$.
2. The mean value for the perimeter lengths of the circles produced by random sectioning of the sphere $= 2\pi \cdot \pi/4 = \pi^2/2$.
3. The mean value of the area of the circles produced by random sectioning of the sphere is equal to the volume of the hemisphere divided by the radius, $= 2\pi/3$.
4. Combining 2 and 3, the mean perimeter per area ratio of the circles will be $\pi^2/2 \cdot 3/2\pi = 3\pi/4$.
5. The surface area of the sphere $= 4$ and the volume $= 4\pi/3$, so that the true surface per volume ratio of the sphere $= 3$.
6. Consequently, $k_{PS} = 3 \cdot 4/3\pi = 4/\pi$.

APPENDIX 2
Different methods of computing k_{WT}, the dimensional addition factor for converting mean profile width to true thickness[58]

1. k_{WT} for flat surface with rotation about a single axis (Figure 9), Method 1 is given by:

$$k_{WT} = \overline{Cos\theta} = \int \frac{Cos\theta\, d\theta}{\theta} = \frac{Sin\theta}{\theta}$$

For values of θ between 0 and 90° ($\pi/2$ radians):

$$\overline{Cos\theta} = \frac{2}{\pi} (Sin\pi/2 - Sin\theta) = \frac{2}{\pi}$$

If the upper limit of θ is taken as 75° (5/12 radians)[51]

$$\overline{Cos\theta} = \frac{2.4}{\pi} (Sin\, 5\pi/12 - Sin\theta) = 0.738$$

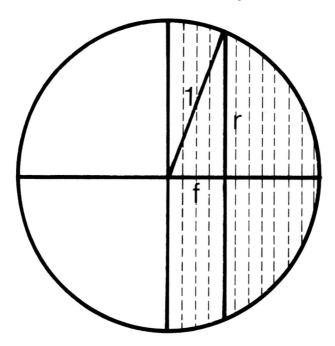

FIGURE 12.   Diagram showing the method of deriving k_{PS} for a sphere. For explanation of symbols see text.

Method 2 is given by:

$$k_{WT} = \frac{1}{Sec\theta} = \int \frac{\theta}{Sec\theta \, d\theta} = \frac{\theta}{Log \, tan \, (\pi/4 + \theta/2)}$$

For values of θ between 0 and 90° this equals zero, but with an upper limit of 75°, $1/\overline{Sec\theta} = 0.646$.

2. k_{WT} for flat surface with rotation about two axes (Figure 13) is given by $1/\overline{Sec\theta}$, weighted for three-dimensional orientation. The section plane PPPP (Figure 13) can rotate about the horizontal axis HH or about the vertical axis VV so that the perpendicular to the section plane (ON) can point in any direction. This is equivalent to N falling on any point on the surface of the sphere in Figure 13. When N lies on the equator of the sphere the section plane is perpendicular to the horizontal tissue slice, $\theta = 0$ and $Sec\theta = 1$. As θ increases, $Sec\theta$ also increases but the probability of a particular angle of section diminishes in proportion to the circumference of the circle parallel to the equator; this circumference is given by $2\pi/RCos\theta$ where R is the radius of the sphere. Consequently, the mean value of $Sec\theta$ averaged for all directions of ON is given by:

$$\overline{Sec\theta} = \frac{2\pi R \int Cos\theta Sec\theta \, d\theta}{2\pi R \int Cos\theta} = \frac{\int d\theta}{\int Cos\theta} = \frac{\theta}{Sin\theta}$$

Therefore, $1/Sec\theta$ averaged for two axes of rotation is the same as $Cos\theta$ averaged for one axis of rotation.

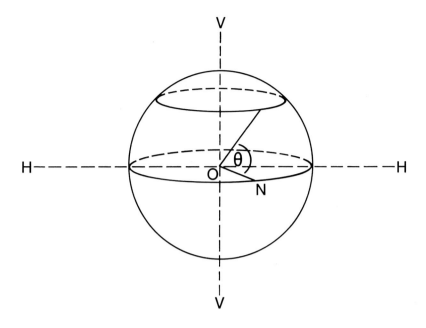

FIGURE 13.   Diagram to show the method of computing k_{wT} for a flat surface with two axes of rotation of section plane. For explanation of symbols see text.

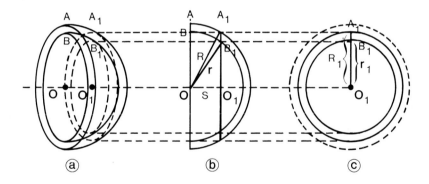

FIGURE 14.   Diagram to show the method of computing k_{wT} for a spherical surface. For explanation of symbols see text. (A) Three-dimensional relationships of concentric hemispheres of center O. (B) Section perpendicular to cut faces of hemispheres. (C) Section parallel to cut faces of hemispheres through A_1 in B.

3. k_{wT} for spherical surface; the three-dimensional relationships are shown diagrammatically in Figure 14. The cut faces of the two hemispheres of center O and upper poles A and B show concentric circles of radius OA $(= R)$ and OB $(= r)$; the true distance between the spheres is AB $(= R - r)$. Figure 14 shows a section perpendicular to the face of the hemispheres. A section plane parallel to the face of the hemispheres at distance $0 - 0_1$ $(= S)$ from the center of the sphere produces two smaller concentric circles (Figure 14) of radius $0_1 - A_1$ $(= R_1)$ and $0_1 - B_1$ $(= r_1)$. From the theorem of Pythagoras applied to Figure 14 the radii of these circles are related to the radii of the spheres as follows:

$$R_1 = \sqrt{R^2 - S^2}$$

and

$$r_1 = \sqrt{r^2 - S^2}$$

The apparent distance between the spheres (or profile width) for a particular section is given by:

$$R_1 - r_1 = \sqrt{R^2 - S^2} - \sqrt{r^2 - S^2}$$

The ratio of apparent to real distance for a particular section is given by:

$$\frac{R_1 - r_1}{R - r} = \frac{\sqrt{R^2 - S^2} - \sqrt{r^2 - S^2}}{R - r}$$

The limiting value for this ratio when $S = r$ is given by:

$$\frac{\sqrt{R^2 - r^2}}{R - r} = \frac{\sqrt{R + r}}{\sqrt{R - r}}$$

An alternative expression for $(R_1 - r_1)$ as a function of S can be derived from the cross-sectional area and mean circumference of the profile in the section (the shaded area in Figure 14).

Area $= \pi(R_1^2 - r_1^2) = \pi(R^2 - S^2 - r^2 + S^2) = \pi(R^2 - r^2)$;

mean circumference $= 1/2 \cdot 2\pi(R_1 + r_1) = \pi(\sqrt{R^2 - S^2} + \sqrt{r^2 - S^2})$

$R_1 - r_1 = $ area/mean circumference.

$$= \frac{(R^2 - r^2)}{\sqrt{R^2 - S^2} + \sqrt{r^2 - S^2}}$$

and

$$\frac{R_1 - r_1}{R - r} = \frac{R + r}{\sqrt{R^2 - S^2} + \sqrt{r^2 - S^2}}$$

The corresponding expression averaged for all sections $(\overline{R_1 - r_1})$ is derived from the cross-sectional area and mean circumference average for the volume corresponding to the shaded areas in Figure 14. The mean values for R_1 and r_1 for all sections $(\overline{R}_1$ and $\overline{r}_1)$ are given by:

$$\overline{R}_1 = \frac{1}{S} \int \sqrt{R^2 - S^2} \, dS$$

$$= \frac{1}{2S} \left(S\sqrt{R^2 - S^2} + R^2 \sin^{-1}(S/R) \right)$$

and

$$\bar{r}_1 = \frac{1}{S} \int \sqrt{r^2 - S^2} \, ds$$

$$= \frac{1}{2S} \left(S \sqrt{r^2 - S^2} + r^2 \sin^{-1} (S/r) \right)$$

The cross-sectional area is independent of S, so that mean area:

$$= \frac{\pi(R^2 - r^2)}{(R - r)}$$

Overall mean circumference:

$$= 1/2 \cdot 2\pi(\bar{R}_1 + \bar{r}_1)$$

The mean vertical distance between the spheres $(\overline{R_1 - r_1})$ is equal to area per overall mean circumference, so that:

$$\overline{R_1 - r_1} = \frac{R^2 - r^2}{(R_1 + r_1)}$$

$$= \frac{2S(R^2 - r^2)}{S \left(\sqrt{R^2 - S^2} + \sqrt{r^2 - S^2} \right) + R^2 \sin^{-1} (S/R) + r^2 \sin^{-1} (S/r)}$$

To obtain k_{WT} this expression must be divided by $(R - r)$ and the reciprocal taken so that:

$$k_{WT} = \frac{S \left(\sqrt{R^2 - S^2} + \sqrt{r^2 - S^2} \right) + R^2 \sin^{-1} (S/R) + r^2 \sin^{-1}(S/r)}{2S(R + r)}$$

REFERENCES

1. Elias, H., Hennig, A., and Schwarz, D. E., Stereology: application to biomedical research, *Physiol. Rev.*, 51, 158, 1971.
2. Elias, H., Three-dimensional structure identified from single sections, *Science,* 174, 999, 1971.
3. Cohen, J. and Harris, W. N., The three-dimensional anatomy of Haversian systems, *J. Bone Jt. Surg.*, 40, 419, 1958.
4. LaCroix, P., The internal remodeling of bone, in *The Biochemistry and Phyiology of Bone,* 2nd ed., Bourne, G. H., Ed., Academic Press, New York, 1971.

5. Jaworski, Z. F., Some morphologic and dynamic aspects of remodeling on the endosteal-cortical and trabecular surfaces, in *Calcified Tissue: Structural, Functional and Metabolic Aspects,* Menczel, J. and Harell, A., Eds., Academic Press, New York, 1971.

6. **Whitehouse, W. J., Dyson, E. D., and Jackson, C. K.,** Scanning electron microscope studies of human trabecular bone, *Nature (London),* 225, 957, 1970.

7. **Whitehouse, W. J.,** Structural variations in spongy bone as revealed by measurements of scanning electron micrographs, in *The Health Effects of Plutonium and Radium,* Jee, W. S. S., Ed., J. W. Press, University of Utah, Salt Lake City, 1976.

8. **Boyde, A., Howell, P. G. T., and Jones, S. J.,** Measurement of lacunar volume in bone using a stereological grid counting method evolved for the scanning electron microscope, *J. Microsc.,* 101, 261, 1974.

9. **Jowsey, J., Kelly, P. J., Riggs, B., Bianco, A. J., Scholz, D. A., and Gershon-Cohen, J.,** Quantitative microradiographic studies of normal and osteoporotic bone, *J. Bone Jt. Surg.,* 47-A, 785, 1965.

10. **Schenk, A.,** Basic stereological principles; Basic symbolism for stereology, in *Proc. of the First Workshop on Bone Morphometry,* Jaworski, Z. F. G., Ed., University of Ottawa Press, 1976, 21, 360.

11. **Freere, R. H. and Weibel, E. R.,** Stereologic techniques in microscopy, *J. R. Microsc. Soc.,* 87, 25, 1967.

12. **Merz, W. A.,** Die Streckenmessung an gerichteten Strukturen in Mikroskop und ihre Anwendung zur Bestimmung von Oberflachen-Volumen-Relationen im Knochengewebe, *Mikoscopie,* 22, 132, 1967.

13. **Martin, R. B.,** Equivalence of Porosity and Specific Surface Measures in One, Two, and Three Dimensions, Proc. Int. Symp. RILEM/IUPAC, Vol. 1, Prague, Sept. 18, 1973.

14. **Weibel, E. R.,** Stereological principles for morphometry in electron microscopic cytology, *Int. Rev. Cytol.,* 26, 235, 1969.

15. **Underwood, E. E.,** *Quantitative Stereology,* Addison-Wesley, Reading, Mass., 1970.

16. **Aherne, W.,** Quantitative methods in histology, *J. Med. Lab. Technol.,* 27, 160, 1970.

17. **Williams, E. D.,** Automated histoquantitation studies of bone, *Proc. R. Soc. Med.,* 65, 539, 1972.

18. **Olah, A. J.,** Influence of microscopic resolution in the estimation of structural parameters in cancellous bone, in *Bone Histomorphometry,* Meunier, P. J., Ed., Armour Montagu, Paris, 1977.

19. **Jaworski, Z. F. G.,** Parameters and indices of bone resorption, in *Bone Histomorphometry,* Meunier, P. J., Ed., Armour Montagu, Paris, 1977.

20. **Wakamatsu, E. and Sissons, H. A.,** The cancellous bone of the iliac crest, *Calcif. Tiss. Res.,* 4, 147, 1969.

21. **Whitehouse, W. J.,** The quantitative morphology of anisotropic trabecular bone, *J. Microsc.,* 101, 153, 1974.

22. **Lloyd, E. and Hodges, D.,** Quantitative characterization of bone: a computer analysis of microradiographs, *Clin. Orthop.,* 78, 230, 1971.

23. **Merz, W. A. and Schenk, R. K.,** Quantitative structural analysis of human cancellous bone, *Acta Anat.,* 75, 54, 1970.

24. **Merz, W. A. and Schenk, R. K.,** A quantitative histological study on bone formation in human cancellous bone, *Acta Anat.,* 76, 1, 1970.

25. **Melsen, F., Melsen, B., Mosekilde, L., and Bergmann, S.,** Histomorphometric analysis of normal bone from the iliac crest, *Acta Pathol. Microbiol. Scand. Sect. A,* 86, 70, 1978.

26. **Meunier, P., Edouard, C., Richard, O., and Laurent, J.,** Histomorphometry of osteoid tissue. The hyperosteoidoses, in *Bone Histomorphometry,* Meunier, P. J., Ed., Armour Montagu, Paris, 1977.

27. **Gunderson, H. J. G.,** Estimators of the number of objects per area unbiased by edge effects, *Microsc. Acta,* 81, 107, 1978.

28. **Sherrard, D. J., Baylink, D. M., Wergedal, J. E., and Maloney, N. A.,** Quantitative histological studies on the pathogenesis of uremic bone disease, *J. Clin. Endocrinol. Metab.,* 29, 119, 1974.

29. **Holmes, A.,** *Petrographic Methods and Calculations,* Murby, London, 1930.

30. **Parfitt, A. M.,** The quantitative approach to bone morphology. A critique of current methods and their interpretation, in *Clinical Aspects of Metabolic Bone Disease, International Congress Series No. 270,* Frame, B., Parfitt, A. M., and Duncan, H., Eds., Excerpta Medica, Amsterdam, 1973, 86.

31. **Whitehouse, W. J.,** Errors in area measurement in thick sections, with special reference to trabecular bone, *J. Microsc.,* 107, 183, 1976.

32. **Frost, H. M.,** Microscopy: depth of focus, optical sectioning and integrating eyepiece measurement, *Henry Ford Hosp. Med. Bull.,* 10, 267, 1962.

33. **Colton, T.,** *Statistics in Medicine,* Little, Brown, Boston, 1974.

34. **Hayslett, H. T.,** *Statistics Made Simple,* Doubleday, New York, 1967.

35. **Chalkley, H. W.,** Method for the quantitative morphologic analysis of tissues, *J. Natl. Cancer Inst.,* 4, 47, 1943.

36. Chayes, F., A simple point counter for thin-section analysis, *Am. Mineral. J.*, 34, 1, 1949.

37. Hally, A. D., A counting method for measuring the volumes of tissue components in microscopical sections, *Q. J. Microsc. Sci.*, 105, 503, 1964.

38. Curtis, A. S. G., Area and volume measurements by random sampling methods, *Med. Biol. Illus.*, 10, 261, 1960.

39. Lennox, B., Observations on the accuracy of point counting including a description of a new graticule, *J. Clin. Pathol.*, 28, 99, 1975.

40. Martin, R. B., The specific surface and porosity of bone, *Crit. Rev. Bioeng.*, in preparation, 1982.

41. Chayes, F., Determination of relative volume by sectional analysis, *Lab. Invest.*, 14, 987, 1965.

42. Mayhew, T. M. and Cruz-Oriv, L.-M., Stereological correction procedures for estimating true volume proportions from biased samples, *J. Microsc.*, 99, 287, 1973.

43. Schwartz, M. P. and Recker, R. R., Bone histomorphometry: applied stereology, *Calcif. Tissue Int.*, 33, 565, 1981.

44. Whitehouse, W. J., A stereological method for calculating internal surface areas in structures which have become isotropic as the result of linear expansion or contraction, *J. Microsc.*, 101, 169, 1974.

45. Rogers, C. A. and Short, R. H. D., Alveolar epithelium in relation to growth in the lung, *Philos. Trans. R. Soc. Lond. Ser. B.*, 235, 35, 1951.

46. Chalkley, H. W., Cornfield, J., and Park, H., A method for estimating volume-surface ratios, *Science*, 110, 295, 1949.

47. Whitehouse, W. J., Scanning electron micrographs of cancellous bone from the human sternum, *J. Pathol.*, 116, 213, 1974.

48. Whitehouse, W. J., Cancellous bone in the anterior part of the iliac crest, *Calcif. Tissue Res.*, 23, 67, 1977.

49. Mednick, L. W., The evolution of the human ilium, *Am. J. Phys. Anthropol.*, 13, 203, 1955.

50. Woods, C. G., Morgan, D. B., Paterson, C. R., and Gossman, H. H., Measurement of osteoid in bone biopsy, *J. Pathol. Bacteriol.*, 95, 441, 1968.

51. Frost, H. M., Histomorphometry of trabecular bone. I. Theoretical correction of appositional rate measurement, in *Bone Histomorphometry*, Meunier, P. J., Ed., Armour Montagu, Paris, 1977.

52. Casley-Smith, J. R. and Davy, P., The estimation of distances between parallel membranes on thick sections, *J. Microsc.*, 114, 249, 1978.

53. Jensen, E. B., Gundersen, H. J. G., and Osterby, R., Determination of membrane thickness distribution from orthogonal intercepts, *J. Microsc.*, 115, 19, 1979.

54. Mayhew, T. M. and Cruz-Orive, L.-M., Caveat on the use of the Delesse principle of areal analysis for estimating component volume densities, *J. Microsc.*, 102, 195, 1974.

55. Williamson, J. R., Vogler, N. J., and Kilo, C., Estimation of vascular basement membrane thickness: theoretical and practical considerations, *Diabetes*, 18, 567, 1969.

56. Weibel, E. R., Principles and methods for the morphometric study of the lung and other organs, *Lab. Invest.*, 12, 131, 1963.

57. Aherne, W., Methods of counting discrete tissue components in microscopical sections, *J. R. Microsc. Soc.*, 87, 493, 1967.

58. Selby, S. M., Ed., *Standard Mathematical Tables*, 15th ed., Chemical Rubber Co., Cleveland, 1967.

59. Parfitt, A. M., unpublished data.

60. Cruz-Orive, L.-M., personal communication.

61. Frost, H. M., personal communication.

Chapter 6

MEASUREMENTS OF AREA, PERIMETER, AND DISTANCE: DETAILS OF DATA COLLECTION IN BONE HISTOMORPHOMETRY

Donald B. Kimmel and Webster S. S. Jee

TABLE OF CONTENTS

I. INTRODUCTION

The purpose of doing bone histomorphometry is to make a quantitative expression of the patterns present in sections of bone. The measurements may be used to derive absolute values of bone formation and bone turnover and are frequently used to test the effects of some treatment on bone metabolism in an experiment where sequential biopsies of patients are performed. They are of greatest value when accompanied by a qualitative evaluation of the pattern present.

In considering the techniques of measuring, one must consider what features describe the appearance of the bone. Bone quantity, bone formation, and bone resorption are three such aspects of normal bone to which the "how much" question applies. In pathologic conditions, the occurrence of woven bone and endosteal fibrosis fits this category. Bone quantity is best understood as the fraction of a particular volume of cancellous bone tissue which is occupied by bone and/or osteoid tissue (as opposed to bone marrow itself). Bone formation is associated with such items as osteoid surface labeled with a tissue-time marker (see Chapters 4 and 7) and distance between a pair of such markers separated in time. Bone resorption is associated with the extent of surface covered with osteoclasts. Bone disease is associated with woven bone and fibrosis as well as nonnormal amounts of bone quantity, bone formation, or bone resorption.

The purpose of this chapter is to describe measuring techniques, the appearance of things to be measured, a rational procedure for data collection, and an idea of the useful limits to which the data may be extended.

II. MEASURING TECHNIQUES

Bone histomorphometry is an application of stereology, the study of three-dimensional structures (bone trabeculae) from two-dimensional samples (bone sections).[1] As explained in Chapter 5 the volume of a given component of an object can be estimated by knowing its area in a two-dimensional projection, while the perimeter or boundary length of the component can be estimated from the two-dimensional perimeter multiplied by $4/\pi$ (Chapter 5). The actual conversion of data to a third dimension is a mathematical procedure which is routine, so long as formulas, units, and counting methods are clearly specified as data are collected and the basic stereological assumptions are valid.

A. Area

The area occupied by a particular component is proportional to the number of regularly arranged test points overlying it (Figure 1). Each point is considered to be representative of the area lying in a square whose center coincides with a test point and whose sides are of length d, the distance between the points.

B. Perimeter

The perimeter of a particular component is proportional to the number of intersections of equidistant parallel lines with the projected perimeter of the component (Figure 2). Each intersection is considered to be representative of the distance d, which surrounds each line for a distance d/2, where d is the distance between the lines.

C. Distance

The distance between two items is proportional to the number of scale unit intervals between them (Figure 3). Each scale unit is representative of the distance d which has

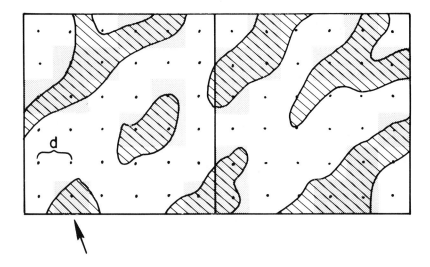

FIGURE 1. Two adjacent fields of 36 test points each are shown as they might appear in a microscope overlying a bone section. Twenty-five of 72 points overlie the cross-hatched feature. Each point actually represents a shaded square (arrow), but it can be seen that a sufficient amount of sampling will minimize any discrepancies. The absolute area occupied by the cross-hatched feature in the two fields is $25d^2$, where d is the distance between the points (Figure 4).

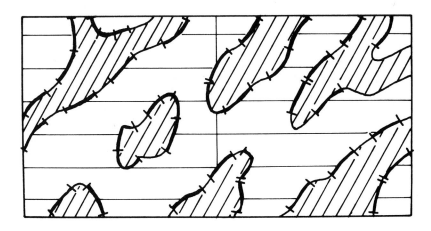

FIGURE 2. Two adjacent fields of six parallel test lines each are shown as they might appear in the microscope overlying a bone section. There are 37 intersections of test lines with the boundary of the cross-hatched feature. The distance between lines is d. Each intersection actually represents this distance d, located for d/2 on either side of the intersection (lower left). It can be seen that only when a boundary is exactly perpendicular to the test lines is d a correct measurement — in all other cases it is an underestimate. The factor $4/\pi$ compensates for this. The absolute perimeter of the cross-hatched feature in the 2 fields is $37 \times d \times 4/\pi$.

been calibrated for the eyepiece micrometer. Calibration of d for a grid or scale is necessary for absolute measurements and conversions. It is found by placing a stage micrometer graduated in 0.1- and 0.01-mm units on the microscope stage, then superimposing the eyepiece reticule with grid or scale over the known units of the micrometer (Figure 4). The distance in millimeters between points, lines, or scale units, (d) is an intrinsic value of each magnification on a microscope (for each grid and scale).

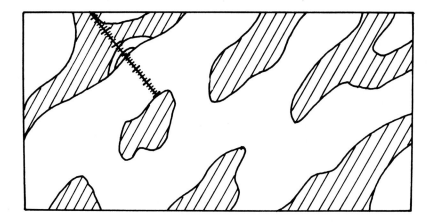

FIGURE 3.   A graduated scale of 30 units length is superimposed over the section as it might appear in a microscope overlying a bone section. The distance between the "tetracycline labels" is about 2 units.

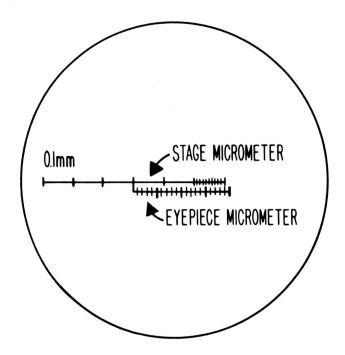

FIGURE 4.   Calibration of scale or grid. In this case, 17 scale units equal 0.270 mm, so scale unit equals 0.016 mm, which is d for this scale at this magnification on this microscope.

D. Grids

A variety of grids is available for use on integrating eyepiece reticules. A typical example, combining points and straight parallel lines is shown (Figure 5). Merz improved the more conventional grids by retaining the 36 test point arrangement, but using semicircle parallel lines instead of straight parallel lines to compensate for the nonrandomly oriented structures which are commonly found in cancellous bone (Figure 6). From this design we have added a centrally located scale micrometer. The grid

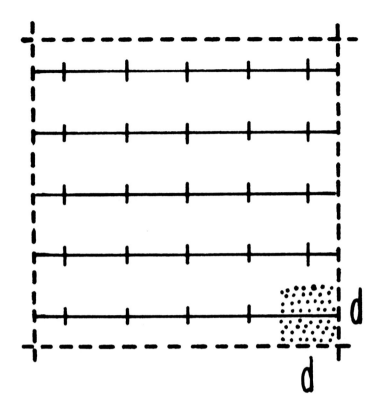

FIGURE 5. A typical grid of six straight parallel lines and 36 regularly arranged test points.

in Figure 7 combines the needs of sampling thoroughness with consideration of grid structure, and should be suitable for all light microscopic bone histomorphometric procedures, short of the automated methods which are described in Chapter 13.

III. STATIC PARAMETERS TO STUDY

A. Bone Quantity

This refers to the fraction of the section which is occupied by either mineralized bone or nonmineralized bone (osteoid). This is assayed in each field by counting the number of test points which lie over bone (h_b) or osteoid (h_o) and dividing by the total number of grid hits in the field.

B. Bone-Forming Surface

On stained histologic sections, this is best studied by recognizing trabecular surface with osteoid. There are flattened bone lining cells ("inactive osteoid") or plump osteoblasts at these surfaces.[2] Great care must be taken to ensure that the terminal lamella (lamina limitans), a thin seam of partially mineralized bone less than 1 μm thick which frequently assumes an osteoid-like stain, is not included in this measurement. Examples of osteoid are presented in Figure 8. The number of intersections of grid parallel lines with osteoid surface is counted in each field.

C. Bone Resorption

On stained histologic sections, this is best studied by recognizing bone surface in the

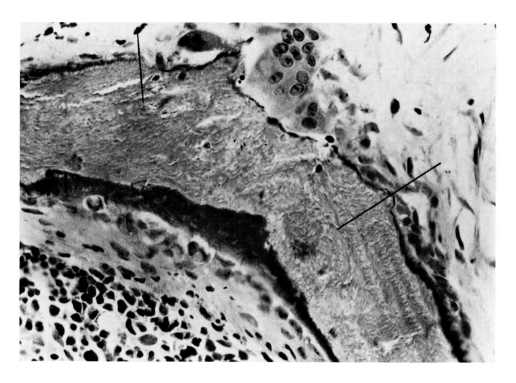

FIGURE 9.   Resorptive surface. In this Goldner-stained section, osteoclasts in eroded areas are seen. The resorptive surface is delineated by the lines. Allowance is made for minor artifacts in specimen fixation and section preparation.

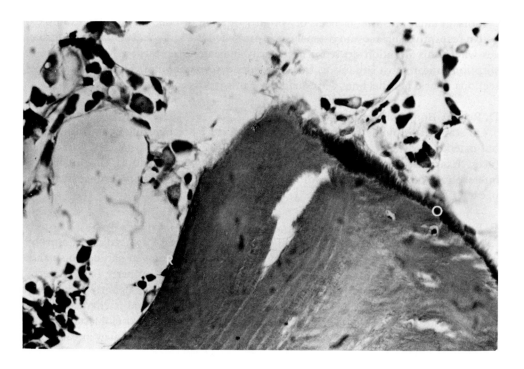

FIGURE 10.   Resting surface and nonforming osteoid surface. The surface labeled R is considered resting, while that labeled O is considered nonforming osteoid. Note the thin area of osteoid-like stain on the resting surface which represents the terminal lamellae and/or condensed stromal tissue.

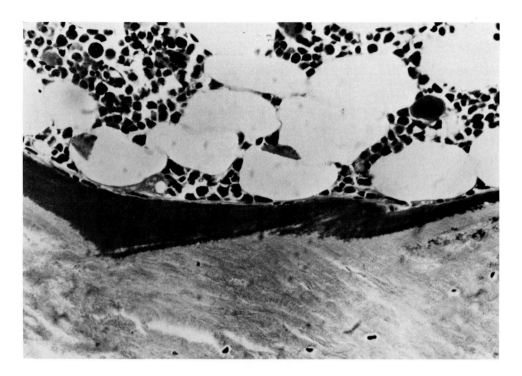

FIGURE 11. Nonforming osteoid surface. The whole surface in this picture is considered nonforming osteoid. Note the flattened lining cells adjacent to it.

F. Woven Bone

In severe hyperparathyroidism, renal osteodystrophy, and Paget's disease, another prominent feature is woven bone and osteoid. This is best recognized in decalcified bone sections viewed with polarizing light, which highlights the orientation of collagen fibers within the bone. In normal, lamellar bone the arrangement of parallel lamellae is clear, but in woven bone the disorganization of collagen is clear (Figure 12). The number of points over woven bone is counted in each case where these diseases are suspected.

IV. DYNAMIC PARAMETERS TO STUDY

In patients who have received a double tetracycline label, an additional valuable parameter may be studied. In unstained sections studied under ultraviolet (UV) light, tetracycline fluorescence is clearly seen. The number of intersections of parallel grid lines with bone surfaces bearing single, double, diffuse, or no tetracycline label is counted. Examples of these are seen in Figures 13 to 15. The double label designation is limited to that surface where a distance between two labels can be established, even when tetracycline is relatively diffusely spread between the two labels. Care must be taken to avoid confusing nonspecific surface fluorescence with true tetracycline fluorescence in the case of single labels. Areas of double label are sampled for interlabel (center-to-center) distance, from which the rate of bone mineral apposition between the times of the tetracycline label is calculated.

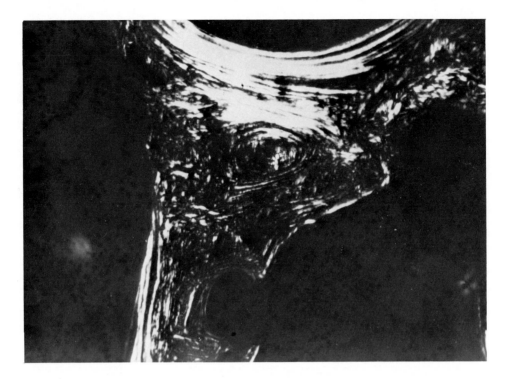

FIGURE 12.   Lamellar and woven bone under polarized light. Birefringent lamellae and disorganized woven areas are contrasted in this section.

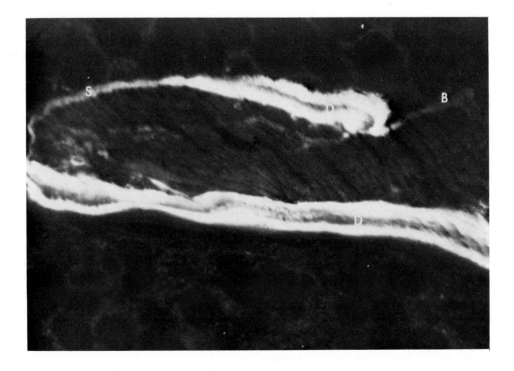

FIGURE 13.   Doubly-labeled tetracycline surface. Two near-ideal examples of double-labeled tetracycline surface areas (D) are presented. Bare (B) and single (S)-labeled surface are also seen.

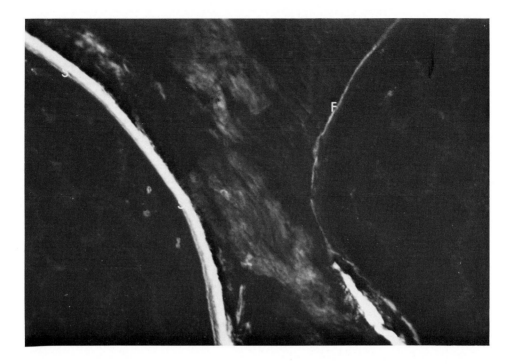

FIGURE 14. Single-labeled tetracycline surface. A long singly labeled tetracycline surface is seen (S). On the right is the phenomenon of surface fluorescence (F) which should not be confused with true tetracycline labeling of mineralizing bone.

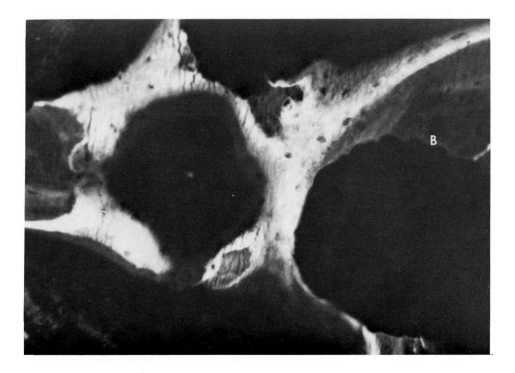

FIGURE 15. Diffusely labeled tetracycline surface. Wide confluent labels (D) are noted here in woven bone. Bare surface (B) is also present.

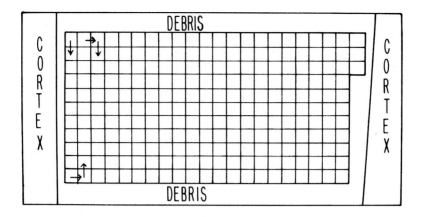

FIGURE 16.   In this representation of a section from a transilial biopsy specimen, the area to be examined is outlined as 234 fields at ×200. The cortex and debris-laden regions are carefully excluded. The arrows indicate a possible means of proceeding through the biopsy. Collection of alternate fields of this section would constitute an adequate sample.

V. TECHNICAL CONSIDERATIONS IN ACHIEVING MAXIMUM PRECISION AND ACCURACY OF BONE HISTOMORPHOMETRY

The accuracy of the measurements in bone histomorphometry is of prime concern in attempting to derive absolute values for bone formation and bone resorption rates. However, the precision in measurement of changes in sequential biopsies taken during an attempt to treat metabolic bone disease is also very important — perhaps even more so. It can be shown that only when the sample size is nearly 3% of all bone in the ilium, can one be certain that it is adequately representative of the entire ilium. Clearly, sampling this much is an unrealistic hope. The biopsy specimen is assumed to be a representative sample of the superior region of the ilium. For bone volume, this is true.

One great challenge lies in adequately sampling the material in the biopsy.[7] Since the outer 1.5 mm of a biopsy specimen is likely to contain crushed distorted bone fragments, it is advisable to enter each embedded specimen at least 3.5 mm during the microtoming procedure before preserving any sections. Since a typical trabecular thickness is 0.1 mm, a marrow space is about 0.25 mm, and a bone remodeling unit may occupy up to 0.4 mm of bone surface, it is reasonable to sample two regions of the specimen separated by at least 0.6 mm which, while geographically close in terms of the whole ilum, are at least composed of different trabeculae and different bone remodeling sites.

Six consecutive 5-μm thick sections and three consecutive 10- to 12-μm thick sections from each region should be prepared on the Jung® microtome and affixed singly to slides. One 5-μm thick section from each site should be stained with toluidine blue, another with Goldner, and the rest kept in reserve. Coverslips are affixed (no stain) to the 10- to 12-μm thick sections with mounting medium which preserves fluorescence in sections.

The analysis must be confined to a specific area of each section. It is wise to isolate the cortical-endosteal surface by at least 0.05 μm and analyze it separately. Further, the areas of distortion and bone fragments found at the periphery must be avoided. In principle, all remaining fields within the sections are studied, moving from one end of the section to the other in the manner indicated in Figure 16. The cortex and the cortical-endosteal areas should be treated separately.

VI. MAGNIFICATION

Another key to precision is proper sampling of the features of sections. There is a special problem in sampling bone surfaces. The features are usually not found uniformly dispersed through biopsies in infinitely small areas, but rather are most often found in groups. For example, sites of osteoid formation (and tetracycline labels) are usually found as surfaces of length 0.1 mm or more and single osteoclasts may be found in 0.02-mm regions along surfaces. Unnecessary and unacceptable errors will be produced if the surface is sampled too infrequently and the type of surface is relatively infrequent. For instance, if the surface were sampled every 0.1 mm and resorption were arranged in 0.02 areas, the probability would be $(0.1-0.02)/0.1$ or 0.08 that one specific area of resorption would not be included in the sample. The probability that two such areas would be missed would be $(0.8)^3$ or 0.512. If these were the only three resorptive sites in a section, there is a better than even chance that the recorded data would show no resorption in the specimen. However, sampling every 0.06 mm at about × 200, improves things such that the chances of missing any one area is only $(0.06$ to $0.02)/0.06$ or 0.67, with the chances of two being missed 0.44, and three being missed 0.29. When only a few such areas are present in a section, it is not sufficient to allow the chance that they may escape detection.

A. Distance Between Double Labels

For measurement of distance between double labels, a good magnification is about × 250, where d for the micrometer is about 0.003 mm. In a typical section of bone, the interlabel distance measures 4 or 5 scale units with a standard deviation (SD) of 30 to 50%. Using similar logic to that used in finding sample size, it can be found that a sample size of 40 from a population with this variance allows better than 50% certainty of detecting a significant ($p < 0.05$) change of 30% in the parameter.[7]

All doubly labeled surface must be weighed equally. If measurements at the rate of four per individual double label are taken as in cortical bone, the data will be more representative of the smaller doubly labeled areas. Instead, double-labeled surface is sampled at a frequency determined by the amount of double-labeled surface in the section. For instance, if 6% of the surface (as found from 900 samples at d = 0.125 mm) were double labeled, that would mean that there was 900 × 0.06 × 0.125 mm = 6.75 mm of surface with double label. If one sampled once per 0.1 mm of double-labeled surface, one would acquire 67 samples, an adequate number.

B. Section Artifacts

Sectioning artifacts are present often enough to merit discussion. With methyl methacrylate embedding, the interior of trabeculae often shows shredding during sectioning (Figure 8), but an estimate of what would have been in the position is not difficult when point counting. Less often, surfaces are distorted or cells are destroyed during sectioning. Every attempt to classify such regions should be made, but an unknown category should be made available when no decision is possible. With sectioning of bioplastic-embedded material, the surfaces are more often distorted, while the interior is intact.

Proper biopsy technique, fixation of the specimen, and staining of sections are aspects which must be mastered, as inconsistencies in these can only subtract from the value of the data which is generated.[1,5,6]

C. Considerations for Pathologic Specimens

In some sections, the number of fields studied will not yield a sufficient number of surface intersections, due to a paucity of trabecular bone. In others, the intracortical distance in the bone core may be as little as 2 to 3 mm. In these cases, either the number of fields must be increased or statistical consideration must be given to the results.

In the analysis of bone from patients where specific disease problems are suspected, counting the unique features can be done. An example is volume of fibrosis in renal osteodystrophy/hyperparathyroidism in the stained sections. Another useful approach in these patients is volume analysis of woven bone done on demineralized hematoxylin and eosin-stained sections under polarized light where points over both woven and total bone would be counted in the same manner as the usual study of stained sections.

VII. EFFECT OF OBSERVER ON VARIABILITY

Since human observers are being used, the effect of judgment and inconsistency of the observer's discrimination is of interest. Human observers have the ability to screen section artifacts and make judgments in borderline situations, something which has not been duplicated by any machine now available. The effect of observer judgment on the collection of histologic data was tested in the author's laboratory (unpublished) in 60 specimens from which 100 fields were counted in a section from each of two regions of the specimen. Prior to the counting, the observers were trained identically and reached an agreement on the identification of the histologic features by viewing sections together on a two-headed microscope.

In Table 1(A), the results of the comparison of different observers' counts for the five histologic features found in adjacent stained sections are shown. There was significant and consistent overestimation by one observer of the nonforming osteoid surface. For all other parameters, there was no significant difference, variance of about 10% being the rule, with R values for the count correlations ranging from 0.84 to 0.99.

Also in Table 1(B), the results of the comparison of the same observer's counts for the five histologic features found in different regions of the biopsy specimen are shown. It appears that the observer was unable to maintain a constant ability to judge osteoid, but remained relatively constant in his ability to identify all other parameters.

The results of Table 1(C) show the comparison of different observers' counts of sections of different regions of the same biopsy. The variance is on the whole somewhat larger; again detection of the osteoid is the largest variable with serious and consistent difference between the two observers.

The parameters which are more accurately defined, such as bone hits, or resorbing surface (only that in the domain of adjacent osteoclasts), seem to be the ones which provide the most reliability when more than one observer is involved. For the well-defined features, an interobserver error of about 10% is expected. For osteoid, the detection of the terminal lamella is such a judgment problem that the value of the parameter is reduced. From the results of Table 1(B) one would speculate that an error of about 10% might result in simple reexamination of a section due to the inevitable sampling of a completely different set of surfaces spaced at 0.06 mm from one another.

VIII. SAMPLE SIZE DETERMINATION IN MEASUREMENT OF BONE HISTOMORPHOMETRIC DATA

Bone histomorphometric data are collected to enable investigators to increase the sensitivity of their experimental assays in finding changes in the condition of bone as a result of some treatment or as an aid in evaluating the extent to which the skeleton of a particular person is diseased. To gain general acceptance, the data should be sub-

Table 1
THE EFFECTS OF OBSERVER JUDGMENT ON A
COLLECTION OF HISTOLOGIC DATA

Parameter	Correlation coefficient	Significant difference in judgment between readings
(A) Same Region — Different Observer		
Bone volume	0.898	No
Osteoid volume	0.987	No
Fractional osteoid	0.896	Yes
Fractional Howship's lacunae	0.771	No
Fractional resorption	0.863	No
(B) Different Region — Same Observer		
Bone volume	0.888	No
Osteoid volume	0.993	No
Fractional osteoid	0.918	Yes
Fractional Howship's lacunae	0.752	No
Fractional resorption	0.802	No
(C) Different Region — Different Observer		
Bone volume	0.812	No
Osteoid volume	0.964	No
Fractional osteoid	0.783	Yes
Fractional Howship's lacunae	0.741	No
Fractional resorption	0.891	No

jected to classical statistical methods and assumptions.[7] In addition, some guidance in sample size selection would help investigators to maximize their efforts in compromising time and extent of data collection. Furthermore, in cases where only a limited amount of samples may be collected from a given specimen, it would be an advantage for investigators to have some idea to what extent their findings in those cases are significant. At least three situations present themselves immediately:

1. A pretreatment biopsy is taken and measured. The investigator desires to know to what extent a posttreatment biopsy must be analyzed to ensure finding a change in a particular bone histomorphometric parameter.
2. A biopsy of a patient is presented and it is desired to know whether the patient falls within the range of a normal group, according to bone histomorphometric standards.
3. A biopsy in which only a limited amount of bone surface is available for sampling is presented. It is desired to know to what extent any derived data represents significant findings.

The statistical task to be done is either of two sorts: (a) to determine the amount of sampling to be done where sufficient bone appears to be available or (b) to determine the significance which can be attributed to data derived from specimens where only limited sampling is possible.

There are three factors which bear on sample size determination: (1) the values which one wishes to detect as different (p_1 and p_2) in the two specimens (their difference is d), (2) the acceptability (a) of finding that the difference occurred by chance and not because there was a real difference, and (3) the surety with which one wishes to detect a difference (P).

A. Assumptions

1. The biopsy specimen which is obtained is representative of the condition of the bone in the whole skeleton or at least in the ilium.
2. The section(s) of the specimen which are chosen for analysis are representative of the biopsy specimen as a whole. They are far enough removed from one another that they do not represent study of the same remodeling sites.
3. All observations, be they samples of categorized grid line-bone surface intersections or a series of digitized points, are mutually independent of one another. In other words, the condition of the last sample taken says nothing about what the next sample may be.

B. Procedure

Standard biopsy methods and sectioning methods are employed to obtain histopathologic sections which are stained by some method (toluidine blue, Goldner,[6] etc.) or left unstained prior to examination under the ultraviolet microscope. All basic bone histomorphometric data concern the fractional portion of the whole sampled population which is of a particular type. This is true for sampling of bone surfaces or points overlying given tissue types.

N is always the sample size. N_1 or N_2 refers to the sample sizes obtained at a first or second biopsy. The value which must be exceeded for particular levels of significance to exist is t_a.

$$d = \arcsin \sqrt{(p_1)} - \arcsin \sqrt{(p_2)}$$

The difference of the two frequencies is expressed as the difference of the square roots of their arcsines expressed in degrees because this is the transformation of the data which is most likely to give a normal distribution.

Where it is desired to know the sample size to collect in advance, the following equations are used:

$$N = \frac{K}{d^2} \tag{1a}$$

where

$$K = 2s^2 (t_a + t_{2(1-p)})^2 \tag{1b}$$

where s^2 is 820.8, the SD of a population which is measured by an infinite amount of sampling of that population; and the constant K for given values of P and a, is given in Table 2 for varying values of P and a. The values of t dictated by the P and a levels selected are t_a and t_2.

Where it is desired to know significance based upon a limited sample size (Table 3), the following equation is used:

$$t_a = \frac{d}{\sqrt{820.8 \, (1/N_1 + 1/N_2)}} \tag{2}$$

Table 2
K AS A FUNCTION OF a
AND P

a	0.05	0.02	0.01
P			
0.8	12887	16482	19174
0.9	17253	21374	24427
0.95	21341	25888	29253

Table 3
LEVELS OF SIGNIFICANCE
FOR VARYING VALUES OF t_a

If $t_a \frac{1}{2} >$	Then $p \frac{1}{4} <$
3.291	0.001
2.576	0.01
2.326	0.02
1.960	0.05

To aid one in relating to the N values derived, a typical transilial biopsy studied at × 100 with an intersample distance of about 75 μm yields about 125 fields. A vertically obtained specimen can be expected to yield two to three times as many fields. In each field, one may obtain 36 volume points and 8 to 10 intersects with bone surface. In fully sampling one section of a typical transilial biopsy specimen, about 4500 volume points and 1200 surface intersects would be encountered.

1. Examples

Example 1 — The first biopsy of a patient with renal osteodystrophy showed that the fractional occurrence of osteoid with osteoblasts was 0.12. It is desired to treat the patient for 6 months, then obtain a second biopsy in which it is hoped that a decrease in such surface to 0.06 will be observed. How many surface samples must be collected to be 90% (P) certain that the results are significant at the 0.02 (a) level? In this case, P is 0.9 and a is 0.02, so the constant (K) from Table 1 is 21374.

$$d = \arcsin \sqrt{0.12} - \arcsin \sqrt{0.06}$$

$$d = \arcsin 0.3464 - \arcsin 0.2449$$

$$d = 20.2679 - 14.1788$$

$$d^2 = 6.0891^2 = 37.0769$$

$$N = \frac{K}{d^2} = \frac{21374}{37.0769} = 576.4$$

N, the number of surface samples which must be collected is 576 for this particular level of sureness and significance. This degree of certainty could be obtained from sampling of about 60 fields, clearly a reasonable thing. For comparison, N would be 347 if only 80% certainty of finding a difference significant at the 0.05 level was required, but would be 788 to be 95% certain of finding a difference significant at the 0.01 level.

Example 2 — A single biopsy of a patient suspected to have incipient osteopenia is presented. The trabecular bone volume is about 0.15, while that for age-matched controls is 0.19. How many point samples of tissue volume within the biopsy must be collected to be 90% certain that the results are significant at the 0.02 level? In this case, P is 0.9 and a is 0.02, so K is again 21374.

$$d = \arcsin \sqrt{0.19} - \arcsin \sqrt{0.15}$$

$$d = \arcsin (0.4359) - \arcsin (0.3873)$$

$$d = 25.8419 - 22.7865$$

$$d^2 = 3.0554^2 = 9.3357$$

$$N = \frac{K}{d^2} = \frac{21374}{9.3357} = 2289.5$$

N, the number of points over tissue volume to be counted is 2289 for this particular level of sureness and significance. This degree of certainty could be obtained by sampling about 60 fields from the specimen, again a realistic possibility. For comparison, N would be only 1380 if 80% certainty of finding a difference significant at the 0.05 level was required, but would be 3133 to be 95% certain of finding a difference significant at the 0.01 level.

Example 3 — A pretreatment biopsy of a patient with hyperparathyroidism shows that surface covered by osteoclasts represents 0.02 of the surface of the biopsy. It is desired to restudy the patient 6 months after subtotal parathyroidectomy, to find whether the frequency of surfaces covered with osteoclasts has decreased to 0.01. Again, 90% surety of finding a difference significant at the 0.02 level is sought. How many surface samples must be collected to accomplish this?

$$d = \arcsin \sqrt{0.02} - \arcsin \sqrt{0.01}$$

$$d = \arcsin (0.1414) - \arcsin (0.1)$$

$$d = 8.3101 - 5.7392$$

$$d^2 = 2.3909^2 = 5.7165$$

$$N = \frac{K}{d^2} = \frac{21374}{5.7165} = 3739.0$$

N, the number of intersections with bone surface which must be sampled is 3739 for this level of certainty and significance. To sample 3739 intersections of grid lines with bone surface would require over 400 fields, probably not a cost effective search. For comparison, N would be only 2256 if 80% certainty of finding a difference significant at the 0.05 level was required, but would be 5122 if 95% certainty of finding a difference significant at the 0.01 level was required.

Example 4 — Two biopsies of a patient with severe osteopenia were taken and measured, the first prior to and a second 1 year after fluoride-vitamin D therapy. It was found that the extent of surface covered with double tetracycline label increased from 0.05 to 0.09 as a result of fluoride treatment. However, in the first biopsy it was possible to measure only 250 surface intersects (N_1), while in the second biopsy it was possible to measure only 400 surface intersects (N_2). Was the increase in double tetracycline-labeled surface significant, and if so, to what extent?

$$t_a = \frac{d}{\sqrt{820.8\,(1/N_1 + 1/N_2)}}$$

$$d = \arcsin\sqrt{0.09} - \arcsin\sqrt{0.05}$$

$$d = \arcsin(0.3000) - \arcsin(0.2236)$$

$$d = 17.4576 - 12.9210 = 4.5366$$

$$t_a = \frac{4.5366}{\sqrt{820.8\,(0.0040 + 0.0025)}} = \frac{4.5536}{\sqrt{5.3352}} = 1.9641$$

In this case, t_a has a value of 1.9641. From Table 3, it can be seen that the measured change is significant at the 0.05 level, but not at the 0.02 level.

C. Independence

The assumption of mutual independence is very critical to this discussion. If it can be clearly shown to be totally false, then this whole approach is in serious question. It is probably a less severe problem where the intersample distance is 50 μm or more. Where samples are collected every 10 μm it is a large problem, because the classification of one point may indeed predict the class into which the next point falls. For instance many osteoid seams are 50 to 100 μm long, so eight or ten points in a row may well be found which lie on osteoid. It is most likely that resorption and formation are linked locally in trabecular bone, so that the finding of an osteoid seam may predict the existence of a nearby osteoclast. For techniques where graphics tablets which sample every 10 μm or so are used, it may be desirable to collect only a random sampling of the digitized points in proper calculation of the frequency of bone histomorphometric parameters. The problem of independence of sample observations requires more investigation.

IX. CONCLUSION

Every possible effort must be expended in the acquisition of an intact biopsy specimen. Artifact-free, uniformly stained sections from two areas located at least 0.6 mm apart must then be employed. Areas, perimeters, and distances are readily measured, using a light microscope and reticule-mounted eyepiece grid. Both normal and pathologic features may be measured. A magnification of about × 200 is appropriate for study of stained sections, about × 100 for studies of surface labeling with tetracycline, and about × 250 for interlabel dstance measurements. The more precisely defined features are more reliable quantitative parameters whether one or more than one observers collect the data. The number of fields to be studied should be determined for each specimen in advance of data collection to limit microscope time to what is essential.

Sample size determination can be done by a straightforward method. Doing this will help investigators limit their data collection efforts to what is statistically useful and avoid situations where only excessive sampling would yield desired significance. The size depends upon the frequency of the items in the biopsy, the desired significance level (l), the size of change to detect (d), and the probability of success which one wishes (P). Attempting to detect significant changes of even 100% in features which have a fractional occurrence of 0.03 or less is probably a poor practice. The degree of significance which may be attributed to data derived from a specimen of limited sample size is easily calculated and provides a meaningful tool in the evaluation of such biopsy material.

REFERENCES

1. **Merz, W. A. and Schenk, R. K.,** A quantitative histologic study on bone formation in human cancellous bone, *Acta Anat.,* 76, 1, 1970.
2. **Pritchard, J. J.,** *The Osteoblast in the Biochemistry and Physiology of Bone,* Bourne, G. H., Ed., Academic Press, New York, 1972.
3. **Hancox, N. M.,** *The Osteoclast in the Biochemistry and Physiology of Bone,* Bourne, G. H., Ed., Academic Press, New York, 1972.
4. **Bordier, P. J. and TunChot, S.,** Quantitative histology of metabolic bone disease, *J. Clin. Endocrin. Metab.,* 1, 197, 1972.
5. **Rasmussen, H. and Bordier, P. J.,** *Physiological and Cellular Basis of Metabolic Bone Disease,* Williams & Wilkins, Baltimore, 1974, 57.
6. **Goldner, J. A.,** A modification of the Masson trichrome technique for routine laboratory purposes, *Am. J. Pathol.,* 14, 237, 1938.
7. **Sokal, R. R. and Rohlf, F. J.,** *Biometry,* W. H. Freeman, San Francisco, 1969, 607.

Chapter 7

BONE HISTOMORPHOMETRY: ANALYSIS OF TRABECULAR BONE DYNAMICS

H. M. Frost

TABLE OF CONTENTS

I. INTRODUCTION

Since the late 1950s a capability has existed to measure bone turnover in compact bone, based on the deposition of tetracycline tissue-time markers.[1] It could not be applied to trabecular bone at the time because of a number of unsolved methodological problems. Thus, one had to settle for measurement of static features that could be obtained by microradiography[2] or light microscopic histomorphometry.[3] Dynamic interpretation of such static features often proved erroneous, however. For example, while increased numbers and extent of osteoblasts might suggest increased bone formation rates, the true change could be exactly the opposite.

Recent solutions to some of the problems involved in analysis of trabecular bone now allow direct measurement of its tissue turnover parameters. This chapter describes an improved version of the original analytic system for such measurements. It will work in man or in animals which possess true bone remodeling, using trabecular bone from any site. The most commonly used site in man is the ilium near the crest.

The techniques of labeling, biopsy, sectioning, and measurement are described in other chapters and will not be detailed here. This chapter outlines the major relationships that can provide usefully accurate numerical values for the rates of bone formation and resorption (expressed with respect to various referents), plus changes in bone balance and values for sigma and additional quantities in trabecular bone. This system can also be applied with minor modifications to cortical bone.

The overall objectives will be to describe first the measurement of bone formation rate per unit of bone tissue volume or surface and per osteoblast; second, the bone resorption rate in the same volume and surface units and per osteoclast and osteoclast nucleus; and finally, the bone balance or rate of change in trabecular bone volume. Values will also be obtained for trabecular bone volume (mineralized and unmineralized), the total trabecular surface area and in the fractions engaged in formation and resorption, the mineral appositional rate (the mean rate of advance of the interface between osteoid tissue and bone), and the length of time required for the typical site of turnover to proceed through its resorption and formation phase, referred to as sigma.

The design of the system affords considerable flexibility and versatility. The primary data are measured in such a way that any expression of activity at any level of biological organization (cell level, tissue level, organ level) can be converted to a corresponding value at any other level, and volume based values can be converted to corresponding surface based values. The analytical system operated on the following ten basic assumptions.

1. The patient or animal was subjected to in vivo labeling with an appropriate agent and schedule as described in Chapter 4.
2. The biopsy specimen (or at least that portion used for measurements) was not fragmented during biopsy or compressed during removal from the trephine
3. No wrinkling or shattering of the specimen occurred during cutting of thin sections
4. Undecalcified sections of 5 to 8 μm in thickness (10 to 20 μm in thickness for fluorescence measurements) are suitably prepared
5. Staining of the sections is of high and uniform quality
6. Area and perimeter measurements are performed at the proper primary magnification with suitable grids[17] as described in Chapters 5 and 6
7. Simple lengths are measured with a calibrated eyepiece micrometer
8. Tetracycline label data are obtained from unstained or lightly stained sections by blue light fluorescence microscopy using adequately designed, collimated, and adjusted equipment[8]
9. Cell counts are performed on properly made and suitably stained sections
10. There is no troublesome observer error in the appropriate classification of tissues or types of surface

Problems of section-section, site-site, and inter- and intraobserver variation are discussed in Chapters 5 and 6, and the stereological problems of deriving three-dimensional quantities from measurements in two-dimensional sections are discussed in Chapter 5.

II. THE ANALYTICAL SYSTEM

The analysis is conveniently arranged into three levels called primary data, first order derived quantities, and second order derived quantities.

The primary data consist of the actual grid and cell counts and simple measurement of lengths performed at the microscope. No coding numbers are provided for these. The first order derived quantities are computed directly from the primary data, some with the aid of additional information such as micrometer or grid constants and marker time intervals. They are given two-digit coding numbers and a symbol in the text and are listed in Tables 1 and 3. The second order derived quantities are computed from the first order quantities and generally relate more clearly to the kinetic features of bone physiology that concern us. They are given three digit coding numbers and a symbol in the text.

The equations are numbered in the order in which they appear in the text. The symbols employed are culled from those used in the past by Frost,[8,12,15] Merz,[17] Schenk,[18-21] Meunier,[22-25] Bordier,[26] Courpron,[27-29] and Bressot[30] but incorporate several recent improvements; they are listed in Tables 1 to 4. The two and three digit code numbers serve for identification, and for addressing in a measurement data reduction program written by the author for the Hewlett-Packard® HP-67 and HP-97 programmable calculators.

Table 1
FIRST ORDER DERIVED QUANTITIES

Code number	Symbol	Name	Dimensions
11	V	Trabecular bone volume	mm³/mm³
12	V_{os}	Osteoid volume	mm³/mm³
13	S_f	Fractional formation surface	mm²/mm²
14	S_r	Fractional resorption surface	mm²/mm²
15	S_{ra}	Fractional active resorption surface	mm²/mm²
16	S_{fa}	Fractional active formation surface	mm²/mm²
17	S_v	Trabecular bone-specific surface	mm²/mm³
18	$\overline{M}$	Appositional rate	mm/year
19	MWT	Mean wall thickness	mm
20	C_{fa}	Osteoblasts per formation perimeter	cells/mm
21	C_{ra}	Osteoclasts per active resorption perimeter	cells/mm
22	N	Nuclei per osteoclast	nuclei/cell

Optional Parameters Without Code Numbers

	V_{osf}	Osteoid volume, fraction	mm³/mm³
	S_{ri}	Fractional inactive resorption surface	mm²/mm²

Table 2
SECOND ORDER DERIVED QUANTITIES

Code number	Symbol	Name	Dimensions
101	F_a	Bone formation rate, active surface referent	mm³/mm²/year
102	F_f	Bone formation rate, formation surface referent	mm³/mm²/year
103	F_s	Bone formation rate, total surface referent	mm³/mm²/year
104	F_v	Bone formation rate, volume referent	mm³/mm³/year
105	F_o	Bone formation rate, osteoblast referent	mm²/cell/year
106	B_v	Bone balance, volume referent	mm³/mm³/year
107	B_s	Bone balance, surface referent	mm³/mm²/year
201	R_s	Bone resorption rate, surface referent	mm³/mm²/year
202	R_v	Bone resorption rate, volume referent	mm³/mm³/year
203	R_r	Bone resorption rate, resorption surface referent	mm³/mm²/year
204	R_a	Bone resorption rate, active surface referent	mms³/mm²/year
205	R_o	Bone resorption rate, osteoclast referent	mm²/cell/year
206	R_{on}	Bone resorption rate, osteoclast nucleus referent	mm²/nucleus/year
207	$\overline{M}_{ra}$	Linear bone resorption rate	mm/year
110	M_f	Radial closure rate	mm/year
210	M_r	Radial resorption rate	mm/year
401	σ_f	Bone formation period	years
402	σ_r	Bone resorption period	years
403	σ	Sigma	years
504	S/V	Trabecular surface-to-volume ratio	mm²/mm³
505	V/S	Trabecular volume-to-surface ratio	mm³/mm²
501	OSW	Osteoid seam width	mm
502	TTI	Trabecular thickness index	
552	S_{fo}	Fractional active osteoid surface	

Table 3
DIMENSIONS — FIRST ORDER DERIVED QUANTITIES

Symbol

V	mm³ Trabecular bone/mm³ bone tissue (trabeculae and marrow)
V_{os}	mm³ Osteoid/mm³ bone tissue (trabeculae and marrow)
S_f	mm² Osteoid surface area/mm² trabecular surface area
S_r	mm² Resorption surface area/mm² trabecular surface area
S_{ra}	mm² Active resorption surface area/mm² trabecular surface area
S_{fa}	mm² Doubly labeled surface area/mm² trabecular surface area
S_v	mm² Trabecular surface area/mm³ bone tissue (trabeculae and marrow)
$\overline{M}$	mm New bone/year
MWT	mm Thickness of completed BMU
C_{fa}	Osteoblasts/mm active forming (doubly labeled) perimeter
C_{ra}	Osteoclasts/mm active resorbing perimeter
N	Nuclei/osteoclast
V_{osf}	mm³ Osteoid/mm³ trabecular bone
S_{ri}	mm² Inactive resorption surface area/mm² trabecular surface area

Table 4
DIMENSIONS — SECOND ORDER DERIVED QUANTITIES

Symbol

F_a	mm³ New bone/mm² active forming (doubly labeled) surface area/year
F_f	mm³ New bone/mm² forming (osteoid) surface area/year
F_s	mm³ New bone/mm² trabecular surface area/year
F_v	mm³ New bone/mm³ preexisting bone/year
F_o	mm² New bone/osteoblast/year
B_v	mm³ Change/mm³ trabecular volume/year
B_s	mm³ Change/mm² trabecular surface/year
R_r	mm³ Bone resorbed/mm² trabecular surface area/year
R_v	mm³ Bone resorbed/mm³ preexisting trabecular volume/year
R_r	mm³ Bone resorbed/mm² resorption surface area/year
R_a	mm³ Bone resorbed/mm² active (osteoclast covered) surface area/year
R_o	mm² Bone resorbed/osteoclast/year
R_{on}	mm² Bone resorbed/osteoclast nucleus/year
$\overline{M}_{ra}$	mm Bone resorbed/year
M_f	mm Bone formed over forming (osteoid) surface/year
M_r	mm Bone resorbed over resorbing (scalloped) surface/year
σ_f	Years required to complete the formation phase
σ_r	Years required to complete the resorption phase
σ	Years required to complete the formation and resorption phases
S/V	mm² Trabecular surface/mm³ trabecular volume
V/S	mm³ Trabecular volume/mm² trabecular surface
OSW	mm Thickness of average osteoid seam
TTI	An index — no dimensions
S_{fo}	mm² Active (doubly labeled) osteoid/mm² forming (osteoid) surface

III. FIRST ORDER DERIVED QUANTITIES

A. Volumes

Volumes are determined using the grid point counting technique. They may be expressed either as decimal fractions, percentages without dimensions, in units such as mm³/mm³, without calibration of the grid, or in absolute units with suitable calibration. Figure 1, illustrates a typical grid overlying a section of bone as it appears under

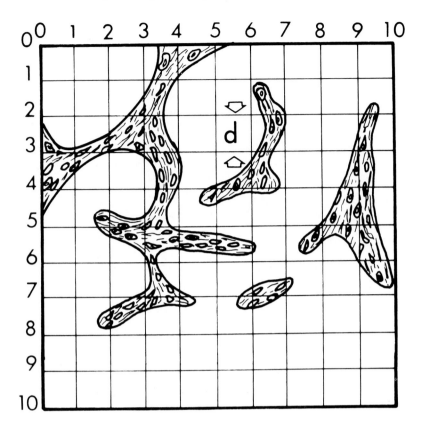

FIGURE 1.   A diagram of a "point count" grid used to measure areas. This one, of a 9 × 9 configuration, has 81 test points each representing the intersection of two orthogonal lines (one ignores the outer border). In this drawing there are 10 clear "hits" and 4 tangent hits (each of the latter counted as ½ hit) for a total of 12. The trabecular area thus equals 12/81 of the grid area. By rearranging the grid in relation to the section one accumulates another "throw", and by moving the slide he both accumulates other throws and enlarges the sample he measures in this way. Typically one repeats this until he accumulates 100 to 200 total hits and measures 5 to 50 mm² of bone. Many different designs of such grids exist, each with its own advantages.

the microscope. Each grid point is defined as the crossing of a horizontal and vertical line. Grid point hits may be tallied on mineralized lamellar bone (h_b), unmineralized lamellar bone (h_o), or on all other tissues combined (h_a).

1. Trabecular Bone Volume, V (#11)

This is the volume occupied by mineralized and unmineralized trabecular bone expressed as a fraction of the volume occupied by marrow plus trabeculae. Woven bone is counted separately or as part of the marrow. It is given by:

$$V = \frac{h_b + h_o}{h_b + h_o + h_a} \tag{1}$$

2. Osteoid Volume, V_{os} (#12)

This is the volume occupied by unmineralized trabecular bone expressed as a fraction of the volume occupied by marrow plus trabeculae. It is given by:

$$V_{os} = \frac{h_o}{h_b + h_o + h_a} \qquad (2)$$

Osteoid volume as a fraction of total trabecular bone volume (V_{osf}) is given by:

$$V_{osf} = \frac{h_o}{h_b + h_o} \qquad (3a)$$

or

$$V_{osf} = \frac{V_{os}}{V} \qquad (3b)$$

These formulas are based on the theorem of DeLesse which states that the ratio of the two areas in a cross section of a biphasic structure is equal to the ratio of the two volumes.[31]

B. Surfaces

Surfaces (seen as perimeters in thin section) are determined using a grid line intersection counting technique. This can be done at the same time as hits are tallied if the grid contains both parallel lines and marked points as in Figure 1. In this discussion use of the horizontal lines will be described, although one may use both the horizontal and vertical lines as described in Chapter 6. Intersections between grid lines and section perimeters are tallied according to various subdivisions of the surface such as osteoid covered (i_o), Howship's Lacunae (i_r), osteoclast covered (i_c), or overlying double labels (i_1). These subdivisions can be expressed as decimal fractions or percentages of the total number of intersections (Σ_i), in units such as mm^2/mm^2 without calibration of the grid, or in absolute units with appropriate calibration (Chapter 4). The total intersections, Σ_i, includes all the above surfaces plus the "inactive" surface where neither osteoid nor Howship's Lacunae are found. The majority of the perimeter in most sections is "inactive".

1. Fractional Formation Surface, S_f (#13)

This is the fraction of the trabecular surface that is covered by osteoid seams. No effort is made to classify the seams on the basis of osteoblast morphology as "active" or "inactive". It is given by:

$$S_f = \frac{i_o}{\Sigma_i} \qquad (4)$$

2. Fractional Resorption Surface, S_r (#14)

This is the fraction of the trabecular surface which is eroded by Howship's lacunae. It is given by:

$$S_r = \frac{i_r}{\Sigma_i} \qquad (5)$$

3. Fractional Active Resorption Surface, S_{ra} (#15)

This is the fraction of trabecular surface beneath and in contact with recognizable osteoclasts (as described in Chapter 6) which almost always lie in scalloped Howship's lacunae. It is given by:

$$S_{ra} = \frac{i_c}{\Sigma_i} \qquad (6)$$

Active resorption surface may also be expressed as a fraction of the resorption surface:

$$\frac{S_{ra}}{S_r} = \frac{i_c}{i_r} \qquad (7)$$

In addition, one may occasionally need the fractional inactive resorption surface (S_{ri}), which is given by:

$$S_{ri} = \frac{i_r - i_c}{\Sigma_i} \qquad (7a)$$

This may also be expressed as a fraction of the total resorption surface:

$$\frac{S_{ri}}{S_r} = \frac{i_r - i_c}{i_r} \qquad (7b)$$

4. Fractional Active Formation Surface, S_{fa} (#16)

This is the fraction of trabecular surface overlying a double tetracycline label. It can be assumed that such surfaces were "active" in the sense that bone was being formed there during the labeling period, but because single-label surfaces are ignored, the true active formation surface is underestimated by an unknown amount (see Chapter 8). It is given by:

$$S_{fa} = \frac{i_1}{\Sigma_i} \qquad (8)$$

The measurement of S_{fa} will nearly always be less than S_f; while S_f includes nearly all forming surfaces (all those with an osteoid seam) some of them will have arrested for some reason at the time the tetracycline was given (before the osteoid was completely mineralized) and thus cannot be considered "active". This arrest may occur normally but is more frequently encountered in disease states.

5. Trabecular Bone-Specific Surface, S_v (#17)

This is the area of trabecular surface per unit volume of bone tissue (bone plus marrow) in mm^3. For this quantity, in contrast to those previously considered, the grid must be calibrated. S_v constitutes the key quantity providing free conversion of any trabecular measurement from a surface referent to a volume refefent and conversely. The justification for converting two-dimensional measurements made on histological sections to three-dimensional expressions is elaborated in Chapter 5. The need for this conversion will become apparent as we proceed. If Σ_h = the total number of grid points examined in all fields, d = the grid constant in millimeters determined using stage micrometer (Chapter 6), then using a grid with horizontal lines only, trabecular bone-specific surface is given by:

$$S_v \, (mm^2/mm^3) = \frac{\Sigma_i}{\Sigma_h \cdot d} \cdot 2 \qquad (9)$$

Corresponding formulas using both horizontal and vertical lines and using other grids are given in Chapter 5. To convert to mm^2/cm^3 one simply multiplies by 1000 (the number of mm^3/cm^3).

C. Appositional Rate, $\overline{M}$ (#18)

This is the thickness of the layer of new mineralized bone laid down per unit time, averaged over all the active bone-forming surfaces, expressed in millimeters per year. The primary measurement is the mean distance between the two bands of fluorescence which are usually parallel or nearly so. The lines are of varying distance apart and varying lengths in any given section, so that the method of sampling must ensure that the measuremerts are representative (Chapter 5 and 6).

The apparent distance between markers does not give the true distance because of the obliquity or malorientation angle, θ, as described in Chapter 5. Varying degrees of obliquity of the plane of section of the biopsy with respect to the plane of the double label cause the apparent separation between the markers to increase, for which an orientation correction factor has been derived. If Σ_x = the sum of all 25 to 100 measurements, in micrometer units, n = total number of measurements, e = the micrometer calibration constant, $\pi/4$ = the currently recommended orientation correction factor (its value = 0.79, compared to the value of 0.74 recommended formerly),[10] and t = the time interval between administration of markers expressed in years, then the appositional rate is given by:

$$\overline{M} = \frac{\Sigma_x \cdot e \cdot \pi/4}{n \cdot t} \qquad (10)$$

As usually measured, $\overline{M}$ applies primarily to bone-forming surfaces that are "ON" or active during the whole marker interval. A related parameter, the radial closure rate, will account later on for "OFF" periods, as well to average activity over time periods much longer than the marker interval.

The writer devised the following sampling procedure to obtain mean values for the appositional rate which represent in their true proportions in the bone the small and the large actively forming surfaces. Using fluorescence microscopy, appropriate sections, a microscope with a mechanical stage, and an eyepiece with a calibrated scale micrometer, proceed thus.

1. With the "X" motion of the mechanical stage scan the section horizontally from one end to the other, then move the "Y" motion say 0.2 mm and repeat the horizontal scan, add another "Y" increment and repeat the "X" scan, and so on. By this means the entire section is scanned, much as the electron beam scans the frame of a television picture tube.
2. Note the exact middle of the micrometer scale and regard the left end of any long-scale division mark there as the "raster point". As the mechanical stage moves the slide, watch this raster point moving across the section and when it intersects a doubly labeled surface stop, rotate the eyepiece to align the scale

perpendicular to the two labels and measure their separation, middle of one to middle of the other, at the locus where the raster point intersected the labels.

3.   Now move on until the next intersection of the raster point with a double label, and so on until the desired number of separate measurements have accumulated (typically n would equal 25 to 100).

D. Miscellaneous

1. Mean Wall Thickness, MWT (#19)

This is the mean thickness of layers of new bone formed at bone-forming sites when the formation phase is completed, or the mean distance between cement lines and the trabecular surfaces of completed structural units. It is needed to compute σ_f as will be described later. If Σ_x = the sum of the micrometer measurements, n = the number of individual measurements, e = the micrometer constant, and $\pi/4$ = the orientation correction factor then the MWT is given by:

$$MWT = \frac{\Sigma_x \cdot e \cdot \pi/4}{n} \tag{11}$$

2. Osteoblasts per Formation Perimeter, C_{fa} (#20)

This is the number of osteoblasts per unit length of osteoid perimeter overlying a double tetracycline label. If x = micrometer divisions along a surface containing both double labels and osteoblasts, n = the number of sites measured, 10 = the number of cells measured at each site, and e = the micrometer constant, then the osteoblasts per unit formation perimeter may be expressed as follows:

$$C_{fa} = \frac{10 \cdot n}{\Sigma_x \cdot e} \tag{12}$$

This measurement and the one to follow are expressed in two dimensions.

3. Osteoclasts per Active Resorption Perimeter, C_{ra} (#21)

This is the number of multinucleated osteoclasts per unit perimeter of active resorption surface. If n = the number of osteoclasts measured, x = the number of micrometer divisions corresponding to the perimeter length of each osteoclast, and e = the micrometer constant then the value can be expressed as:

$$C_{ra} = \frac{n}{\Sigma_x \cdot e} \tag{13}$$

Note that this is a one-dimensional linear measurement.

4. Nuclei per Osteoclast, N_r (#22)

This is the number of nuclei seen in the typical osteoclast in thin section. If x = the number of nuclei within a cell and n = the number of osteoclasts examined, then the number of nuclei per osteoclast can be expressed as:

$$N_r = \frac{\Sigma_x}{n} \tag{14}$$

The following sequence is suggested in order to save time at the microscope for the above parameters. The data for V, V_{os}, S_f, S_r, S_{ra}, S_{ri}, and S_v may be obtained in a single "pass" through approximately 100 fields with a light microscope at ×100 to ×200 in several sections stained appropriately. One simply moves from field to feld tallying grid points on mineral and on osteoid, parallel line intersections with osteoid surface, osteoclasts in Howship's lacunae, Howship's lacunae without osteoclasts, and resting surfaces (neither Howship's lacunae or osteoid) and the number of fields examined. The fractional active formation surface, S_{fa}, should be determined alone on a single "pass" with a fluorescence microscope on a 10-μm section with magnification at ×100 to ×200 as described earlier, tallying parallel line intersects with double-label perimeter, and with all other perimeters and number of fields examined. The appositional rate measurement, $\overline{M}$, should be made on the same sections in another single "pass". The osteoclasts per active resorption perimeter, C_{ra}, and nuclei per osteoclast, N_r, can be made on a fourth "pass", and the MWT and osteoblasts per formation perimeter, C_{fa}, on a fifth "pass". The entire group of measurements might take about 6 hr and the calculations for all the primary and second order parameters another hour or more. These times can be reduced and accuracy increased by tallying the data into a Hewlett Packard® HP-67 or HP-97 programmable calculator using a program written by the author. The raw data may be stored on magnetic cards for use in these calculators. Data reduction may also be accomplished by writing programs for any of a number of programmable desk calculators or microcomputers now commercially available.

IV. SECOND ORDER DERIVED QUANTITIES

Tables 2 and 4 list the 24 second-order derived quantities and their three-digit code numbers which are described in this section. They are computed from data already on hand and no further measurements need be made. There are four sub-groups: formation, resorption, sigma, and miscellaneous (though still important). Each can be expressed in terms of different referents which serve different functions in expressing the biologic activity involved. The reasons for these will become clear as one proceeds.

A. Bone Formation
1. Bone Formation Rate, Active Surface Referent, F_a (#101)
This is the volume of mineralized new bone per unit area of active bone-forming surface per unit of time in $mm^3/mm^2/year$. It is given by:

$$F_a(mm^3/mm^2/year) = \overline{M}\ (mm/year) \qquad (15)$$

Note that F_a and $\overline{M}$ are numerically equal but of different dimensions, having different referents and expressing different biologic functions.

2. Bone Formation Rate, Formation Surface Referent, F_f (#102)
This is the volume of mineralized new bone formed per unit area of osteoid surface per unit time in $mm^3/mm^2/year$. It represents bone formation averaged over the entire bone-forming surface whether or not it was active at the time of labeling and biopsy. It is given by:

$$F_f = \frac{\overline{M} \cdot S_{fa}}{S_f} \quad \text{or} \quad \frac{\#18 \cdot \#16}{\#13} \qquad (16)$$

3. Bone Formation Rate, Total Surface Referent, *F*ₛ (#103)

This is the volume of mineralized new bone made per unit of total trabecular surface per unit time in mm³/mm²/year. It is given by:

$$F_s = \overline{M} \cdot S_{fa} \quad \text{or} \quad \#18 \cdot \#16 \tag{17}$$

4. Bone Formation Rate, Volume Referent, *F*ᵥ (#104)

This is the volume of mineralized new bone made per unit volume of preexisting bone per unit of time in mm³/mm³/year. It is given by:

$$F_v = S_v \cdot \overline{M} \cdot S_{fa} \quad \text{or} \quad \#17 \cdot \#18 \cdot \#16 \tag{18}$$

5. Bone Formation Rate, Osteoblast Referent, *F*ₒ (#105)

This is the area of mineralized new bone made per average osteoblast during unit time in mm²/cell/year. It is given by:

$$F_o = \frac{\overline{M}}{C_{fa}} \quad \text{or} \quad \frac{\#18}{\#20} \tag{19}$$

This is an index of osteoblast work efficiency, but ignores whatever fraction of the osteoblasts are overlying surfaces where mineralization is not occurring at the time of labeling and biopsy. It is an area and not a volume measurement, because C_{fa} is not extrapolated to the trabecular surface in three dimensions.

B. Bone Resorption

Bone resorption rates can be measured by histomorphometric techniques using a strategy devised by the author in 1963.[5] The key datum allowing this is measurement of bone balance.

All bone formed over some past time (F) less all bone resorbed over the same time (R) must equal the change in amount present (B) over the same time or: F − R = ΔB, which on rearrangement gives:

$$R = F - \Delta B \tag{20}$$

F can be obtained from the tetracycline double-label data, and ΔB by measurement of bone volume in two sequential biopsies from identical contralateral sites in the iliac crest taken several months apart (or in serial rib biopsies) provided the patient remains in a steady state without therapeutic or other perturbations of bone cell function.

Two methods of obtaining bone balance will be presented. Nonhistomorphometric techniques, such as measurement of external mineral balance, serial radiogrammetry, serial photon absorptiometry, or serial whole body neutron activation analysis, are less noisy but measure balance integrated over the entire skeleton, of which only 20% is trabecular bone. Histomorphometric methods have the advantage that balance is measured at the same skeletal sampling site as formation rate, so that the calculated resorption data apply to the osteoclasts and resorption surfaces viewed under the microscope.

Trabecular bone volume, V (#11) has considerable variability depending on the size of the biopsy sample and site-to-site variation in the skeleton (Chapter 6). Consequently, accurate estimates of bone balance require that homogeneous groups of patients be serially biopsied. Three points deserve special emphasis:

1. No type of resorption surface is measured to obtain the bone resorption rate.
2. The bone resorption rate cannot be measured by histomorphometric means unless the change in amount of bone per unit time is also measured.
3. The bone resorption rate cannot be deduced from static histomorphometric measurements.

1. Bone Balance, Volume Referent, B_v (#106)

This is the volume of bone gained or lost per unit volume of preexisting bone during unit time in mm³/mm³/year. It is given by:

$$B_v = \frac{V_2 - V_1}{V_1 \cdot (t_2 - t_1)} \qquad (21)$$

V_1 and V_2 are values for trabecular bone volume in the first and second biopsies and t_2 to t_1 is the time interval between the biopsies in years.

2. Bone Balance, Surface Referent, B_s (#107)

This is the volume of bone gained or lost per unit surface of trabecular bone per unit time in mm³/mm²/year. It is given by:

$$B_s = \frac{V_2 - V_1}{S_{V_1} \cdot V_1 \cdot (t_2 - t_1)} \qquad (22)$$

where S_{v1} is the trabecular specific surface (#17) in the first biopsy.

It is now possible to calculate a variety of derived quantities relating to resorption.

3. Bone Resorption Rate, Surface Referent, R_s (#201)

This is the amount of bone resorbed per unit surface of trabecular bone per unit time in mm³/mm²/year. It is given by:

$$R_s = F_{s_2} - B_s \qquad (23)$$

where F_{s2} is the bone formation rate total surface referent from the second biopsy. This can be combined with Equation 23 to give:

$$R_s = F_{s_2} - \left[\frac{V_2 - V_1}{S_{V_1} \cdot V_1 \cdot (t_2 - t_1)} \right] \qquad (24)$$

If B_s is measured by nonhistomorphometric means it must be expressed as a fractional decrease or increase during the time of observation, and F_{s2} must be multiplied by this fraction.

4. Bone Resorption Rate, Volume Referent, R_v (#202)

This is the volume of bone resorbed per unit volume of preexisting bone during unit time. It is given by:

$$R_v = F_{v_2} - B_v \qquad (25)$$

where F_{v2} = bone formation rate, volume referent, from the second biopsy in mm^3/mm^3/year. This can be combined with Equation 22 to give:

$$R_v = F_{v_2} - \left[\frac{V_2 - V_1}{V_1 (t_2 - t_1)} \right] \qquad (26)$$

For these measurements to be useful mean values must be obtained on at least 20 subjects, so that the estimate of B_v or B_s will have less variance than the precision of the histomorphometric measurements. If the change in bone balance is smaller than the precision of measurement and is less than 10% of the bone formation rate, then the best estimate of bone resorption rate by histomorphometric means is the bone formation rate from a single biopsy or:

$$R_s \simeq F_s \qquad (27)$$

To determine the effect of a drug on bone balance by histomorphometry, two serial biopsies are required before beginning treatment and two again during treatment. In most instances this would result in excessive discomfort for the patient, excessive delay in obtaining results, and considerable expense. Consequently, the best strategy for most human work will be to accept the bone formation rate as the best estimate of bone resorption in a single biopsy, and to measure bone balance independently by nonhistomorphometric means.

5. Bone Resorption Rate, Resorption Surface Referent, R_r (#203)

This is the volume of bone resorbed per unit area of resorption surface (active + inactive) per unit time in mm^3/mm^2/year. It is given by:

$$R_r = \frac{R_s}{S_r} \quad \text{or} \quad R_r = \frac{\#201}{\#14} \qquad (28)$$

6. Bone Resorption Rate, Active Surface Referent, R_a (#204)

This is the volume of bone resorbed per unit area of active resorption surface per unit time in mm^3/mm^2/year. It is given by:

$$R_a \simeq \frac{R_s}{S_{ra}} \quad \text{or} \quad R_a \simeq \frac{\#205}{\#15} \qquad (29a)$$

or assuming the bone balance is equal or close to zero,

$$R_a \simeq \frac{F_s}{S_{ra}} \quad \text{or} \quad R_a \simeq \frac{\#103}{\#15} \qquad (29b)$$

This quantity is a measure of collective osteoclast work efficiency.

7. Bone Resorption Rate, Osteoclast Referent, R_o (#205)

This is the area of bone resorbed per osteoclast in mm^2/cell/year. It is given by:

$$R_o = \frac{R_a}{C_{ra}} \quad \text{or} \quad R_o = \frac{\#204}{\#21} \tag{30}$$

As for bone formation rate, osteoblast referent, this quantity is not expressed as a volume because C_{ra} is not extrapolated to the trabecular surface in three dimensions (although it can be by anyone who understands the geometry and algebra involved). It is an index of individual osteoclast work efficiency.

8. Bone Resorption Rate, Osteoclast Nucleus Referent, R_{on} (#206)

This is the area of bone resorbed per osteoclast nucleus in unit time in mm^2/osteoclast nucleus/year. It is given by:

$$R_{on} = \frac{R_o}{N_r} \quad \text{or} \quad R_{on} = \frac{\#205}{\#22} \tag{31}$$

It is another index of individual osteoclast work efficiency.

9. Linear Bone Resorption Rate, M_{ra} (#207)

This is the linear thickness of bone resorbed beneath osteoclasts during unit time. It is given by:

$$M_{ra} = R_a \quad \text{or} \quad M_{ra} = \#204 \tag{32}$$

Note that M_{ra} and R_a are numerically equal but are in different dimensions because each have different referents and express different biologic functions, analogous to the relationship between F_a and $\overline{M}$. M_{ra} represents the resorption counterpart of the appositional rate $(\overline{M})$ and is a useful index of cell level resorption activity.

C. Quantities Related to Bone Tissue Turnover and Duration of Cell Level Activity

The bone turnover rate is of value in understanding bone physiology in health and disease and in comparing histomorphometric with other methods of measuring turnover such as by radioactive calcium kinetics. Recent information indicates that both of these methods of measuring total bone turnover give similar results, about 10% of the entire skeleton per year in the adult human. If bone balance is zero, turnover is equal to the bone formation (or resorption) rate. If this is not so, the simplest approach (others exist and have their uses) is to define bone turnover in mm^3/mm^2/year as the arithmetic mean of the resorption and formation rates, or:

$$V_t = 1/2(F_v + R_v) \tag{33}$$

The life span of Basic Multicellular Units (BMU), referred to as sigma, is inversely proportional to its work efficiency. Alterations in sigma, usually prolongations, accompany many metabolic bone diseases. Sigma is probably the most important datum in understanding and working with bone behavior in health and disease.[6] Sigma can be subdivided into shorter periods corresponding to successive phases of the remodeling sequence. The bone-forming and resorbing surfaces alternately turn on and off so that each includes a fraction that is inactive at the time of the biopsy. These fractions must be included in the computation of sigma, since the total time taken to construct a new BMU must include the durations of both the off and on states.

1. Radial Closure Rate, M_f (#110)

This is the appositional rate averaged over the entire bone-forming surface in mm/year. It is given by:

$$M_f = \frac{\overline{M} \cdot S_{fa}}{S_f} \quad \text{or} \quad \frac{\#18 \cdot \#16}{\#13} \tag{34}$$

M_f is numerically equal to F_f but expressed in different units because of the different referents and different biologic function which it represents. M_f includes the "OFF" as well as the "ON" times of the typical bone-forming center.

2. Radial Resorption Rate, M_r (#210)

This is the mean thickness of the layer of old bone eroded during unit time averaged over the whole resorption surface (including both active and inactive parts) in mm/year. It is analagous to the radial closure rate (M_f) just defined. It is given by:

$$M_r = R_r \quad \text{or} \quad M_r = \#203 \tag{35}$$

As for previous pairs of related quantities ($\overline{M} = F_a$, $M_f = F_f$, $M_{ra} = R_a$), M_r and R_r are numerically equal but in different dimensions. M_r is the key datum needed to define σ_r; the duration of the resorption phase of the remodeling cycle or the mean time taken to erode a Howship's lacunae, is described below.

3. The Bone Formation Period, σ_f (#401)

This is the period of time in years required to complete the formation phase of the average BMU. It is given by:

$$\sigma_f = \frac{MWT}{M_f} \quad \text{or} \quad \sigma_f = \frac{\#19}{\#110} \tag{36}$$

4. The Bone Resorption Period, σ_r (#402)

This is the period of time in years required to complete the resorption phase of an average BMU. It is given by:

$$\sigma_r = \frac{MWT}{M_r} \quad \text{or} \quad \sigma_r = \frac{\#19}{\#210} \tag{37}$$

In practice it is usually calculated in a different manner. From Equations 37 and 35 it follows that:

$$\sigma_r = \frac{\sigma_f \cdot M_f}{M_r} \tag{37a}$$

This can be combined with Equations 17, 35, 36, and 29a to give:

$$\sigma_r = \sigma_f \cdot \frac{S_r}{S_f} \tag{38}$$

These calculations assume in part that the bone resorption rate is approximately equal to the bone formation rate, that the patient was in a steady state at the time of labeling and biopsy, and that predominently remodeling rather than modeling activity is represented in the measurements.

5. Sigma, σ (#403)

This is the period of time required to complete both the resorption and formation phases of an average BMU, including "ON" and "OFF" states. It is given by:

$$\sigma = \sigma_f + \sigma_r \tag{39}$$

D. Miscellaneous Quantities

1. Trabecular Surface/Volume Ratio, S/V (#504)

This is the ratio of the trabecular surface area to its volume in mm²/mm³. It is given by:

$$S/V = \frac{S_v}{V} \quad \text{or} \quad S/V = \frac{\#17}{\#11} \tag{40}$$

2. Trabecular Volume/Surface Ratio, V/S (#505)

This is the reciprocal of S/V, and in units of mm³/mm² is given by:

$$V/S = \frac{V}{S_v} \quad \text{or} \quad V/S = \frac{1}{S/V} \tag{41}$$

To illustrate the interconvertibility between surface and volume referents Equation 18 can be combined with Equation 17 to give:

$$F_v = \frac{S_v}{V} \cdot F_s \tag{42}$$

In other words, volume-based bone formation (F_v) = surface-based bone formation (F_s) multiplied by the surface-to-volume ratio (S_v/V). Any quantity in the surface referent can be converted to volume referent by multiplying by S/V, and any quantity in volume referent converted to surface referent by multiplying by V/S.

Another use for V/S is in the computation of bone balance-surface referent. Equation 23 may be rewritten:

$$B_s = \frac{1}{(t_2 - t_1)} \cdot \frac{V_2}{S_{v_1}} - \frac{V_1}{S_{v_1}}$$

Over relatively short time intervals such as 3 months, S_v(trabecular bone-specific surface) will usually not change significantly, so that S_{v_2} is approximately equal to S_{v_1}. From this it follows that:

$$B_s = \frac{1}{(t_2 - t_1)} \cdot V/S_2 - V/S_1 \tag{43}$$

a cell-level referent) makes new bone at about two thirds normal speed. Note that now we not only have dimensions, space, and time as referents, we also have levels of biological organization: cell, tissue, and organ, to join with volume and surface space dimensions and time.

This example requires that we invoke even more referents to extract from the raw numbers any biological meaning lying in the situation from which they came. For example these investigators (real ones but for obvious reasons not identified) assumed the dynamic data found in their patients represented the states which made their skeletons osteopenic. But in fact these patients suffered from osteogenesis imperfecta and their bone tissue volume deficits arose as a failure to accumulate normal amounts of bone *during growth,* not as a pathological loss in adult life of previously normal bone stores.

As others in this book affirm, accumulation of bone during growth depends upon growth and modeling activities while losses in adult humans depend upon BMU-based remodeling. The one equals apples and pears, the other walnuts, so what one learns of the latter by kinetic analysis reveals absolutely nothing useful about the former. And in other words, because the study group consisted of adults (which itself represented another referent, and a chosen rather than an accidental one) the data could not possibly shed any light upon the childhood cause or development of their disease.

Now as referents we find growth and modeling systems, and separately, remodeling systems, i.e., *kinds* of integrated actitivies which exist in the company of the other referents already named. As already implied, the above examples abstract real studies published on subjects with osteogenesis imperfecta.

In one final example, organ culture workers have repeatedly shown in their systems that calcitonin reduces resorption of isolated fetal mouse and rat calvaria. This has led to a number of human treatment trials of senile and postmenopausal osteoporosis which all failed. The failures were easily predictable in advance (and some of the sigma group become somewhat unpopular in some quarters by making exactly those predictions) *solely on the basis of the choices of referents involved,* as follows.

Adult acquired osteoporoses derive from BMU malfunctions since adults comprise remodeling-only sytems (a referent). The calvaria, however, comprise modeling-only systems (a different referent).

The osteoporoses in question present bone volume deficits only where bone touches marrow, but not on periosteal bone surfaces (a juxtapositional or interactive referent), while the calvaria in question have no marrow. The osteoporoses arose primarily on cortical-endosteal and trabecular envelopes while the calvaria have only periosteal surfaces, no endosteal cavity at all. The failures of those trials verified the accuracy of the above predictions.

In conclusion, the choice of referent can make the whole difference between the ability to find meaning in an experiment or group of data, and total failure. Problems relating to choices of referent remain widespread in the skeletal field although histomorphometrists seem to do better in that regard than most others, perhaps because that work forcibly confronts one with those problems and requires their resolution.

V. CONCLUDING REMARKS

The foregoing material presents in simple algebraic and arithmetical form an interlocking, integrated system of core relationships which allow selected measurements, whether made at the light microscope or otherwise, to provide both static and kinetic data concerning trabecular bone tissue turnover parameters and their determining tissue mechanisms. The system can provide dynamic, quantitative, and usefully accurate data for activities from the cell nucleus up to the intact organ, for formation, resorption, and bone balance, and for varied static compositional and dimensional properties as well. Many further parameters can be derived from the present primary data of the system and, likely, in the future it may become necessary to add further primary data to those accounted for herein. After all, this active field has evolved rapidly, and it seems as likely that it has now attained its zenith in that respect as it does that the U.S. government will set about this year paying off its national debt.

It becomes appropriate in these closing remarks to look briefly at two matters pertinent both to this chapter and to the entire book: the relationship of histomorphometry to "BMU theory" and the evolution of the latter.

As for the former, the mechanical histomorphometric procedures performed at the microscope have very limited value and usefulness by themselves. What has made them extraordinarily useful in this field in the past 15 years and likely will for the next 15 stems not from those procedures, as elegant in their conception as some of them may seem. Rather it stems from a new and far better understanding of the intraskeletal tissue mechanisms which, by responding to factors arising outside the skeleton, directly cause most of its malfunctions and diseases. That new understanding has arisen since about 1963, the writer calls it BMU theory, and with respect to the concerns of this book *it serves the function of telling the histomorphometrist what to look at and measure in some disease to find meaningful information and answers instead of the irrelevant or the trivial.* Effective use of bone histomorphometric expertise per se therefore implies an adequate understanding of how the skeletal system actually works, for without that understanding guiding them the tools of histomorphometry become expensive toys. Some proof that the wedding of that methodology and theory does have the power and other attributes implied above lies in the fact that each of the various individuals who comprise the small clan which has used BMU theory since 1966 has as a partial result become something of a legend in the field in his own time (while no "hers" fly with that clan at the moment — at least known to the writer — at least one does seem to be sprouting wings).

As for BMU theory, its history is being reviewed elsewhere in greater detail but it began between 1963 and 1966 in the writer's laboratory at Henry Ford Hospital with the discovery that the skeleton has previously unknown and unsuspected analogs of the renal nephron, analogs equally as important and as directly responsible for its diseases, and governed by collective game rules equally unique but far removed from those which apply to individual cells, as proves true again of the the nephron in relation to the kidney. In various chapters of this book some of the names we gave those analogs appear time and again: envelopes, BMU, BSU, modeling, remodeling. Their spinoffs include unfamiliar concepts such as appositional rate, radial resorption rate, envelope-specific behavior, sigma, transients and steady states, referents, ON-OFF, activation frequencies (i.e., birth rates), and so on.

The peculiar and time-dependent nature of many such phenomena required very highly specialized techniques to make them visible and self evident, techniques which had the unusual property of recording temporally serial in vivo events in a single ledger in such a way that a later reading ex vivo displayed the events at one time in proper sequence to the observer's eyes.

The writer helped to build such techniques out of the raw lumber of tissue-time markers, undecalcified sections, the light microscope, stereologic and biologic principles, and mathematics. The information exposed by those techniques was totally new and strange at the time to the field at large. It required new concepts and a new jargon to handle it efficiently, both of which helped to make it controversial and hard for others to understand. The overwhelming majority of people in the field who perused the earlier publications coming out of that work had neither personal access to those special techniques nor to others who did, and so they had to take embryo BMU theory and the underlying physiology on faith, reserve judgment or contest it — or even come to Detroit to see for themselves. A few did actually come, they saw and at least in part for that reason each is one of those referred to earlier who has become something of a legend in his own time.

But they were few in number, and overwhelmingly outnumbered before 1980, so for over 15 years and simply by default BMU theory remained the special property of a mere handful, the sigma bone group.

Now the landscape visible to this writer changes. Earlier controversies over the worth of BMU theory begin to subside while most of its one-time students and apprentices now instruct others including their former teacher, and a growing number of scientists both improve that theory and the histomorphometric technology wedded to it, and use it as a tool to mine pathophysiologic gold and diamonds such as only dreamed of or not even conceived of 15 years ago.

REFERENCES

1. **Milch, R. A., Rall, D. P., and Tobie, J. E.,** Bone localization of the tetracyclines, *J. Natl. Cancer Inst.,* 19, 87, 1957.
2. **Jowsey, J., Kelly, P. J., Riggs, B. L., Bianco, A. J., Scholz, D. D., and Gershon-Cohen, J.,** Quantitative microradiographic studies of normal and osteoporotic bone, *J. Bone Jt. Surg.,* 47A, 785, 1965.
3. **Merz, W. A. and Schenk, R. K.,** Quantitative structural analysis of human cancellous bone, *Acta Anat.,* 75, 54, 1970.
4. **Frost, H. M.,** *Bone Remodeling Dynamics,* Charles C Thomas, Springfield, Ill., 1963.
5. **Frost, H. M.,** *Mathematical Elements of Lamellar Bone Remodeling,* Charles C Thomas, Springfield, Ill., 1964.
6. **Frost, H. M.,** *Bone Remodeling and Its Relation to Metabolic Bone Disease,* Charles C Thomas, Springfield, Ill., 1973.
7. **Frost, H. M.,** *Orthopaedic Biomechanics,* Charles C Thomas, Springfield, Ill., 1973.
8. **Frost, H. M.,** Tetracycline based analysis of bone dynamics, *Calcif. Tissue Res.,* 3, 211, 1969.
9. **Frost, H. M.,** Bone histomorphometry: errors due to malorientation of the section plane, in *Second International Workshop on Bone Morphometry,* Meunier, P., Ed., University of Claude Bernard, Lyon, 1977, 69.
10. **Frost, H. M.,** Bone histomorphometry: theoretical correction of appositional rate measurements in trabecular bone, in *Second International Workshop on Bone Morphometry,* Meunier, P., Ed., University of Claude Bernard, Lyon, 1977, 361.
11. **Frost, H. M.,** Bone histomorphometry: empirical correction of appositional rate measurements in trabecular bone, in *Second International Workshop on Bone Morphometry,* Meunier, P., Ed., University of Claude Bernard, Lyon, 1977, 371.
12. **Frost, H. M.,** Bone histomorphometry: a method of analysis of trabecular bone dynamics, in *Second International Workshop on Bone Morphometry,* Meunier, P., Ed., University of Claude Bernard, Lyon, 1977, 455.
13. **Frost, H. M.,** Skeletal manifestations of disease, *Clin. Orthop. Rel. Res.,* 49, 3, 1966.
14. **Frost, H. M.,** The numerical and statistical approach, in *First Workshop on Bone Morphometry,* Jaworski, Z. F. G., Ed., University of Ottawa Press, Ottawa, Canada, 1976, 24.
15. **Frost, H. M.,** Measuring bone dynamics: the property called sigma, in *First Workshop on Bone Morphometry,* Jaworski, Z. F. G., Ed., University of Ottawa Press, Ottawa, Canada, 1976, 246.
16. **Frost, H. M.,** The origin and nature of transients in human bone remodeling dynamics, in *Clinical Aspects of Metabolic Bone Disease,* Frame, B., Parfitt, A. M., and Duncan, H., Eds., Excerpta Medica, New York, 1973, 124.
17. **Merz, W. A.,** Die Streckeumessung an gerichteten structuren in Midroskop und ihrre Anwendung zur Bestimmung von Oberflachen volumen, relationen in knochengewebe, *Mikroskopie,* 22, 132, 1967.

18. **Schenk, R. K., Merz, W. A., and Muller, A.** Quantitative histological study on bone resorption in human cancellous bone, *Acta Anat.*, 74, 44, 1969.

19. **Schenk, R. K., Olah, A. J., and Merz, W. A.**, Bone cell counts, in *Clinical Aspects of Metabolic Bone Disease,* Frame, B., Parfitt, A. M., and Duncan, H., Eds., Excerpta Medica, New York, 1973, 193.

20. **Schenk, R. K.**, Basic stereological principles, in *Frist Workshop on Bone Morphometry,* Jaworski, Z. F. G., University of Ottawa Press, Ottawa, Canada, 1976, 21.

21. **Schenk, R. K.**, Histological estimates of bone resorption surface perimeter in iliac bone, in *First Workshop on Bone Morphometry,* Jaworski, Z. F. G., Ed., University of Ottawa Press, Ottawa, Canada, 1976, 153.

22. **Meunier, P.**, La Dynamique du Remaniement Osseux, Ph.D. Thesis, University of Claude Bernard, Lyon, 1968, 5.

23. **Meunier, P., Edouard, E., Richard, D., and Laurent, J.**, Histomorphometry of osteoid tissue. The hyperosteoidoses, in *Second International Workshop on Bone Morphometry,* Meunier, P., Ed., University of Claude Bernard, Lyon, 1977, 249.

24. **Meunier, P., Aaron, J., Edouard, C., and Vignon, G.**, Osteoporose et involution adipeuse de la population cellulaire de la moelle. Etude quantitative de 51 biopsies iliaques, *Presse Med.,* 78, 531, 1970.

25. **Meunier, P., Aaron, J., Edouard, C., and Vignon, G.**, Ostéoporose et involution adipeuse de la population cellulaire de la moelle. Etude quantitative de 51 biopsies iliaques, *Presse Med.,* 78, 531, 1970.

26. **Bordier, P., Matrajt, H., Miravet, L., and Hioco, D.**, Measure histologique de la résorption osseuse dans l'osteoporose: étude préliminarie, Calc. Tiss. Collect. Coll. Univ. Liege, 1965.

27. **Courpron, P.**, Données Histologiques Quantitatives sur le Vieillissement Osseux Humain, Ph.D. Thesis, University of Claude Bernard, Lyon, 1972, 1.

28. **Courpron, P.**, Amount of bone in iliac crest biopsy. Significance of the trabecular bone volume. Its values in normal and pathological conditions, in *Second International Workshop on Morphometry,* Meunier, P., Ed., University of Claude Bernard, Lyon 1977, 39.

29. **Courpron, P., Giuorx, J. M., Bringuier, J. P., and Meunier, P.**, Histomorphometrie de l'os spongieux iliaque. Influence des techniques de préparation et de lecture des coupes histologiques sur la détermination du volume trabeculaire osseux iliaque, *Lyon Med.,* 232, 515, 1974.

30. **Bressot, C., Courpron, P., Edouard, C., and Meunier, P.**, *Histomorphometrie des Ostéopathies Endocriniennes,* University of Claude Bernard Press, Lyon, 1976, 1.

31. **DeLesse, A.**, Procédé mécanique pour déterminer la composition des roches, *Ann. Mines,* 13, 379, 1848.

Chapter 8

BONE HISTOMORPHOMETRY: CORRECTION OF THE LABELING "ESCAPE ERROR"

H. M. Frost

TABLE OF CONTENTS

I. INTRODUCTION

When living subjects receive two tetracycline bone tissue-time markers spaced in time so as to label bone formation sites (BMU) with paired fluorescent markers, substantial fractions of the bone-forming surface can display only one marker. One factor that accounts for this is that some new bone-forming systems begin to form between the two markers, thereby taking the second marker but "escaping" the first, while others complete formation (or at least calcification of new matrix) between the two markers, thus "escaping" the second (see Chapter 4).

Another factor that may account for some of the singly labeled surface is the presence of osteoid surface (forming surface) which is "turned off" yet still takes label.[1] The existence of these singly labeled "resting" osteoid surfaces forms a subject of some current controversy. If one could be certain that resting osteoid surface would never take a single label, then the analysis contained in this chapter would be unnecessary. But at the present time this issue remains unresolved. Therefore, the analysis contained in these pages is deemed helpful, both in understanding the problem and in pointing the direction for further research.

The labeling "escape phenomenon" will cause "escape errors". This chapter describes a way to quantitate and to correct those errors without making additional or new kinds of measurements (although one must follow one of the writer's double labeling analytical protocols).[2-5] It involves a transformation from a time domain to a space domain first circulated to the sigma group in 1978.

II. THE NATURE OF THE ESCAPE EFFECT AND ITS ERROR

The "ladder graph" in Figure 1 depicts the BMU events in a bone sample. Time proceeds to the right. The dimensionless vertical axis demonstrates each BMU separately from its temporally overlapping neighbors. The horizontal rungs represent individual BMU in their formation phase (the resorption phase has been left out). Bone-forming activity in a given BMU begins on the left line and proceeds horizontally to its conclusion on the right, taking a total time period signified by sigma fa (σ_{fa}).*

Just as one reads this page, following the ladder down and to the right traces the sample's future while following it up and to the left traces its past. The oblique line of origin on the left contains the array of births of new BMU and on the right their array of terminations.

A "flash" tetracycline marker, M_1, would label all BMU actively forming bone at that instant. The vertical M_1 intersections with the horizontal "rungs" of the ladder indicate exactly that. A second "flash" marker, M_2, given at a later time would label all bone-forming systems active at that second and later moment. M_1 and M_2 comprise a double label.

The intersections of the two marker lines with the termination line show about four systems labeled so close to the termination of their bone-forming activities that they escaped the second marker. Likewise where the two marker lines intersect the origin line, about four other BMU began forming after deposition of the first marker so they escaped it. In the interval between these two extremes 14 BMU took both markers. Thus of the 22 labeled BMU, 57% took both, 43% only one.

Figure 2 upper right rediagrams the situation. The shaded rectangle embraces those BMU which took both markers. The triangular regions embrace additional BMU that

---

* Sigma in this instance is written as σ_{fa}. It is the bone-forming period uncorrected for any of the forming surfaces in the "OFF" state and is to be distinguished from σ_f (Chapter 7) which takes into account such "OFF" surfaces.

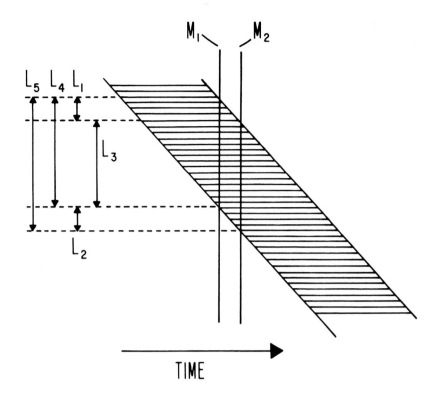

FIGURE 1. "Ladder graph" depicting the BMU events in a bone sample.

took only single markers. Those in the upper triangle took only the first marker, those below only the second.

Now with everything else constant but in a different subject whose σ_{fa} is one third of the previous value, as in Figure 2 lower left, the relative fractions of doubly and singly labeled BMU have changed so that only 30% of the active surface shows double labels, 70% of it single ones. If histomorphometric analysis assumed that only double labels identified instantaneously active bone-forming surface (an assumption widely accepted), the formation and resorption rates and sigma values computed for this sample would be more than 200% in error.

Having shown the nature of the problem the text next describes how to find the instantaneously active surface without measuring singly labeled surface and explains why it follows that procedure.

III. CORRECTING THE ESCAPE ERROR

A. Cumulative Fractions of Active Surface Doubly and Singly Labeled

In Figure 1 the vertical lines on the left "map" the progression in time of the events on the ladder to the right.* Thus L_1 corresponds to and maps on the vertical axis the time between the first and second markers, and also the extent of that part of the surface singly labeled with M_1 only. The line, L_2 maps the same time interval and also the extent of surface singly labeled by M_2 only. The line, L_4, maps in an identical sense the value of σ_{fa} as well as the size of the doubly labeled surface plus half the singly labeled surface.

* Technically the graph also transforms time domain events to surface domain events.

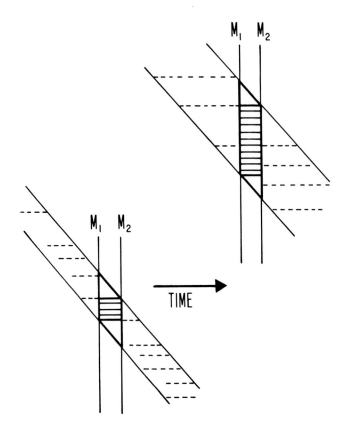

FIGURE 2.   Rediagram of same BMU events in a bone sample.

A number of useful identities can be written to describe what is displayed in Figure 1. In the following discussion the equals sign with a period over (i.e., $\doteq$) means "corresponds to and proportionally equals".

L_1, L_2, L_3, L_4, and L_5 correspond proportionally to real time periods and $M_1 - M_2$ = MI which is the marker interval.

$$L_1 = L_2 \doteq MI \tag{1}$$

$$L_4 \doteq \sigma_{fa} \tag{2}$$

$$L_5 = \sigma_{fa} + MI \tag{3}$$

$$L_3 = \sigma_{fa} - MI \tag{4}$$

It is important that σ_{fa} here signifies the BMU formation period uncorrected for any "OFF" or "resting" osteoid and is calculated as

$$\sigma_{fa} = \frac{MWT}{\overline{M}} \tag{5}$$

where MWT = the mean wall thickness and $\overline{M}$ = the appositional rate, both obtained from measurements at the microscope (see Chapter 7).

Another concept that must be understood is the "cumulative active forming surface''. This is the forming surface seen in the microscopic section as the doubly labeled surface plus the entire singly labeled surface that is so labeled because of the label escape phenomenon. It excludes any singly labeled surface that might lie on resting osteoid or that appears singly labeled due to unresolvable double labels.

Let DL signify the fraction of the total cumulative active forming surface doubly labeled and SL the fraction singly labeled.

Then:

$$DL \doteq \frac{L_3}{L_5}$$

$$\therefore \quad \text{(from Equation 3, 4)} \quad DL \doteq \frac{\sigma_{fa} - MI}{\sigma_{fa} + MI} \tag{6}$$

$$\text{and} \quad SL \doteq \frac{L_1 + L_2}{L_5}$$

$$\therefore \quad \text{(from Equation 1, 3)} \quad SL \doteq \frac{2MI}{\sigma_{fa} + MI} \tag{7}$$

$$\text{Note that } DL + SL \doteq \frac{L_1 + L_2 + L_3}{L_5} = 1 \tag{8}$$

$$\text{and} \quad \frac{SL}{DL} \doteq \frac{2MI}{\sigma_{fa} - MI} \tag{9}$$

One more datum required for the analysis is the fraction of total trabecular surface with double label. This is obtained by direct measurement at the microscope and is expressed as S_{fa} (see Chapter 7).

B. The Instantaneously Active Surface

The object of this analysis is to determine the instantaneously active forming surface (S_{fa}') without depending on direct measurement of any singly labeled surface (note the prime in the preceding symbol.) It can be expressed as

$$S_{fa}' \doteq L_4 \tag{10}$$

but L_4 must be converted somehow from its time units (remember it is equal to σ_{fa}) to its proportional trabecular surface units. This may be done as follows. Let DL_f equal the fraction of the total trabecular surface that is doubly labeled and SL_f equal the fraction of total trabecular surface that is singly labeled due to the escape phenomenon. Then:

$$DL_f = S_{fa} \tag{11}$$

and further,

$$\frac{SL}{DL} = \frac{SL_f}{DL_f} \tag{12}$$

The only unknown term in this equation is SL_f, the fraction of total trabecular surface that is singly labeled due to the escape phenomenon. By rearranging this equation to solve for the unknown we have

$$SL_f = \frac{SL}{DL} \cdot DL_f \qquad (13)$$

$$\therefore \text{ (from Equations 9 and 11)} \cdot SL_f = \frac{2MI}{\sigma_{fa} - MI} \cdot S_{fa} \qquad (14)$$

Now, we can express $S_{fa}{}'$ in another manner as

$$S_{fa}{}' \doteq L_4 \doteq DL_f + 1/2\, SL_f \qquad (15)$$

$$\therefore \text{ (from Equations 11 and 14) } S_{fa}{}' = S_{fa} \left(1 + \frac{MI}{\sigma_{fa} - MI} \right) \qquad (16)$$

This expresses $S_{fa}{}'$ in terms of the primary data.

All the terms in this equation are now known and none of the singly labeled surfaces in the microscopic section have been measured in order to arrive at a value for $S_{fa}{}'$.

In the analytical system in Chapter 7, one may simply substitute $S_{fa}{}'$ as derived in this text for the measured value of S_{fa}.

The concepts, "instantaneously active forming surface" and "cumulatively active forming surface" should be further elucidated. The former is that fraction of trabecular surface engaged in forming bone at any given instant in time, i.e., that fraction of trabecular surface which takes a label at any one point in time because it is active or "ON". If one could administer two tetracycline labels infinitely close together in time and could still resolve them under the microscope, the entire instantaneously active forming surface would contain double label. The instantaneously active forming surface, $S_{fa}{}'$, is the datum required to accurately calculate all bone formation and resorption rate parameters for a given specimen, the former, however, only above the cell level of organization.

On the other hand, the "cumulatively active forming surface" is that fraction of the trabecular surface taking two labels plus all surface which labeled singly due to the escape phenomenon.

As Figure 2 shows, prolonging the marker interval relative to the value of σ_{fa} augments the singly labeled surface and diminishes the doubly labeled surface. When the marker interval exceeds the value of σ_{fa} no double labels at all can appear.

C. The Marker Interval: σ_{fa} Ratio

Equation 16 may be rewritten as follows:

$$S_{fa}{}' = S_{fa} \left(1 + \frac{MI/\sigma_{fa}}{1 - MI/\sigma_{fa}} \right)$$

This expresses the correction factor needed to derive $S_{fa}{}'$ as a function of the ratio of the marker interval to σ_{fa}. Designating this ratio as R we may write:

$$DL = \frac{1 - R}{1 + R}$$

$$SL = \frac{2R}{1 + R}$$

$$SL/DL = \frac{2R}{1 - R}$$

$$\text{and } \frac{S_{fa}{}'}{S_{fa}} = 1 + \left(\frac{R}{1 - R}\right)$$

In Table 1 the left hand column lists possible values for R,r (the MI:σ_{fa} ratio). Columns 2 and 3 provide corresponding values for DL and SL and the right hand column shows the corresponding correction factors needed to obtain the instantaneously active surface from the doubly labeled surface.

Reflection upon the data in Table 1 will reveal the following:

1. As σ_{fa} increases, the escape errors decrease but not in proportion, and as the marker interval decreases the escape errors decrease also not in proportion.
2. Given normal adult trabecular bone σ_{fa} values in the 30 to 60 day range, then a 20-day marker interval (a commonly chosen time span) leads to a 41 to 285% escape error in histomorphometrically derived values of $S_{fa}{}'$ and the bone formation (and resorption) rates. A 40-day marker interval generates hopelessly large errors.
3. When the MI equals or exceeds σ_{fa} no double labeling at all can occur so no useful kinetic data whatsoever can be obtained from such a sample by histomorphometric means. When the MI approaches but still remains below the value of σ_{fa}, the correction factors become unwieldy and prudence suggests questioning their reliability, which leads to a major statement. In designing experiments try to keep the MI/σ_{fa} ratio as small as feasible, at least below 0.5 where the escape error equals 100% of $S_{fa}{}'$ and preferably below 0.2 when the error equals 25% of $S_{fa}{}'$.

Shortening the marker intervals cannot eliminate escape errors or reduce them to trivial levels. Given the usual appositional rate of between 1 and 2 μm/day, the effect of section obliquity, the thickness of the tetracycline bands, and fuzziness of their edges, one cannot go below a marker interval of about 5 to 10 days in most human studies and still be able to resolve all the labels well enough in the microscope to obtain an interlabel width measurement. Consequently, the escape error problem cannot be eliminated.

In summation then: *the marker interval: σ_{fa} ratio specifically determines the magnitude of the labeling escape error.* That may prove true only in part, however, as indicated shortly.

D. Bone Formation and Resorption Rates

True bone formation rate values (in all referents except the cellular level) exceed those computed by previous formulas[2-5] which assumed that active surface constituted only doubly labeled surface. The formulas remain correct, but the data were not accurate. More accurate data can be supplied by applying the contents of this text. The correct values will increase proportionally to the increase in the true active bone-forming surface relative to the doubly labeled part of that surface. Bone resorption rates

Table 1
INSTANTANEOUSLY ACTIVE SURFACE (S_{fa}'): CORRECTION FACTORS FOR THE ESCAPE EFFECT

MI/σ_{fa}	DL	SL	$S_{fa}' = S_{fa} \times$ the following
0.01	0.980	0.020	1.01
0.03	0.942	0.058	1.03
0.05	0.905	0.095	1.05
0.05	0.905	0.095	1.05
0.07	0.869	0.131	1.08
0.10	0.818	0.182	1.11
0.13	0.770	0.230	1.15
0.16	0.724	0.276	1.19
0.19	0.724	0.276	1.23
0.24	0.613	0.387	1.32
0.29	0.550	0.450	1.41
0.34	0.493	0.507	1.51
0.39	0.439	0.561	1.65
0.45	0.379	0.621	1.82
0.50	0.333	0.667	2.00
0.55	0.290	0.710	2.22
0.60	0.250	0.750	2.5
0.65	0.212	0.788	2.85
0.70	0.176	0.824	3.34
0.80	0.111	0.889	5.00
0.90	0.053	0.947	9.93
0.99	0.005	0.995	100.5
1.00	0	1.00	Infinity

Note: MI = marker intervals in days; σ_{fa} = sigma in days assuming no "OFF" periods (σ_{fa} = MWT/M); DL = fraction of the cumulatively active bone-forming surface taking double labels; SL = fraction of the cumulatively active bone-forming surface taking only single labels because of the escape effect. For a given MI/σ_f ratio (listed in the first column), multiply the actually measured doubly labeled surface in the biopsy, S_{fa}, by the correction factor in the same row, far right hand column, to find the true instantaneously active bone-forming surface, S_{fa}'. Use this corrected value (instead of S_{fa}) to compute formation rates, BMU sigma values, etc.

also will prove proportionally higher than previously computed, and at all levels of organization including the cellular one. Since annual normal adult human trabecular bone tissue turnover is probably at least triple previous estimates (i.e., $\simeq 50\%$ instead of $\simeq 15\%$ annually), yet the imbalance that sheds bone in normal adult life retains its original value ($\simeq 1\%$/year), the coupling between the resorption and formation activities implied by the BMU concept appears even more efficient than before.

IV. UNCERTAINTIES REGARDING THE LABEL ESCAPE PHENOMENON

Two major uncertainties in this area at the present time are (1) can "resting" osteoid surface appear as a singly labeled surface? and (2) how much of the singly labeled surface actually represents double labels too close together to resolve?

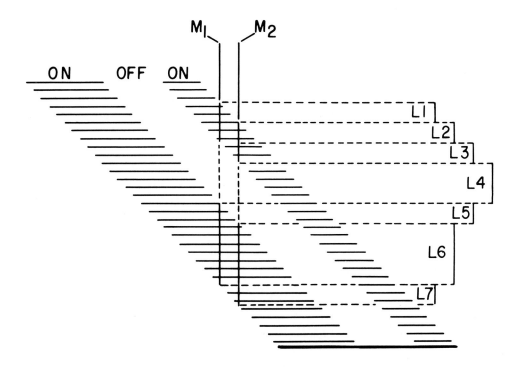

FIGURE 3. M_1 and M_2 are "flash" markers of tetracycline in a hypothetical system of BMU. The system is drawn in the same manner as Figure 1 only an interruption in bone formation is occurring in each BMU, signified by "OFF". L_1 through L_7 are lengths of time or trabecular surface and are marked as follows: L_1 — M_1 only (singly labeled); L_2 — M_1 + M_2 (doubly labeled); L_3 — M_2 (singly labeled); L_4 — neither (no label); L_5 — M_1 (singly labeled); L_6 — M_1 + M_2 (doubly labeled); L_7 — M_2 (singly labeled). (Courtesy of A. M. Parfitt).

Application of this model to predict the size of the cumulatively active forming surface in human samples may help solve such problems. At present little or no solid data support the hypothesis that resting osteoid can label singly, and the idea rests upon observations the writer made many years ago.

Nevertheless the method for finding S_{fa}' presented here may be more accurate than simply assuming that S_{fa}' equals all doubly labeled surface plus one half of the singly labeled surface as suggested by Melsen.[6] If there really is "bogus" single labeling due to singly labeled resting osteoid and unresolvable double label then Melsen's method will overestimate S_{fa}'. It seems clear that if the fractional doubly labeled surface is used to represent S_{fa}' then it will be underestimated.

Another concern centers on the effects of the temporary pauses in the formation phase of BMU originally suggested by the writer.[7] Figure 3 illustrates another ladder graph with the addition of a period of downtime or "OFF" during the course of formation in each BMU. To simplify the situation, the downtime is assumed to be in the same relative position and of the same duration throughout.

In this hypothetical situation there will be single labels produced by the escape phenomenon related to the downtime as well as the beginning and the end of formation. For example, L_3 represents a period when only the second label will take because the first label was given during the switched off period, but formation switches on before the second label. Likewise, L_5 represents taking of the first label followed by a switch off before the second label. These may be instances where a single label is produced by a period of interruption of bone formation but in each case the bone surface was normally active at the time the label was given, and these are not examples of "resting" osteoid taking a single label.

In the diagram, S_{fa}' is equivalent to the doubly labeled surface ($L_2 + L_6$) plus one half of the surface which is singly labeled due to the escape phenomenon at the beginning and end of BMU formation ($L_1 + L_7$)/2, plus one half of the surface singly labeled due to the escape phenomenon at the beginning and end of the downtime ($L_3 + L_5$)/2. In this case, the proposal to use one half the singly labeled surface is correct despite the added complication of the downtime.

Here too, however, the magnitude of the effect varies directly with the duration of the downtime, and for downtimes lasting 10% or less of σ_f any errors created thereby should prove tolerable.

More work is needed to determine the best way to measure and calculate S_{fa}', and we must determine whether "resting" osteoid can take a single label. If it can, then problems arise not only in determining S_{fa}' but in measuring the appositional rate as well. For example, if the first of two labels is taken up by a quiescent surface during downtime and bone formation subsequently resumes, the first label would remain *in situ,* while the second would be taken up. The distance between them, however, would be shorter than it should be given the appositional rate in question.

Resolution of these problems awaits further research.

ACKNOWLEDGMENTS

The writer is indebted and grateful to Dr. R. R. Recker for providing professionally executed drawings in Figures 1 and 2 and to A. M. Parfitt as well for Figure 3 and the ideas it addresses. They have both added materially to the ideas of the original manuscripts sent to the bone group sigma members, ideas which stemmed directly from comments and thoughts of Drs. F. Melsen and P. J. Meunier presented at the University of Utah Bone Workshop at Sun Valley in 1976, and further comments in 1979 by Dr. J. Smith of the University of Utah.

REFERENCES

1. **Frost, H. M.,** Observations on osteoid seams: the existence of a resting state, *Henry Ford Hosp. Med. Bull.,* 8, 220, 1961.
2. **Frost, H. M.,** *Bone Remodeling and its Relationship to Metabolic Bone Disease,* Charles C Thomas, Springfield, Ill., 1973.
3. **Frost, H. M.,** A method of analysis of trabecular bone dynamics, in *Bone Histomorphometry,* Meunier, P., Ed., Armour-Montague, Paris, 1977, 445.
4. **Frost, H. M.,** *Mathematical Elements of Lamellar Bone Remodeling,* Charles C Thomas, Springfield, Ill., 1964.
5. **Frost, H. M.,** Tetracycline based analysis of bone remodeling, *Calcif. Tissue Res.,* 3, 211, 1969.
6. **Melsen, F.,** Tetracycline double-labeling of iliac trabecular bone in 41 normal adults, *Calif. Tissue Res.,* 26, 99, 1978.

Chapter 9

THE PHYSIOLOGIC AND CLINICAL SIGNIFICANCE OF BONE HISTOMORPHOMETRIC DATA

A. M. Parfitt

TABLE OF CONTENTS

I. INTRODUCTION

Proponents of quantitative methods sometimes refer disparagingly to traditional microscopic pathology as little more than pattern recognition, but a poorly understood and possibly irrelevant measurement is no substitute for clear thinking and informed judgment. A set of numbers can be treated as a pattern as readily as a set of visual images; indeed, this is how physicians commonly interpret numerical data. Samples of bone tissue are usually examined to assist in the diagnosis or treatment of metabolic bone disease. For this purpose a purely qualitative appraisal may be entirely adequate,[1] but to enable concepts of pathogenesis to be formulated and tested, the quantitative approach is essential. If the results of bone histomorphometry are to be more than a number pattern associated with a particular disease or treatment effect, their interpretation must be based on a detailed grasp of the structure and function of bone and its cells.

II. THE MICROANATOMY AND PHYSIOLOGY OF BONE IN RELATION TO BONE HISTOMORPHOMETRY

A. General Description of the Function and Structure of Bone[2-7]

Bone is a specialized connective tissue in which the matrix of collagen fibers and ground substance is impregnated with a solid mineral. Bone is the main constituent of the bones, which are the individual organs of the skeletal system. This system has both biomechanical and metabolic functions. The hardness and rigidity of bone enable the skeleton to maintain the shape of the body, to protect the soft tissues, to provide a framework for the bone marrow, and to transmit the force of muscular contraction from one part of the body to another. The mineral content of bone contributes to the regulation of extracellular fluid (ECF) composition, particularly the ionized calcium concentration. The biomechanical competence of the bones is maintained by continuous replacement or turnover of whole volumes of bone tissue. This process is carried out by the osteoclasts and osteoblasts on bone surfaces which, together with their precursor cells, make up the remodeling system. The solid phase of bone mineral is in

Table 1
QUANTITATIVE FEATURES OF BONE

Whole bone	Cortical	Trabecular
Fractional volume (V_v) (mm³/mm³)	0.95	0.20
Surface/tissue volume (S_v) (mm²/mm³)	2.4	4.0
Surface/bone volume (S/V) (mm²/mm³)	2.5	20
Total bone volume (mm³)	$1.4 \cdot 10^6$	$0.35 \cdot 10^6$
Total tissue volume (mm³)	$1.5 \cdot 10^6$	$1.75 \cdot 10^6$
Total internal surface (mm²)	$3.5 \cdot 10^6$	$7.0 \cdot 10^6$
Bone structural units		
Length (mm)	2.5	1.0
Circumference or width (mm)	0.6	0.6
Wall thickness (mm)	0.075	0.040
Volume of contained bone (mm³)	0.065	0.025
Number/mm³ tissue volume	14	8
Number/mm³ bone volume	15	40
Total number in skeleton	$21 \cdot 10^6$	$14 \cdot 10^6$

Note: Comparison of cortical and trabecular bone. Representative quantitative data relating to whole bone and bone structural units. Cortical bone data derived mainly from rib and not applicable to ilium. Osteons are assumed to have a mean cement line diameter of 0.19 mm and a mean Haversian canal diameter of 0.04 mm. Trabecular bone data derived mainly from ilium. For further details see text.

Table 2
QUANTITATIVE FEATURES OF BONE[a]

Dry fat free weight (kg)	4
Total bone volume (mm³)	$1.75 \cdot 10^6$
Total tissue volume (mm³)	$3.25 \cdot 10^6$
Total bone calcium (g)	1050
Total internal surface (mm²)[b]	$11.0 \cdot 10^6$
Mean S_v (mm²/mm³)	3.4
Mean S/V (mm²/mm³)	6.3
Total number of BSU	$35 \cdot 10^6$

[a] Some representative quantitative characteristics of the whole skeleton.
[b] Includes $0.5 \cdot 10^6$ for cortical-endosteal surface.

contact with the bone ECF, which is separated from the systemic ECF by the flat cells which line all bone surfaces. Distributed throughout the bone are osteocytes lying within lentil-shaped lacunae, joined to each other and to the cells on the surface by filamentous prolongations of the cell running within canaliculae. The bone ECF extends throught the lacunar-canalicular system, and as on the surface, lies between the solid bone and the cells. Apart from the losses or gains of bone tissue in bulk, all transfers of ions between bone mineral and systemic ECF take place between these surface cells and osteocytes or through the surface cells, which collectively make up the homeostatic system.

It is convenient to distinguish two types of bone structure in the adult skeleton-cortical bone (or compacta) and trabecular bone (or spongiosa). The porosity, or soft tissue content, is usually less than 10% by volume in cortical bone and more than 75% by volume in trabecular bone but, particularly in the aging skeleton, all degrees of porosity may be found. About 80% of total bone mass is cortical bone, which forms

Table 3
SURFACE AND VOLUME MEASUREMENTS

	Cortical	Trabecular	Combined total
Resorbing surface			
Percent of total surface	0.6	1.2	1.0
Per unit bone volume (mm²/mm³)	0.015	0.24	0.06
Total (m²)	0.021	0.084	0.105
Forming surface			
Percent of total surface	3.0	6.0	5.0
Per unit bone volume (mm²/mm³)	0.075	1.2	0.3
Total (m²)	0.105	0.42	0.525
Inert surface			
Percent of total surface	96.4	92.8	94.0
Per unit bone volume (mm²/mm³)	2.41	18.56	5.64
Total (m²)	3.37	6.16	10.01
Mineralized bone volume			
Percent of total volume	99.9	99.2	99.75
mm³	$1.399 \cdot 10^6$	$0.347 \cdot 10^6$	$1.746 \cdot 10^6$
Osteoid volume			
Percent of total volume	0.1	0.8	0.24
cm³	1.4	2.8	4.2

Note: Representative static surface and volume measurements related to bone remodeling in cortical bone, trabecular bone, and whole skeleton. Values for cortical bone are based on data from the rib; values for trabecular bone are about one third those found in the ilium.

the outer wall of all bones, but the bulk of cortical bone is in the shafts of the long bones. The remaining 20% of total bone mass is trabecular bone, containing either hematopoietic or fatty marrow; it is found mainly in the bones of the axial skeleton and in the ends of the long bones. The plates and bars of trabecular bone are continuous with the inner side of the cortex. Some structural characteristics of cortical and trabecular bone are compared in Table 1, and some values pertaining to the whole skeleton are given in Table 2.

1. Bone Surfaces

Each bone has four surfaces — periosteal, intracortical or Haversian, inner cortical or cortical-endosteal, and trabecular endosteal, the latter three being in continuity. All bone remodeling and turnover takes place at bone surfaces and each of these surfaces is always in one of three functional states — forming, resorbing, or quiescent. Bone-forming surfaces are covered by osteoid seams and osteoblasts, and bone-resorbing surfaces are scalloped by Howship's lacunae containing osteoclasts and poorly characterized mononuclear cells. Approximate values for the extent of surface undergoing resorption and formation are given as fractions of the total surface, as amounts per unit of bone volume, and as totals for the whole skeleton for both cortical and trabecular bone in Table 3. At quiescent surfaces the most superficial part of the bone is formed by a thin layer of unmineralized but electron-dense connective tissue, the lamina limitans,[6] which in trabecular bone is sometimes confused with a thin osteoid seam.[8] This is covered by a continuous layer of flat lining cells with osteogenic potential, sometimes referred to as "resting osteoblasts" or "surface osteocytes", which

FIGURE 1. Cross-section of cortical bone to show two kinds of cement line-reversal lines which circumscribe a completed osteon and arrest lines which lie within a completed osteon. (Redrawn from Jaffe[192]).

separates the bone from the adjoining soft tissues.[9] On the inside of the bones they form the endosteum and its prolongations which line the Haversian canals and subendosteal spaces. The endosteum can be regarded as the outer layer of the adjacent bone marrow as well as the inner layer of the bone. On the outside of the bones the osteogenic cells form the inner layer of the periosteum; the outer layer of the periosteum is dense fibrous tissue continuous with tendon and ligament insertions and with the joint capsule.

A surface which is closed and so unbounded and divides space into an inside and an outside may be referred to as an envelope.[10] The periosteal envelope encloses all of the hard and soft tissues of a single bone, and the endosteal envelope, with its three subdivisions (trabecular, inner cortical, and Haversian) encloses all of the soft tissues within the bone with the exception of the osteocytes and their processes. Bone tissue is therefore inside the periosteal envelope and outside the endosteal envelope. The behavior of bone cells and their responses to mechanical, chemical, and hormonal stimuli differ significantly between the periosteal and endosteal envelopes. This is obvious during growth, but even in the adult skeleton there may simultaneously be a net gain of bone adjacent to the periosteum and a net loss of bone adjacent to the endosteum. The distribution of bone loss between the three subdivisions of the endosteal envelope is also different at different ages, in different bones, and in different disease states.

2. Bone Structural Units

Histologists have long recognized that the skeleton is constructed of many small elements of bone made at different times;[11] these will be referred to as bone structural units (BSU).[12] As bricks are held together by mortar, so the BSU are held together by an electron-dense but collagen-free connective tissue, which is visible in the two-dimensional world of the histologic section as a cement line — in three-dimensional reality, a cement plane or surface (Figure 1). The strength of bone is increased by this method of construction but slippage at the cement plane can be produced by prolonged mechanical stress.[13] Almost all cement lines are reversal lines, recognized by their irregular scalloped appearance, acid phosphatase staining,[14] and discontinuity of canaliculae.[15] They mark the furthest extent of a previous episode of bone resorption and are formed during the quiescent interval prior to commencement of bone formation. A few cement lines are arrest lines, recognized by their smooth surface, lack of acid phosphatase staining, and continuity of canaliculae. These are formed during a period of temporary interruption of bone formation and indicate that the BSU was completed in two or more separate periods rather than continuously. If bone formation had not resumed, the arrest line would have become the lamina limitans on the new bone surface. The number of osteons with arrest lines increases with age.[16] Smooth cement lines are also found at the periphery of primary osteons, which formed during the growth in width of the long bones. This takes place by subperiosteal apposition, enclosing longitudinal vessels around which concentric lamellae are subsequently formed.

The characteristic structure of adult cortical bone is the secondary osteon or Haversian system, a cylinder about 200 μm in diameter roughly parallel to the long axis of the bone, with a central canal about 40 μm in diameter. Within the canal run blood vessels, lymphatics, nerves, and connective tissue which are all continuous with those of the bone marrow and the periosteum. The total length of a single canal from periosteum to endosteum is about 10,000 μm or 1 cm.[17] The Haversian canals are connected to each other by transverse Volkmann's canals and in places either divide or reunite to form a branching network. The mean length of a single BSU is conveniently taken as the mean distance between branch points, which is about 4 to 5 mm in the dog[18] and approximately 2.5 mm in man,[19] so that the mean volume of a human cortical BSU is about 0.065 mm^3 (Table 1). Osteons form approximately two thirds of the total cortical bone volume, a proportion which falls with age,[20] the remainder being made up of interstitial bone which represents the remnants of previous generations of osteons, and the subperiosteal and subendosteal circumferential lamellae. The walls of the osteon consist of concentric lamellae about 7 μm thick[21] between which the osteocytes are arranged circumferentially. The osteocytes of adjacent rings are staggered, so that the radially oriented canaliculae connect osteocytes which are 1 to 3 lamellae apart. The lamellae are distinguishable because of alternation of the main direction along which the collagen fibers are oriented; one pair of adjacent lamellae gives one pair of bright and dark lines of birefringence when examined by polarized light.

In the trabecular bone the BSU are flattened, as if an osteon had been slit open and unrolled, and lie roughly parallel to the plane of the trabecular plates. The BSU which form the trabecular surface are shaped somewhat like shallow segments of a cylinder of radius about 600 μm, each segment being about 50 μm in depth at the center and about 1 mm in length.[22] The mean depth (or wall thickness) measured at four equidistant points in each BSU and corrected for section obliquity was 39 μm in normal subjects;[28] the mean volume of a trabecular surface BSU is therefore about 0.025 mm^3 (Table 1). Most of the surface is concave towards the marrow spaces, although the central area of large plates may be almost flat, and the bars and the edges of the plates are convex. As in cortical bone there is a lamellar structure with alternating collagen fiber orientation, and a similar distribution of osteocytes and their connections.

3. Surface and Volume Data

Because bone is of variable density the amount of bone in the body is best defined on the basis of volume rather than mass. The total volume of bone tissue, or absolute bone volume, is the total volume of the skeleton less the volume of the various soft tissues within the endosteal envelope. Strictly speaking, the volume of the lacunar-canicular system should also be excluded, but this is not possible with the usual ways that volume and density are measured. Based on a mean value for dry, fat-free skeletal weight of about 4 kg for black and white male and female cadavers[24,25] and a bone tissue density of 2.3,[26] a representative mean value for the total volume of bone tissue is 1750 cm^3 (Table 2). With a calcium content of 0.6 g/cm^3, this corresponds to a mean total bone calcium of 1050 g. Of the total volume, 80% of (1400 cm^3) is cortical bone and 20% (350 cm^3) is trabecular bone. These volumes are equivalent to a total number of cortical BSU (15/mm^3) of $21 \cdot 10^6$, and a total number of trabecular BSU (40/mm^3) of $14 \cdot 10^6$, with a combined total for the whole skeleton of $35 \cdot 10^6$ (Table 2). In fact, there will be about two thirds of this number of complete BSU and many incomplete fragments remaining from previous BSU which have been partially remodeled. The mean internal surface-to-volume ratio (S/V) for cortical bone is about 2.5 mm^2/mm^3 and for trabecular bone about 20 mm^2/mm^3.[19,27] The total surface areas (in mm^2) for the subdivisions of the endosteal envelope are $7 \cdot 10^6$ for the trabecular, $0.5 \cdot 10^6$ for the inner cortical, and $3.5 \cdot 10^6$ for the Haversian for a combined total of $11 \cdot 10^6$ mm² or 11 m². The mean surface area of a trabecular BSU is 0.6 mm² so that more than three quarters ($11.7 \cdot 10^6$) of the trabecular BSU occupy surface positions. The total endosteal surface area of 11 m² compares with about 0.5 m² for the periosteal envelope, about 7 m² for the total lacunar and canalicular surface,[26] and about 50,000 m² for the interfibrillary microcanalicular surface.[28] The relationships between bone volume, porosity, and surface area will be further discussed in a later section.

B. The Cellular Basis of Bone Remodeling[2-6]

1. Bone Resorption

The cells mainly responsible for bone resorption are the osteoclasts — large and generally multinucleated cells found within Howship's lacunae. When examined by electron microscopy the osteoclasts have a central ruffled border with many cytoplasmic extensions which appear to infiltrate the disintegrating bone surface. This is surrounded by a clear zone where the cell is temporarily sealed to the surface; the ratio of ruffled border to clear zone is an indication of the activity of the cell.[29] In tissue or organ culture the osteoclast is mobile and can resorb bone over a domain of larger area than is in contact with the cell at one time.[6] On the endosteal surface of the long bones in the growing rabbit the extent of the osteoclast domain is about 2.5 times the contact area.[30] The corresponding ratio of osteoclast domain to contact area in adult mammalian bone is unknown, but indirect evidence suggests values of 1 to 2 for cortical bone in the dog and 2 to 3 for trabecular bone in man; this will be discussed in more detail later.

During bone resorption the solid mineral is made soluble, the glycosaminoglycans of the ground substance are depolymerized, and the collagen fibrils are disaggregated and their tropocollagen molecules broken down into small peptides. At the light microscopic level these processes occur so close together in time that the bone appears to be resorbed as a whole, but it is likely that the mineral and ground substance are removed before collagen can be digested. The latter process may be carried out by mononuclear cells (such as monocytes or macrophages) which secrete collagenase, rather than by osteoclasts,[31] and may continue after the osteoclasts have moved along or away from the surface.

An important but unexplained peculiarity of bone resorption is the usual avoidance of unmineralized surfaces. Although resorption of osteoid has occasionally been observed in tissue or organ culture, this only rarely occurs in vivo. Since the postulated collagenase-secreting cells do not resorb osteoid, the signal to these cells would have to involve the prior appearance of osteoclasts. Because implants of decalcified bone are readily resorbed it is some peculiarity of bone matrix which has never been mineralized, rather than simple absence of mineral, which inhibits the appearance of osteoclasts. Furthermore, the avoidance of osteoid is confined to the initiation of resorption. Once the process is underway, all bone in the path of the osteoclasts is resorbed irrespective of its degree of mineralization.

Immediately beneath a surface undergoing vigorous resorption, such as the endosteal surface of a growing long bone, the osteocyte lacunae in the bone which is about to be resorbed may enlarge, but with this exception osteocytes do not participate in normal bone resorption or bone turnover.[32] During pathologic bone resorption, such as adjacent to rheumatoid pannus, the osteocytes may produce extensive perilacunar demineralization; this may render the bone susceptible to attack by collagenase-secreting mononuclear phagocytic cells so that resorption may be accomplished without the intervention of osteoclasts.[33] The bone beneath the walls of the lacunae (perilacunar bone) is less dense and more permeable than the bulk of interlacunar bone, and probably function as a short-term mineral reservoir.[34,35] This specialized bone may be subject to periodic removal and replacement — so call osteocytic mini-remodeling[36] — but the existence of this process is disputed;[37] more will be said on this subject in a later section.

2. Bone Formation

In contrast to bone resorption, bone formation occurs in two stages, matrix formation and mineralization, which are separated both in time and in space. The cells responsible for bone formation are the osteoblasts. These are columnar or cuboidal cells with basophilic cytoplasm arranged in a continuous epithelial-like palisade. They lie immediately adjacent to a layer of as yet unmineralized matrix forming an osteoid seam, the average thickness of which is normally about 8 to 10 μm. Matrix formation determines the volume of the newly formed bone but not its density; it involves the biosynthesis of collagen and of the proteoglycans and glycoproteins and other components of the ground substance. Mineralization increases the density of the newly formed bone but does not alter the volume, since water is displaced by solid mineral; it involves the initial deposition of amorphous tricalcium phosphate which is slowly converted into crystals of hydroxyapatite. The boundary between mineralized bone and osteoid, or the osteoid-bone interface, is usually blurred, with a transitional region about 3 μm wide known as the zone of demarcation[11] or "ligne frontiere".[38] Three characteristics of this zone indicate that mineralization was proceeding at a particular seam at the time of a biopsy. First, there are high concentrations of phospholipid, zinc, and a variety of other substances indicated by their characteristic histochemical properties.[39-41] Second, there is a line of confluent granules about 1 μm in diameter known as the mineralization front[36] or phosphate ridge[40] shown by toluidine blue or other appropriate staining; the granules probably consist of aggregates of amorphous mineral or of calcium-phospholipid-phosphate complexes which precede the formation of amorphous mineral.[42] Finally, tetracycline and other bone-seeking substances such as radiocalcium are deposited.[6] During the calcification of cartilage and of the woven bone which forms the primary spongiosa in enchondral ossification, the cells (chondroblasts and osteoblasts) give rise to small membrane-bound matrix vesicles which can actively accumulate calcium and phosphate ions.[43] These vesicles have not been demonstrated in adult lamellar bone formation, but they or their remnants may be responsible for some of the microscopic characteristics of the zone of demarcation.

Matrix formation takes place at the interface between osteoblasts and osteoid, whereas mineralization takes place at the interface between osteoid and mineralized bone; these two processes are thus separated both in space and in time. Separation in space is represented by the thickness of the osteoid seam, and the corresponding separation in time results from the delay, normally about 10 to 20 days, between the deposition of matrix and its subsequent mineralization. During this period, known as the mineralization lag time,[44] the matrix undergoes a variety of changes collectively known as maturation, which prepare it for mineralization.[45] These include an increase in the cross links between the polypeptide chains of the collagen molecules, which decrease the randomness of collagen fiber orientation,[40] binding of phospholipids to collagen[40,46] and of calcium to noncollagenous matrix proteins,[47] accumulation of silicon[48] and zinc,[49] and an initial increase in glycosaminoglycan content with a precipitous decline just before the onset of mineralization.[50] Corresponding to these maturational changes, several subdivisions of the osteoid seam can be recognized by various histochemical procedures[41,51] which show differences between the osteoid immediately adjacent to the osteoblasts and the osteoid adjacent to the zone of demarcation.

In the appositional formation of lamellar bone, two different aspects of mineralization must be distinguished[52] — the rate at which the zone of demarcation advances (or

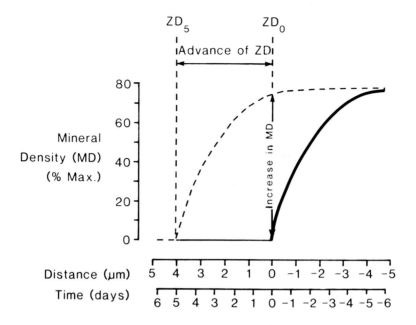

FIGURE 2. Distinction between two aspects of mineralization. In a plane perpendicular to a bone-forming surface, mineral density (MD) as percent of maximum is plotted as a function of distance from the location of the zone of demarcation at time zero (ZD_0) and of time. ZD is the junction between osteoid to the left and mineralized bone to the right; correspondence between distance and time scales depends on movement of ZD from right to left at a constant rate, assumed here to be 0.8 μm/day. Curved solid line shows MD at different locations at time zero. Curved dotted line shows MD at different locations 5 days later during which ZD has advanced 4 μm to ZD_s, and mineral density at ZD_0 has increased to 75% of maximum. The rate of advance of ZD, or mineral appositional rate ("horizontal mineralization"), and rate of mineral accumulation ("vertical mineralization") can vary independently, although they usually change in the same direction.

the mineral apposition rate) and the rate at which mineralization proceeds after it is initiated (Figure 2). In an individual moiety of bone matrix, mineral accumulation as a function of time is a continuous process but is conveniently subdivided into different stages (Figure 3). During the first few hours there is a rapid increase both in density and in Ca/P ratio,[52] referred to as primary mineralization. Mineral accumulation then continues at an exponentially decreasing rate with little further change in Ca/P ratio; this is referred to as secondary mineralization. In the rat, about 90% of maximum possible mineral density is reached within 4 to 5 days.[52] In man, only about 70% of maximum is reached in the first few days and 90 to 95% of maximum after several months.[6,11] Primary mineralization is under the control of the local cells, both the osteoblasts on the surface and the osteoblasts which have become buried within the osteoid. The latter cells, although commonly referred to as osteoid osteocytes, would more appropriately be termed mineralizing osteoblasts. Secondary mineralization continues at an increasing distance from the surface; it is probably governed by the chemical composition of the fluid phase of bone and the extent to which the mineral is accessible to diffusion.[53] Further increase in mineralization beyond 95% of maximum is prevented in some way by the osteocytes and occurs only if these cells die — the phenomenon of micropetrosis.[54] The lower density of recently deposited bone mineral is the basis for its permeability to basic fuchsin[55] and for the identification of bone-forming surfaces by microradiography.[56]

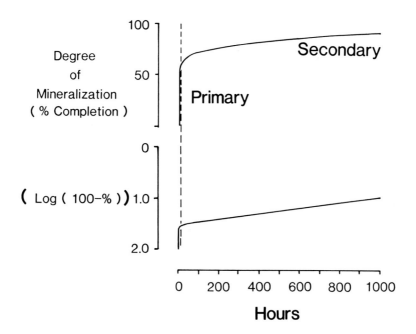

FIGURE 3.   Mineralization of an individual moiety of bone as a function of time. Mineral density (expressed as % of maximum) plotted on linear scale in upper panel and logarithmic scale in lower panel. The rapid early phase (primary mineralization) is based on electron probe microanalysis in the rat[52] and the slow late phase (secondary mineralization) is based on densitometric analysis of microradiographs in dog and man.[6,11]

Much of the knowledge of osteoblast function and bone formation has been obtained by the use of in vivo markers, especially the tetracyclines. Other markers are available for use in experimental animals, but all except for tetracycline are too toxic for use in human subjects. If one of these substances is trapped at the mineralization front it remains in the same place for the lifetime of that moiety of bone and is detectable by its fluorescence.[57] Tetracycline is incorporated into bone in a similar manner to the formation of a hot spot after radiocalcium administration. When the blood level of either tetracycline or radiocalcium is high, all free bone surfaces which are accessible to the circulation, including the borders of osteocyte lacunae, become labeled; this is a passive process which is demonstrable in vitro as well as in vivo.[58,59] Radiocalcium exchanges with stable calcium in the surface compartment and tetracycline forms a chelate with calcium ions. Both of these processes are reversible, so that as the blood level falls, both radiocalcium and tetracycline escape from the surfaces back into the circulation except where they are permanently fixed at sites of bone formation. Fixation of radiocalcium occurs solely because of trapping beneath a layer of newly deposited bone mineral so that outward diffusion is blocked.[6] Fixation of tetracycline also occurs by this mechanism, but unlike radiocalcium, tetracycline is irreversibly deposited at sites of bone formation in vitro as well as in vivo.[59,60] Consequently, there must be an additional purely physiochemical mechanism such as preferential binding to collagen[61] or more likely to the amorphous or immature mineral at the zone of demarcation.[62]

In the rat, tetracycline is fixed soon after the onset of mineralization when the density is still less than 20% of maximum[63] and the Ca/P ratio is still low.[52] Much the same was found in the rabbit,[64] with the additional observation of a high tetracycline concentration in labeled bone 5 min after administration. The chemical reactivity of

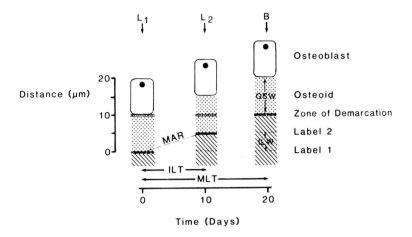

FIGURE 4. Interpretation of in vivo tetracycline labeling. Panels depict successive locations of a bone-forming surface with (from above down) osteoblast, osteoid seam, zone of demarcation (ZD), and mineralized bone. Locations shown at time of administration of first label (L_1), of second label (L_2), and at time of biopsy (B). ILT = interlabel time, here 10 days. Consider the fate of an individual moiety of bone matrix just deposited adjacent to the osteoblast in L_1. With time, the osteoblast recedes upwards and ZD approaches from below. The time taken for ZD to reach the moiety of matrix is the mineralization lag time (MLT), here 20 days, during which the matrix undergoes a variety of maturational changes. Interlabel width (ILW) divided by ILT is the mineral appositional rate (MAR). If osteoid seam width (OSW) remains constant, the matrix appositional rate is the same as the mineral appositional rate and OSW = MAR · MLT.

bone mineral and its affinity for tetracycline are much greater when it has just been deposited than when it is more than a few days old.[65] The precise physicochemical basis of this change in behavior is not known, but it is probably related to the transformation from amorphous to crystalline mineral. Since the existence of amorphous mineral depends on recently active mineralization, the significance of tetracycline fixation is much the same whether it occurs by trapping under newly deposited crystalline mineral[57] or by preferential binding to amorphous mineral,[62] but failure of fixation may result from a defect in either mechanism, as will be discussed further in a later section. Because tetracycline diffuses into recently deposited low-density bone[44] there is a minimum width of fluorescence known as the zone of instantaneous labeling (ZIL), which is normally about 4 μm in width, both in man[66] and in the rat.[44,63] Any delay in mineral accumulation will increase the thickness of low-density bone and so increase the permeability to tetracycline and the width of the ZIL. There is also a wider ZIL when there is a greater proportion of amorphous mineral as in renal osteodystrophy[62] but this also could reflect greater permeability rather than preferential binding. Whatever the mechanism of fixation, the mean distance between bands of fluorescence produced by tetracycline administration at known times enables the mineral appositional rate (MiAR) and several other derived quantities to be measured, thus introducing the critical dimension of time to the understanding of bone remodeling (Figure 4).

C. The Supracellular Organization of Bone Remodeling
1. Bone Remodeling Units (BRU)

Lamellar bone formation in the adult human skeleton is purely a replacement mechanism — it occurs only at locations where bone resorption has recently been completed.[67] Precursor cells, probably both local and blood borne, proliferate into a group of newly formed osteoclasts which work together as a unit; they excavate the bone to

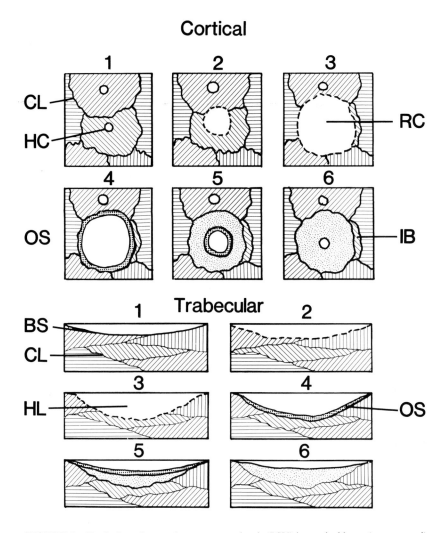

FIGURE 5. Evolution of a new bone structural unit (BSU) in cortical bone (upper panel) and trabecular bone (lower panel). The original BSU (stage 1) are demarcated by cement lines (CL). In cortical bone the resorptive process (stages 2 and 3) enlarges the Haversian canal (HC) and removes most of one and part of four other BSU to form a resorption cavity (RC). In trabecular bone the resorptive process erodes from the bone surface (BS) and removes part of three BSU to form a Howship's Lacuna (HL). The formation process converts the resorption perimeter (interrupted line) to a new cement line (stage 4) within which an osteoid seam (OS) is laid down and the new BSU is progressively reconstructed (stages 5 and 6). IB = interstitial bone.

a depth which varies only between narrow limits and thus determine the location, shape, and size of the future BSU. When they have completed their task the osteoclasts disappear and there is a quiescent interval or reversal phase[68] of variable duration during which the irregular surface of the resorption space is smoothed off and the cement line is laid down.[69] There next appears a coordinated unit of osteoblasts derived from a probably different precursor cell pool, whose task is to replace as exactly as possible the recently removed bone and so to rebuild a new BSU (Figure 5). Bone turnover thus occurs in anatomically discrete foci within which remodeling activity lasts for about 4 to 8 months. The specialized groups of cells with a finite life span which accomplish one quantum of bone turnover in such a focus (erosion and refill of one cavity) will be referred to as bone remodeling units (BRU); some characteristics

Table 4
BONE REMODELING UNITS (BRU) AND BONE
REMODELING CYCLES (BRC) — QUANTITATIVE
CHARACTERISTICS

	Cortical	Trabecular
Length (mm)	2.5	0.6
Mean circumference or width (mm)	0.36	0.6
Surface area (mm²)	0.9	0.36
Volume of missing bone (mm³)	0.02	0.005
Number BRU/mm² bone surface	0.08	0.2
Number BRU/mm³ tissue volume	0.19	0.8
Number BRU/mm³ bone volume	0.20	4.0
Total number BRU in skeleton	$0.28 \cdot 10^6$	$1.4 \cdot 10^6$
Total number BRC in skeleton	$0.7 \cdot 10^6$	$1.4 \cdot 10^6$
Volume of missing bone (mm³/mm³)	0.004	0.02
Total remodeling space (mm³)	$5.6 \cdot 10^3$	$7.1 \cdot 10^3$
Calcium equivalent (g)	3.4	4.2

	Whole skeleton
Total number of BRU	$1.68 \cdot 10^6$
Total number of BRC	$2.1 \cdot 10^6$
Total remodeling space (cm³)	12.6 (0.7%)
Calcium equivalent (g)	7.6

of cortical and trabecular BRU are compared in Table 4. The BRU travel through the cortex or across the trabecular surface to carry out bone remodeling in an ordered and predictable manner; each lasts for about the same period of time and replaces about the same volume of old bone by new. Activation of precursor cell proliferation, osteoclastic resorption, and osteoblastic formation follow one another in an unvarying succession symbolized by A → R → F,[70] although the mechanism of the close spatial and temporal coupling between these processes is still unknown.[6]

The structural order of bone results from the coordination of cellular activity within each BRU. Because of the longitudinal orientation of osteons in the cortex, they rarely depart by more than 10° from being perpendicular to a standard cross section and parallel to a standard longitudinal section. Consequently, most current knowledge of the sequence of events within individual BRU has been derived from the study of cortical bone, especially that of the rib.[71] Although this is now rarely used as a site of biopsy, the details of cortical remodeling underlie many of the concepts which apply also to the trabecular bone of the ilium.

2. Cortical Remodeling

In cortical bone the osteoclast unit arises within a Haversian or Volkmann's canal and forms a cavity which is extended longitudinally through the cortex to produce a resorption tunnel. Behind the advancing resorption front the tunnel is refilled from the circumference towards the center by osteoblastic apposition of new bone on the surface, a process which continues until a normal Haversian canal diameter has been attained. In a fully mature cortical BRU the osteoclasts at the front are arranged like a burr, about 400 μm long and 200 μm wide at the base (Figure 6). This is usually termed the "cutting cone"[40] although it is more often shaped like a paraboloid or hemisphere. From the few published illustrations, most of the cutting cone surface appears to be lined by osteoclasts so that the osteoclasts domain seems to be no more than 1.5 times the contact area, but there are no published data on the fraction of resorption surface covered by osteoclasts in the cortex of long bones in man or in animals with adult bone remodeling. Behind the cutting cone is a belt of variable width

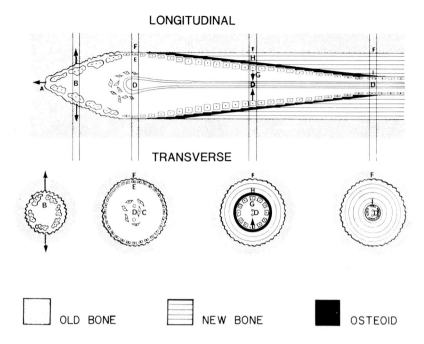

LONGITUDINAL

TRANSVERSE

☐ OLD BONE   ☰ NEW BONE   ■ OSTEOID

FIGURE 6.   BRU in cortical bone shown in longitudinal section (upper panel). Corresponding transverse sections (lower panel) show BRC at different stages of development. A — apex of advancing osteoclastic front moving from right to left; B — multinucleated osteoclasts enlarging a resorption space (they are shown advancing centrifugally but other directions of movement are possible — see Figure 10; C — spindle-shaped mitotically active precursor cells; D — capillary loop; E — mononuclear cells lining quiescent reversal zone between resorption and formation; F — cement line separating new bone from old; G — osteoblasts advancing centripetally during radial closure; H — osteoid seam between osteoblasts and new bone; I — flattened cells lining canal of completed Haversian system. (Slightly modified from Parfitt[5]).

where resorption has been completed but formation not yet begun (the quiescent or reversal zone), followed by the so-called "closing cone", an elongated cavity lined by osteoblasts and osteoid seam which continues into the Haversian canal behind.[40]

The dimensions of the cutting cone have been established by direct measurement in longitudinal sections (Table 5) but the dimensions of the closing cone are less certain. The course of most osteons is slightly oblique so that longitudinal sections rarely include the entire length of the closing cone. In the dog, reconstruction from serial autoradiographs showed that the "hot spots" of ^{45}Ca uptake seen in two-dimensional cross section correspond in three dimensions to "hot rods", which had a mean length of about 2200 μm and a maximum length of 6000 μm.[76] Similar three-dimensional reconstruction after tetracycline labeling gave a mean length of 1750 μm and a maximum length of 7250 μm.[77] An indirect method of estimating closing cone length (which will be described later) also gave values of approximately 2000 μm in dog and monkey (Table 5). From these dimensions the mean volume of bone removed to make room for a mature BRU is about 0.025 mm²; most cortical BRU are fully mature so that the corresponding mean volume for both evolving and mature BRU is about 0.02 mm³. The number of BRU seen in cross section varies widely at different ages, but a representative normal value for adults is 0.5/mm². With a mean total length of 2.5 mm, this corresponds to 0.2/mm³ bone volume so that the total number of cortical BRU in the skeleton is about 280,000 (Table 4). The combined volume of all BRU, representing the total volume of temporarily missing bone or remodeling space (12), is about 0.004

Table 5
CORTICAL BONE IN DIFFERENT SPECIES

	Dog	Monkey	Man
Length of resorption (cutting) cone (LRC) (μm)	343[72]	478[a]	409[73]
Width of resorption (cutting) cone (WRC) (μm)	120[72]	101[a]	188[73]
Longitudinal erosion rate (LER) (μm/day)	39.2[72]	36.7[a]	(27.0)[b]
Radial erosion rate (RER) (μm/day)	6.9[c]	6.2[d]	(6.2)[d]
Duration of resorption (σ_R) (days)	8.8[e]	13.0[e]	(15.1)[e]
Longitudinal formation rate° (LFR) (μm/day)	44.4[72]	58.1[a]	—
Duration of formation (σ_F) (days)	63[74]	55[a]	76[75]
Calculated length of closing cone (LFC) (μm)	2470[f]	2018[f]	(2052)[f]
Total length of BRU (LBRU) (μm)	2813[g]	2496[g]	(2461)[g]
Ideal A_r/A_f	0.139[h]	0.237[h]	(0.199)[h]
Actual A_r/A_f	0.287[74]	0.510[a]	1.17[75]

Note: Comparison of some three-dimensional geometrical characteristics of cortical bone remodeling units (BRU) in different species. Numbers in parentheses are computed on the assumption that the radial erosion rate in man is the same as in the monkey.

[a] Data of Oliver and Crouch.[194]
[b] Calculated as LRC/WRC · RER.
[c] Calculated as ½ WRC/LRC · LER.
[d] Assumed to be same as in monkey.
[e] Calculated as LRC/LER.
[f] Calculated as σ_R · LER.
[g] Calculated as LRC + LFC.
[h] Calculated as LRC/LFC.

mm³/mm³. This is 0.4% of cortical bone volume or 5.6 cm³ for the whole skeleton. Much higher values for both total number of BRU and total volume of remodeling space are found in children and in diseases in which bone turnover is increased.

a. Two Types of Cortical BRU and Their Evolution

The precise mode of evolution of a cortical BRU in time and space is still uncertain. Observations in young dogs after double tetracycline labeling indicate that a complete BRU, both cutting and closing cones, moves *en bloc* through the cortex.[78] The rate of advance (longitudinal erosion rate or LER) is about 40 μm/day,[72] from the dimensions of the cutting cone the rate of centrifugal resorption (radial erosion rate or RER) is about 7 μm/day (Figure 7), and the time taken to complete resorption at one level (σ_r) is about 9 days. In our laboratory similar values of 36.7 μm/day for LER and 6.2 μm/ day for RER have been found in the monkey (Table 5); σ_r is longer than in the dog because the diameter of the resorption spaces is larger. The radial erosion rate is calculated on the assumption that resorption begins at the center, so that the radial distance traveled by the resorption front is one half the width of the cutting cone at its base. If the resorption space was formed by enlargement of an existing Haversian canal, the distance traveled by the resorption front would have been 15 to 20 μm less, and the calculated rate correspondingly slower. The radial erosion rate is an index of the bone resorbing activity of the individual cell. In view of the great similarity between cortical remodeling in man and monkey[79] it is assumed in the absence of evidence to the contrary that this important quantity is the same in man as in the monkey. The calculated longitudinal erosion rate corresponding to the cutting cone dimensions shown is 27 μm/day, and the calculated value for σ_r is 15 days.

The data assembled by Johnson[40] indicated much greater longitudinal erosion rates on the order of 100 to 300 μm/day under a variety of circumstances. Johnson suggested

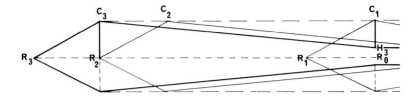

FIGURE 7.   Geometric relationships in Haversian remodeling. R1, R2, and R3 are successive positions of the apex at the resorption front that is moving from right to left. C1, C2, and C3 are the corresponding successive positions of the cement line determined by the furthest point of radial excursion during resorption. H3 is the boundary of the Haversian canal. During the time taken for the apex to advance longitudinally from R1 to R3, which is the length of the forming or closing cone (LFC), the formation front has moved centripetally from C1 to H3, which is the wall thickness of the completed osteon. The rate of longitudinal advance or longitudinal erosion rate (LER) is given by LFC/σ_f, where σ_f is the duration of radial closure. Similarly, during the time taken for the apex to advance from R2 to R3, which is the length of the resorption or cutting cone (LRC), the resorption front has moved centrifugally from R2 to C3, which is the mean radius of the cement line (CLR). The radial erosion rate (RER) is given by LER · CLR/LRC. These calculations assume that there is no delay between completion of resorption and commencement of formation, but inclusion of a constant delay does not alter the relationships between the rates of longitudinal resorption, radial resorption, and radial closure which determine the slopes of the cutting and closing cones.

that these rates were applicable to normal human cortical remodeling, and that radial excavation at one level was completed within 24 hr. If the rate directly determined in the dog and monkey applies throughout the development of a complete BRU from the onset of activation of osteoclast proliferation, then structures at all stages of development short of a complete BRU should be found with equal frequency (Figure 8). However, incomplete BRU are only rarely observed. For example, very few hot rods shorter than 500 μm in length were found in 50-μm serial sections.[76] In longitudinal sections of dog and monkey rib not a single example of the hypothetical intermediate structures shown in Figure 9 could be found despite an extensive search.[194] In many instances, both in published illustrations and in our own observations, the osteoid seams and tetracycline labels seen in longitudinal section appear to be cylindrical rather than conical in shape, indicating that refilling of a resorption space began more or less simultaneously over an extended distance. These data suggest that once the resorption cavity has been started it progresses with explosive rapidity, reaching a length of 1 to 2 mm within a few days, the short duration accounting for the rarity with which such cavities are found in longitudinal sections. Formation then begins along the entire length of the cavity and resorption either stops or continues at the rate previously described for a variable time and distance, corresponding to the varying lengths of a completed BSU. Consistent with this interpretation are the observations that both in the dog[72] and in the monkey (Table 5), the rate of advance of the forward edge of the osteoid seam just initiated (the longitudinal formation rate) was significantly faster than the longitudinal erosion rate, suggesting that longitudinal advance of the BRU was slowing down.

To summarize the preceding discussion, the apparent conflict between the direct measurements of longitudinal erosion rate by tetracycline labeling[72] and the indirect measurements reported by Johnson[40] can be resolved if the life history of a single cortical BRU is divided into three separate stages — evolution, advancement, and termination (Figure 9). During evolution resorption occurs alone without formation, during advancement both resorption and formation occur together, and during termination formation occurs alone without resorption. The stages of evolution and termination occur in all BRU, but the stage of advancement is of variable duration and may be omitted entirely. This explains why the mean total length of a cortical BRU is about the same as the mean total length of a cortical BSU. The frequency

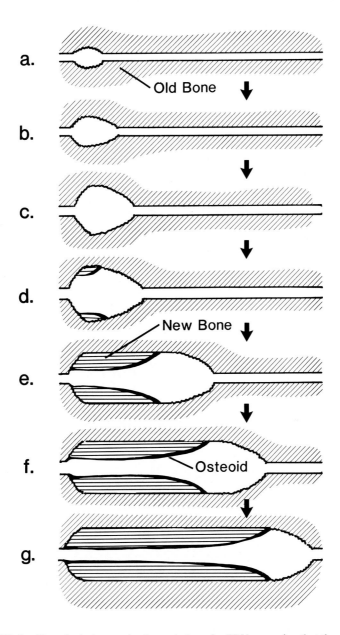

FIGURE 8. Hypothetical stages in the evolution of a BRU, assuming that the relationships between longitudinal, radial erosion, and radial closure is the same as depicted in Figures 6 and 7 for a mature or completed BSU. In longitudinal sections these intervening stages should collectively be at least as common as completed BRU but in practice are rarely if ever seen, suggesting that this is not the usual mode of evolution.

distribution of Haversian canal lengths[19] is skewed in a very similar manner to the frequency distribution of hot rod lengths,[76] with a mode in the region of 2 to 3 mm, and maximum length extending up to 6 or 7 mm in both cases (Figure 8). This indicates that most BRU do not advance further once their evolution is complete, but that some advance for distances up to 3 to 4 times their own length.

There are thus two types of BRU (Figure 10). Both types begin with a stage of evolution lasting only a few days, during which the resorption cavity is expanded rapidly to its full length and width; σ_r during this stage is probably in the range of 3 to 5

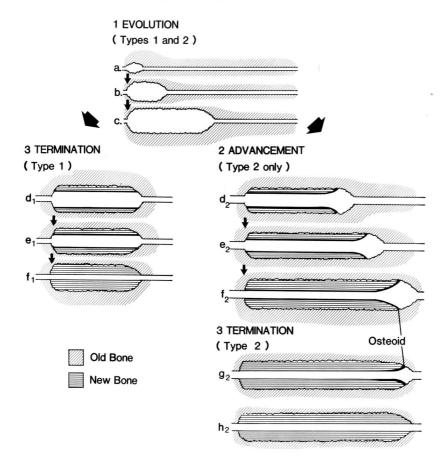

FIGURE 9.   Hypothetical stages in life history of two types of cortical BRU which enable conflicting estimates of the rate and duration of resorption to be reconciled.[40,72] Type 1 BRU proceed directly from evolution (stage 1) to termination (stage 3), whereas Type 2 BRU have an intervening stage of advancement (stage 2). See text for further details.

days. In Type I BRU resorption then ceases and the cavity is refilled along its whole length as a closing cylinder. In Type II BRU longitudinal advancement continues at a much slower rate and the entire BRU advances *en bloc* as previously described. Because direct determination of the longitudinal erosion rate involves measuring the distance traveled between two tetracycline labels, only Type II BRU during their phase of longitudinal advance would be included in the measurements; σ_r during this stage is in the range of 10 to 20 days. After a variable time and distance resorption ceases and the cavity is refilled as a closing cone.

Equally uncertain is the relationship between the apparent movement of the complete BRU and the movement of individual cells. The precise three-dimensional path of a single osteoclast is unknown, but their movement within the cutting cone is usually resolved into longitudinal and radial components. It is necessary to distinguish between two different methods of evolution and longitudinal advance (Figure 11). If the resorption tunnel is extended along an existing Haversian canal there need be no longitudinal movement of osteoclasts at all. Enlargement of the Haversian canal at any cross-sectional level could be accomplished by osteoclasts which arise at that point from the precursor cells, either those already present in the connective tissue or new cells brought in via the circulation. The apparent movement of the BRU would then be like a peristaltic wave; the rate of advance would be a function of the rate at which the

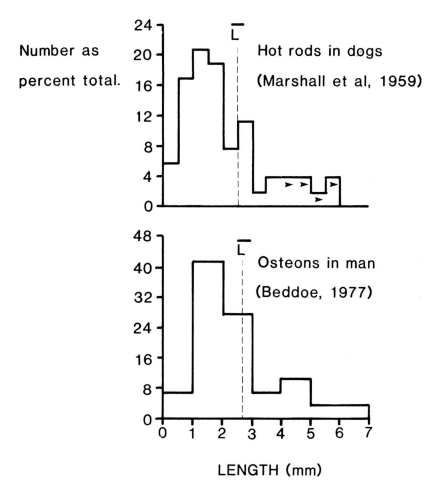

FIGURE 10. Frequency distribution of lengths of active osteons ("hot rods") measured to the nearest 0.5 mm and inactive (completed osteons) measured to the nearest 1.0 mm, expressed as % of total number measured.[19,76] $\overline{L}$ = mean length; → denotes minimum length where actual length could not be determined.

activation stimulus was transmitted along the canal, not on the erosive capacity of the osteoclasts, which would be exclusively in a radial direction. However, if the resorption cavity arises within a Volkmann's canal or diverges from an existing Haversian canal, it must tunnel into solid bone and some longitudinal movement of individual osteoclasts would have to occur. All osteoclasts in a cutting cone could advance longitudinally at the same rate, maintaining their relative positions (Figure 12). Alternatively, newly recruited osteoclasts might advance longitudinally at the point of the burr (or apex of the cone) for a short distance and then turn to erode radially or obliquely, making room for new osteoclasts in the center (Figure 12). Neither of these possibilities is supported by direct evidence, but the second maintains greater consistency between the two methods of advance. Tunneling through solid bone involves the creation of new vascular channels and therefore the formation of a new blood vessels and new connective tissue, but nothing is known about how this occurs. It has been suggested that all secondary osteons are made during the growth period, and that adult remodeling consists entirely of the replacement of existing osteons. However, this cannot be true in man because the number of vascular channels per unit cross-sectional area of cortical bone increases progressively with age.[80]

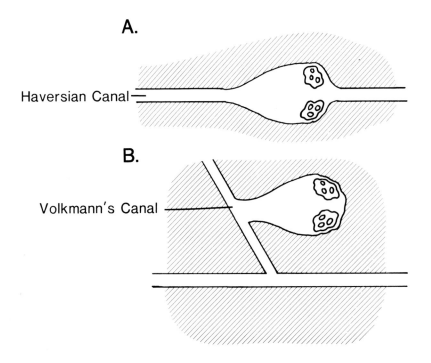

FIGURE 11.   Two modes of origin of a new BRU. (A) From an existing Haversian canal, not requiring longitudinal erosion through new bone or creation of new blood vessels; (B) from a Volkmann's canal, requiring both longitudinal erosion through new bone and creation of new blood vessels.

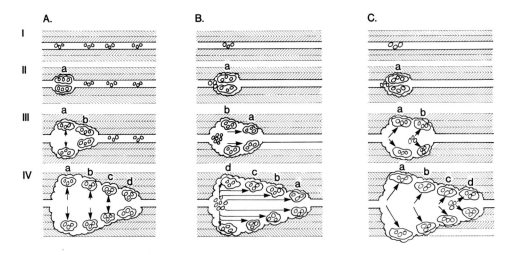

FIGURE 12.   Possible paths of movement of individual osteoclasts in cortical BRU. A — predominantly transverse movement perpendicular to the long axis of the BRU. If the BRU travels along an existing Haversian canal (Figure 11A) no movement of precursor cells is needed and exclusively transverse movement is possible; if the BRU tunnels through new bone (Figure 11B), precursor cells must move and transverse movement of osteoclasts must be preceded by longitudinal movement for a short distance. Note that in a completed cutting cone the youngest osteoclasts are at the apex and closest to the center, whereas the oldest osteoclasts are furthest behind and furthest from the center. B — predominantly longitudinal movement, which must be preceded by transverse movement for a short distance. Note that the oldest osteoclasts are at the apex of the cone and closest to the center and the youngest osteoclasts are further behind and furthest from the center. All osteoclasts can originate from a common precursor pool. C — predominantly oblique movement. The disposition of osteoclasts in a complete cutting cone with respect to age is the same as in B, but the location of the precursor cells in relation to the apex of the cone is different.

There is also considerable variation in the duration of the reversal phase between completion of resorption and commencement of formation, and hence in the length of the quiescent zone between the cutting and closing cones. In the young dog, refilling normally follows immediately behind excavation[78] but in man the coupling may be less close. Johnson observed that during adolescence the length of the quiescent segment could exceed 1000 μm.[40] Regions corresponding to periods of temporary arrest of formation may be found along the closing cone and remodeling may occur at multiple sites in the same osteon traveling either in the same direction or in opposite directions to form a double ended system.[81]

b. Cross-Sectional Analysis of Bone Formation

The relationship between the two-dimensional events seen in cross sections and the three-dimensional events of an evolving BRU is clearer for bone formation than for bone resorption. At the onset of radial closure the osteoblasts are tall and columnar in shape and closely packed together. In human bone the density is about 4500/mm^2[82] so that the area of bone covered by one osteoblast — its secretory territory — is about 220 μm^2. In the calvarium of the rat the osteoblasts are packed even tighter with a secretory territory of about 160 μm^2.[83] The cells are also intensely basophilic, a histochemical characteristic which correlates with the presence on electron microscopy of the endoplasmic reticulum, and so is an indication of the activity of the cell in protein synthesis.[84] Matrix formation is rapid and the osteoid seam quickly reaches a thickness of 15 to 20 μm before mineralization begins. As radial closure proceeds the osteoblasts get smaller in volume and flatter in shape. Each cell covers a progressively larger area and there is a concurrent reduction in basophilia; there are parallel changes in the size and shape of the osteocytes formed as some osteoblasts are entombed in the new bone.[85] These structural changes in the osteoblast are accompanied by corresponding functional changes manifested by a reduction in osteoid seam thickness and in mineral appositional rate.[85] At the termination of bone formation first matrix synthesis and then mineralization cease, and the cells on the surface become indistinguishable from the flat cells which line mature Haversian canals. Observations in several species have indicated an exponential decline in the rate of mineralization as radial closure approaches completion[85-87] but in part this is explained by a faster initial rate in larger than in smaller cavities.[88] When osteons of the same initial size are compared, the kinetics of osteon closure may be triphasic — an initial rapid fall in rate, maintenance of the same rate throughout the middle two thirds, and a further fall in rate in the terminal phase.[88]

Johnson[40] proposed a radically different concept of bone formation during osteon closure. He suggested that there are cycles of alternating matrix synthesis and mineralization, each cycle carried out by a single group of cells which disintegrate when mineralization is completed. In each cycle, seam thickness would increase from zero to a maximum value, matrix synthesis would then cease, and when matrix maturation was complete the whole of the seam would rapidly mineralize, the entire sequence being repeated 6 or 7 times. According to this model, mineralization would be in abeyance for a substantial fraction of the total duration of radial closure, but when serial tetracycline labels are given to a young dog most osteons show the same number of concentric fluorescent bands, indicating that mineralization was active over the entire circumference of that osteon at the time that each label was given.[89] Similar observations have been made in the rat and rabbit, in which many sites of bone growth show all of 5 or more tetracycline labels given at 12 to 48 hourly intervals.[90,91] Such short-term labeling has also revealed temporary interruptions in mineralization and fluctuation in the appositional rate in the normal rabbit, with a period of about 48 hr.[91] These results may have been influenced by parathyroid stimulation resulting from depression

Table 6
CELL BALANCE DURING FORMATION OF A BSU

	Cortical	Trabecular
Initial cement surface area (mm²)	1.5	0.62[a]
Total number of osteoblasts[b]	6750	2800
Final BSU surface area (mm²)	0.3[c]	0.6
Total number of lining cells[d]	700	1400
BSU bone volume (mm³)	0.065	0.025
Total number of osteocytes[e]	1300	500
Total number of cells in completed BSU	2000	1900
Loss of cells/completed BSU	4750	900
Loss (%)	70	32

[a]   During resorption the local surface normally increases by less than
5%, disregarding the irregularities visible with high magnification.
[b]   Assumes density of 4500/mm².
[c]   Haversian canal area.
[d]   Assumes density of 2300/mm².
[e]   Assumes density of 20,000/mm³.

of ionized calcium by the high dose of tetracycline used, but a similar protocol in the rat revealed no short-term periodicity.[91] In enchondral ossification in the rat, there is a circadian rhythm (24-hr period) with matrix synthesis being more rapid during the day and mineralization more rapid during the night.[92] In the human rib the frequency distribution of osteoid seam perimeter lengths showed 7 maxima, suggesting a much longer term fluctuation in mineral appositional rate with a period of about 10 to 12 days.[93] This is consistent with Johnson's concept of successive surges of osteoblast activity, but within each surge matrix synthesis and mineralization would be synchronous, not alternating.

The periodicity of bone formation may be related to the loss of cells which must occur during osteon closure, since there are many fewer cells in a completed BSU than were initially present at the onset of bone formation.[40] Assuming a cement line diameter of 190 μm, MWT of 75 μm, Haversian canal diameter of 40 μm, and total length of 2500 μm, the total area of cement surface in a complete BRU prior to bone formation would be 1.5 mm² with a cell density of 4500/mm². The surface density of lining cells has not been measured directly, but the length of the cell profiles in two-dimensional histologic sections (= 20 μm) suggests that the surface area has increased at least twofold so that the total number of lining cells would be no more than 700. Estimates of the volume density of osteocytes in cortical bone have varied between 12,000[94] and 20,000/mm³.[95] With a mean BSU bone volume calculated from the previous dimensional of 0.065 mm³, even the largest of these estimates gives the total number of osteocytes in the completed BSU as only 1300. Thus the total number of cells in the completed BSU (about 2000) is less than a third of the number of osteoblasts present at the onset of bone formation (Table 6). When and how the other cells disappear is completely unknown. Johnson suggested that the cells disintegrate after completion of each surge of bone formation and are replaced by new cells recruited from precursors in the adjacent connective tissue.[40] However, this would increase even further the number of cells whose disappearance must be explained. It seems more likely that after each surge the torch is handed on to other cells which were present at the outset but which had hitherto been less active.

As a rule, once the formation of a new osteon has begun it proceeds to completion within a predictable time, but occasionally there is complete arrest of bone formation to produce a resting seam.[96] This occurs as a normal phenomenon of aging, in several kinds of metabolic bone disease, and diseases such as cardiac failure.[97] A resting seam is characterized by complete absence of both tetracycline uptake and fuchsin-permeable low-density bone and is covered by flat lining cells. This indicates that primary mineralization has been in abeyance long enough for secondary mineralization to have progressed to more than 80% of maximum in the bone adjacent to the osteoid,[98] which probably requires at least a month. In most resting seams bone formation is eventually reactivated, in which case the completed osteon will contain an arrest line as a permanent record of what took place,[16] but some resting seams may remain in that state indefinitely.

c. Bone Remodeling Cycles and Bone Dynamics

Although there is considerable uncertainty concerning the three-dimensional organization of a cortical BRU, the sequence of events at a single location is more stereotyped (Figure 6). With respect to a point of reference just outside a recently completed BSU there would have occurred in succession the appearance of new osteoclasts, centrifugal resorption radially towards the reference point, cessation of resorption and the laying down of the cement line, and a quiescent interval of variable duration followed by centripetal formation away from the reference point towards the Haversian canal. These events occurring at a single cross-sectional location are conveniently referred to as a remodeling cycle whose successive stages correspond to the appearances in histologic cross sections of resorption spaces of various size lined by osteoclasts, resorption spaces lined by the transitional cells of the reversal phase, and osteoid seams corresponding to different degrees of radial closure. The total duration of a remodeling cycle (σ_{RC}) is made up of the successive durations of resorption ($\sigma_{RC(r)}$), quiescence ($\sigma_{RC(q)}$), and formation ($\sigma_{RC(f)}$). Variations in the three-dimensional organization of the remodeling unit primarily reflect variations in the temporal coordination between adjacent remodeling cycles. In Type I BRU the resorptive stages of adjacent cycles are slightly out of phase, the duration of the quiescent interval declines progressively from the earliest to the most recent remodeling cycle, and the formation stages are in phase. In Type II BRU, during *en bloc* longitudinal advancement each cycle is out of phase with the cycle immediately preceding by the same amount for both resorption and formation. In each remodeling cycle there occurs a succession of cellular events in the same place but at different times, whereas in each remodeling unit these same events are occurring in different places but at the same time.

i. Calculation of Sigma and Activation Frequency

The total duration of radial osteon closure (σ_f) can be measured by the in vivo tetracycline labeling technique. When tetracycline is given for two periods of a few days, 10 days apart, about 75% of the osteoid seams seen in cross sections of cortical bone show two fluorescent bands. About 20% will show only a single band because bone formation was initiated or completed between the two labels, and about 5% will show no fluorescence because the second label was given during the initial mineralization lag time, that is after the onset of matrix synthesis but before the onset of mineralization. In a few normal individuals and more commonly in patients with bone disease, the extent of missing label is too great for this to be the only explanation, indicating periodic cessation or slowing down in mineralization. Consequently, completion of radial closure will take longer than if the appositional rate measured in doubly labeled osteons continued without interruption. In order to estimate σ_f a corrected mineral appositional rate or radial closure rate (M_f) must be calculated from the expression:

$$M_f = MiAr \cdot f_L \qquad (1)$$

Table 7
REPRESENTATIVE VALUES FOR BONE REMODELING

Quantity	Cortical	Trabecular	Combined
Mineral apposition rate (M;μm/day)	1.0	0.75	0.84
Corrected apposition rate (M$_f$;μm/day)	0.8	0.6	0.67
Mean wall thickness (MWT;μm)	75	40	51[a]
Duration of formation ($\sigma_{RC(f)}$; day)	94	67	76
Number of osteoid seams (A$_f$/mm^3)	0.3	3.3	0.98
Birth rate of osteoid seams (μ_f; mm^3/year)	1.16	18	4.5
Total birth rate (/hr)	180	720	900
Bone formation rate (F; mm^3/mm^3/year)	0.03	0.26	0.076
Bone turnover rate (%/year)	3	26	7.6
Total turnover (cm^3/day)	0.115	0.25	0.365
Total turnover (mg Ca/day)	70	150	220

Note: Representative values for dynamics of bone remodeling and turnover in cortical bone, trabecular bone, and whole skeletons. A$_f$ in trabecular bone calculated on assumption that each new unit of osteoid seam is of 0.36 mm^2 surface area, and that mean fractional surface values for whole body are one third of those found in the ilium.

[a] This is the mean for all new BSU made in 1 year, not the mean for the whole skeleton at one time.

where f$_L$ is the fraction of the individual osteoid seams which are labeled. Different methods of calculating f$_L$ are discussed in a later section. In the steady state the mean value for M$_f$ is equal to the mean matrix appositional rate (MaAR), and so is an index of the bone-forming activity of the individual cell, the counterpart of the radial erosion rate. The relationships between M$_f$, the matrix and mineral appositional rates, and osteoid seam width will be discussed in more detail in a later section.

σ_f depends on the MWT of new osteons and the radial closure rate as previously defined, or in symbols:

$$\sigma_f \text{ (days)} = \frac{MWT(\mu m)}{M_f(\mu m/day)} \qquad (2)$$

Calculated in this way, σ_f includes both the initial mineralization lag time and the time required to complete mineralization (σ_{Mi}). Representative values are 75 μm for MWT, 0.80 μm/day for M$_f$, and so about 3 months for σ_f (Table 7). This relationship can be used to estimate the birth rate of new remodeling cycles or activation frequency symbolized by μ_{RC}. An ideal remodeling cycle would occur in an osteon segment only one cell in thickness, but in view of the usual impossibility of counting cells in the longitudinal direction the remodeling cycle is assigned an arbitrary length of 1 mm, so that measurements made in cross sections with dimensions of mm^{-2} can be assigned dimension of mm^{-3} without change in the numerical values. An arbitrary length of 20 μm would be more consistent with the real size of the cell, but this would merely complicate the subsequent calculations to no purpose. With this convention, the number of bone remodeling cycles (BRC) in a unit of volume of cortical bone is 2.5 times the number of BRU. If the seams observed in cross section are treated as a population of individuals with a finite life span, then the mean seam number (A$_f$) is equal to the mean birth rate (μ_f) times the mean life span (σ_f), or in symbols:

$$A_f \text{ (/mm}^3\text{)} = \mu_f \text{ (/mm}^3\text{/day)} \cdot \sigma_f \text{ (days)} \qquad (3)$$

Combining this with Equation 2 gives $A_f = \mu_f \cdot MWT/M_f$ which on rearrangement yields:

$$\mu_f \,(/mm^3/day) = \frac{A_f(/mm^3) \cdot M_f(\mu m/day)}{MWT(\mu m)} \qquad (4)$$

This defines the birth rate of new units of osteoid seam of 1-mm length and 0.6-mm mean perimeter/mm^3 of cortical bone (μ_f); in a steady state this must be equal to the birth rate of new BRC, or activation frequency (μ_{RC}). In practice the year is a more convenient unit of time for expressing μ than the day.

ii. Bone Formation Rate and Bone Turnover

It is now possible to derive an important relationship between the activation frequency and the volume-based bone formation rate (F_v). This is defined as the volume of new bone formed in unit time per unit volume of existing bone and so is an index of bone formation at the level of the tissue rather than of the individual cell. It is equal to the birth rate of BRC multiplied by the amount of new bone made in each BRC; with the same dimensional convention this amount in mm^3 is the mean osteon cross-sectional area (MOA) in mm^2 multiplied by 1 mm. In symbols this relationship is

$$F_v \,(mm^3/mm^3/year) = \mu_f \,(/mm^3/year) \cdot MOA \,(mm^2) \cdot 1(mm) \qquad (5)$$

The mean cross-sectional area of completed osteons declines by about 30% between 20 and 70 years[20,99] but does not change significantly over time periods shorter than a decade and differs little between one individual and another. For example, in cross sections of rib cortical bone from 60 normal persons, the mean diameter of completed osteons had a coefficient of variation (CV) of only 6.5%[195] whereas the CV of the activation frequency determined by the tetracycline method is about 50%, almost 8 times the CV of the osteonal diameter.[75] It therefore follows that variation in F_v (the bone formation rate) is almost entirely due to variation in μ_f and hence in μ_{RC} — the rate at which new BRC are initiated.

The relationship between bone resorption, bone formation, and bone turnover must now be defined more clearly. In the broadest sense turnover implies renewal or replacement, either of a cell population or of a tissue or one of its components. If bone resorption and formation are equal, bone volume (local or whole body) does not change with time; the bone turnover rate is equal to both resorption rate and formation rate and is equivalent to a fractional rate constant with dimensions T^{-1}. If bone resorption and formation are not equal, so that bone volume does change with time, the correct expression for bone turnover has never been defined. If the focus of interest is on the rate of disappearance from the skeleton of bone-seeking substances such as radium or strontium, then it is convenient to equate bone turnover with bone resorption. However, in accordance with the concept of turnover as tissue replacement, it will here be defined as the fractional volume of bone completely replaced in unit time. If bone formation is equal to or less than bone resorption, then turnover equals the bone formation rate. However, if bone formation exceeds bone resorption, then turnover is equal to the bone resorption rate. Total remodeling activity will then consist of bone turnover with an additional component of either surplus bone resorption or surplus bone formation; in essence, turnover is defined as either total resorption or formation, whichever is the smaller. A simpler but less rigorous approach is to define turnover as the arithmetic mean of resorption and formation as in Chapter 5. In practice, the values corresponding to these definitions and the even simpler definition of turnover as equal to the bone formation rate almost never differ by more than 10%. The more rigorous definitions can only be applied if local bone mass is measured in conjunction with histomorphometric estimates of bone turnover, or if external balance is measured in conjunction with kinetic estimates.

iii. Work vs. Time and Levels of Organization

The total amount of bone resorbed within a single BRC depends on the rate of advance (or velocity) and duration of activity of a set of osteoclasts, which together determine the distance which the resorption front will travel through the bone. Similarly, the total amount of bone formed within a single BRC depends on the velocity and duration of activity of a corresponding set of osteoblasts, which together determine how far the formation front will travel through the bone. The difference between these quantities, or terminal balance, represents the difference in size between the new BSU and the cavity in which it is built.[67] For both osteoclasts and osteoblasts, velocity, time, and distance are interrelated variables. Normally, the distance traveled (a measure of the work to be completed) and the rate of advance (a measure of individual cell vigor) are primary or independent variables, and time is the dependent variable, although in some disease states time may become a limiting factor. This is because bone cells are work oriented rather than time oriented — they are programmed to carry out a particular amount of work (however long it takes), rather than to work only for a particular period of time (however little has been accomplished).

It is for this reason that, for both resorption and formation, tissue level and cell level activity can vary independently, so that the rate of bone turnover is not necessarily related to the rate at which individual osteoclasts and osteoblasts do their work. When bone is described as turning over more quickly or more slowly it is tempting to visualize the individual cells (on which turnover depends) working more quickly or more slowly, but this is to confuse the amount of work performed (which rarely varies by more than 20%) with the time taken to complete it, which can be prolonged tenfold or more. Because of the relative constancy of the mean dimensions of an osteon in cross section, if individual osteoblasts work more slowly, the period of time for which the seam remains in existence (and hence the number of seams seen in cross section) will increase in inverse proportion to the decline in rate. Eventually, a new steady state will be reached in which a larger number of more lethargic cells are making the same total amount of bone as before.

As well as the cell and tissue level, bone remodeling must be considered at the level of the whole body. Thus, tissue level values for A_f, μ_f, and F_v are all expressed per mm^3 of bone. If they are multiplied by total bone volume in mm^3, corresponding whole-body values would be obtained; analogous calculations apply to bone resorption. It is the whole-body values which are related to bone turnover determined by radiocalcium kinetics and to indirect biochemical indices of turnover such as urinary total hydroxyproline excretion and alkaline phosphatase. If total bone volume is decreased, these whole-body measurements may be normal even though tissue level bone turnover is increased. Unfortunately, tissue level turnover varies substantially from site to site but has been measured only in rib and ilium, so that appropriate whole-body mean values for A_f, μ_f, and F_v are unknown. Correlations have been established between local histomorphometric data and biochemical indices and kinetic estimates of whole-body turnover,[100] but whether the latter are in good or poor agreement with the histomorphometric estimates depends on the assumptions which are made. This issue will be discussed further after trabecular bone remodeling has been examined.

iv. Relationship of Cross-Sectional to Longitudinal Events

The activation frequency as defined (the frequency with which new BRC are created per unit cross-sectional area of 1-mm thickness) is thus the primary determinant of the

Table 8
SIGMA IN CORTICAL BONE: CROSS-
SECTIONAL AND LONGITUDINAL VALUES

	Cross-sectional remodeling cycle (σ_{RC} — days)	Longitudinal remodeling unit (σ_{RU} — days)
Resorption (evolution)	5	5
Resorption (advancement)	15	165
Quiescence	(0—20)	(0—20)
Formation	90	260
Total	95—125	265—285

Note: Comparison of cross-sectional and longitudinal values for sigma (σ) in cortical bone. The time taken to create and refill a resorption cavity at a single cross-sectional location is σ_{RC}, and the time taken to rebuild a complete new osteon of 5 mm in length is σ_{RU}.

bone resorption and formation rates and hence of the rate of bone turnover. This cross-sectional concept must now be related more clearly to the three-dimensional geometry of the remodeling unit.[6] Since remodeling cycles are created in succession as the remodeling unit evolves or advances longitudinally through the bone, their rate of creation will be a function of the longitudinal erosion rate. Doubling this rate without any other change must double the distance which the resorption front advances during osteon closure, and so will double the length of the closing cone; this in turn will double the number of seams seen in cross section and so double the calculated activation frequency. The cross-sectionally defined activation frequency will also be dependent on the total duration for which the BRU remains in existence. If this is doubled for example, then again the number of seams seen in cross section will be doubled. However, in contrast to the radial erosion rate which is a function of the activity of individual osteoclasts, the rate of longitudinal advance and the duration of continued advance of the BRU are both dependent on continued recruitment of new osteoclasts. It follows therefore that the cross-sectionally defined activation frequency will be a true reflection of the rate of precursor cell proliferation and differentiation into osteoclasts, which is the primary determinant of the rate of bone turnover. It will not, however, be a measure of the birth rate of new remodeling units, since it is not possible from cross-sectional measurements alone to resolve the activation frequency so measured into its three components of the initiation of new BRU, their rate of longitudinal advance, and the duration for which they persist as discrete anatomical entities.

The distinction between the two-dimensional cross-sectional concepts and the three-dimensional longitudinal concepts applies also to sigma. As normally defined, σ_{RC} with its subdivisions is the time taken for completion of one remodeling cycle in cross section. During the stage of rapid evolution $\sigma_{RC(r)}$ is probably much shorter, but this is not normally amenable to measurement. But the concept of sigma also applies to the total life span of the BRU in three dimensions, symbolized by σ_{RU} (Table 8). In Type I BRU, σ_{RU} is only slightly longer than σ_{RC} but in Type II BRU, σ_{RU} may be up to 3 to 4 times as long as σ_{RC}. For example, the time taken to build a new BSU of 5-mm length would be about 265 days or nearly 9 months. This would be made up of 5 days to create a resorption cavity 1 mm in length, 160 days to advance a further 4 mm at a rate of 25 μm/day, and another 100 days to complete radial closure after cessation of longitudinal advance. During the total of 265 days, resorption would continue for 165 days ($\sigma_{RU(r)}$) and formation would continue for 260 days ($\sigma_{RU(f)}$).

d. Effective vs. Arrested Resorption

A rarely stressed but important aspect of cortical remodeling concerns the occurrence of resorption spaces which are not part of the normal remodeling sequence. The number of resorption spaces counted in cross sections is usually greater than would be expected from the three-dimensional geometry of the BRU. If random cross sections are obtained, the ratio of resorption spaces to osteoid seams (A_r/A_f) should be the same as the ratio of the length of the cutting cone to the length of the closing cone, which is approximately 400/2000 or 0.2. However, the observed ratio in human rib is frequency 1.0 or higher.[101] This problem was further investigated by devising an indirect method for estimating the length of the closing cone; from Figure 7 it is evident that this must be equal to $\sigma_{RC(f)}$ determined in cross section multiplied by LER determined in longitudinal section. In both dog and monkey the observed A_r/A_f ratio is more than twice as high as the predicted value, and in man about six times as high (Table 5). The difference is made up of small resorption spaces which in longitudinal section appear as localized dilatations of a Haversian canal in which no bone remodeling activity is evident. These are 100 to 200 μm in length and 50 to 80 μm in diameter, or 10 to 30 μm wider than the Haversian canal.[73] Such cavities in the very elderly have been ascribed to a process of physical debridement not dependent on cellular intervention, akin to weathering of stone.[102] However, since they are found at all ages it seems more likely that they result from a previous attempt to create a new BRU which was aborted at an early stage.[40,73] This phenomenon must be distinguished from the occurrence of zonal osteons in which only the inner third or half has been replaced by new bone of lower mineral density, but leaving a smooth-walled Haversian canal.[16]

If all resorption spaces were produced by random cross sections through evolving (stage 1) BRU or cutting cones, the theoretical frequency distribution would show a preponderance of large spaces and very few small ones (Figures 13 and 14a). But the observed frequency distribution in man shows just the opposite, with a large excess of small resorption spaces consistent with a bimodal distribution (Figure 14b). The surplus of small spaces is sufficient to account for the discrepancies observed in cross section and supports the previous interpretation that resorption spaces are of two fundamentally different types (Figure 14c). Arrested resorption spaces represent sites of permanent interruption of the normal remodeling sequence; they are lined by flat cells indistinguishable from those which line mature Haversian canals. By contrast, effective resorption spaces represent sections through cutting cones and are lined by osteoclasts. However, not all resorption spaces lacking osteoclasts represent abortive BRU. Some may be part of an osteoclast domain, and others correspond to the period of quiescence or reversal phase of variable duration between completion of resorption and commencement of formation. There is no information on the relative prevalence of the three different types of nonosteoclast-lined resorption spaces. The diameter of a quiescent resorption space would be the same as that of the future osteon, but the cells corresponding to the reversal phase are poorly characterized. At the present stage of knowledge it is not possible to distinguish in cross section between a small quiescent resorption space and a large arrested resorption space. The distinction between osteoclast domain, quiescent resorption, and arrested resorption is also important for the indirect estimation of $\sigma_{RC(r)}$. From the previous discussion and from the geometrical relationships illustrated in Figure 7, it is evident that in a steady state the ratio of the durations of radial resorption and radial closure are the same as the ratio of the lengths of the cutting cone and closing cones, or in symbols:

$$\sigma_{RC(r)} = A_r/A_f \cdot \sigma_{RC(f)} \qquad (6)$$

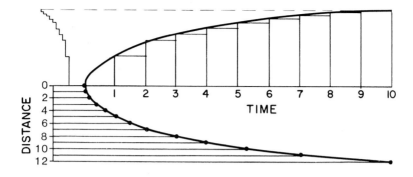

FIGURE 13. Kinetics of resorption in idealized cutting cone. If longitudinal advance occurs at a constant rate, equally spaced vertical subdivisions correspond to equal intervals of time. Note that horizontal subdivision of the total radial excursion into equal intervals of distance from the center corresponds to successively longer intervals of time, indicating that radial erosion is most rapid at the outset and gets progressively slower as radial excavation is completed. If transverse sections are distributed at random along the length of the cutting cone, the frequency distribution of resorption space sizes will be in proportion to the corresponding time intervals. Since each successive increment of distance corresponds to a longer interval of time than the preceding increment, the prevalence of resorption spaces will increase as their radius increases.

For valid estimates of $\sigma_{RC(r)}$ to be obtained from this equation it is essential that A_r is based only on sections which are through effective resorption spaces and so within an osteoclast domain, and that both quiescent and arrested resorption are excluded. As already indicated there is no certain method of making these distinctions, so that all previously published indirect estimates of $\sigma_{RC(r)}$ based on this principle are incorrect. But if A_r was determined solely on the presence of osteoclasts, $\sigma_{RC(r)}$ would only be moderately underestimated. By a similar argument if the number of quiescent resorption spaces (A_q) could be accurately determined, then the duration of the reversal phase ($\sigma_{RC(q)}$) could be estimated from the relationship $\sigma_{RC(q)} = A_a/A_f \cdot \sigma_{RC(f)}$.

The characteristics of the three types of resorption space are summarized in Table 9. The data suggest that less than 2% of activation episodes are aborted, but that arrested resorption spaces are much more prevalent than those which are part of the normal remodeling sequence because they persist until the bone containing them is remodeled. It is likely that arrested resorption spaces contribute significantly to the modest increase in intracortical porosity which occurs with age.

3. Trabecular Remodeling

Because of its solidity, formation of new cortical bone can only occur at sites where bone has been removed. This physical constraint does not apply to the trabecular or cortical endosteal surface, but there is considerable evidence that here also bone formation is coupled spatially and temporally to bone resorption. Examination of the trabecular surface reveals only cavities in the various stages of excavation and repair. Except for callus formation around an occasional microfracture,[103] there are no localized excrescences resulting from bone formation *de novo*, although a smooth curvature could be preserved if formation occurred over the whole of a concave surface. More compelling evidence is that (as in cortical bone) almost all cement lines have the scalloped configuration of reversal lines.[104] This indicates that, on the trabecular surface as well as within the cortex, formation normally occurs only at sites of previous resorption. But it says nothing about the time relationship between these processes, and the three-dimensional spatial relationship between sites of resorption and formation cannot be discerned in standard histologic sections.

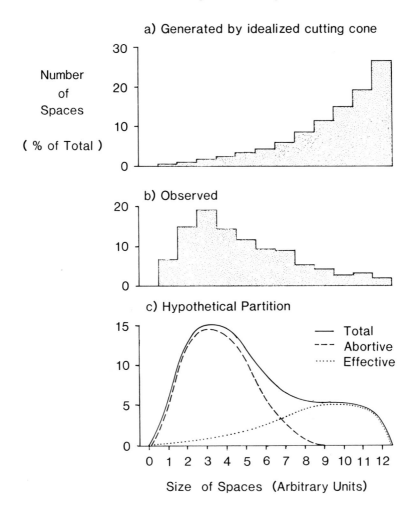

FIGURE 14.   Frequency distribution of resorption space sizes; number of spaces is expressed as % of the total number of observations and size of spaces in arbitrary units with 12 subdivisions. a — Theoretical frequency distribution generated by idealized cutting cone depicted in Figure 11. b — Actual distribution constructed from measurements made on 3207 resorption spaces in human cortical bone.[193] c — Hypothetical partition of observed frequency distribution into two populations, one comprising a large number of small abortive resorption spaces and the other a small number of large effective resorption spaces which are part of a complete remodeling sequence. The frequency distribution of the latter corresponds to panel (a) but with a range of maximum sizes rather than a single maximum size.

a. Trabecular BRU

When thick slabs of trabecular bone are examined with the stereo microscope, a group of osteoclasts may be spread out over a front up to 600 μm wide and 50 to 100 μm deep,[12] and osteoid seams are seen as approximately rectangular patches like shingles, about 600 μm wide and between 500 and 1000 μm long. But whether the osteoclasts and osteoblasts present at different places at the same time are involved in rebuilding the same BSU (as they may be in cortical bone) is usually unknown, although juxtapositions consistent with this interpretation may be observed. A trabecular BRU exists as a sequence of cellular events in time, but not necessarily as a discrete anatomical structure in space. Formation might not begin until resorption of the whole cavity for a future BSU is completed, which could occur very rapidly as in a Type I BRU in

Table 9
TYPES OF RESORPTION SPACES IN CROSS SECTIONS OF CORTICAL BONE

| | Effective (Part of A → R → F) | | Abortive (A → R) |
	Active	Quiescent	
Width (μm)	50—250	150—250	50—200
Lining cell	Osteoclast	Mononuclear[a]	Flat
Prevalence (% total)	20[a]		80
Duration (years)	0.1[a]		20
Incidence[b] (% total)	98[a]		2

Note: In hand-ground sections the type of lining cell cannot be determined, and there are no data on the relative prevalence of active and quiescent spaces. The duration of abortive spaces is presumed to be the same as the mean life span of cortical bone with a turnover of 5%/year. The duration of resorption spaces that is part of the normal remodeling sequence is an estimated value for $\sigma_{RC(r)}$ and $\sigma_{RC(q)}$. [a]Variable and poorly characterized. [b]Relative incidence equals relative prevalence (¼) divided by relative duration (1/200).

[a] Applies to both active and quiescent.

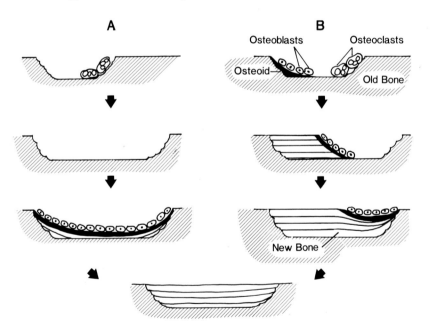

FIGURE 15. Types of trabecular BRU. (A) Resorption completed before formation begins, as in Type 1 cortical BRU (Figure 10); (B) resorption followed closely by formation, as in Type 2 cortical BRU during advancement (Figure 10).

cortical bone. Alternatively formation might begin while resorption of the same cavity was still continuing nearby, as in Type II BRU during their stage of advancement (Figure 15). Conceivably both patterns of evolution might occur in different circumstances.

Nevertheless it is conceptually helpful to partition the observed resorption and formation surfaces as if they were associated in spatially discrete BRU. With conventional stains only about 0.5% of the iliac trabecular surface appears to be in contact with osteoclasts,[105] but using more specific histochemical identification a value of 1.4% was found.[106] If the extent of osteoclast domain on the trabecular bone surface of the ilium is related to the total osteoid surface in the same ratio as the cutting cone length to closing cone length in cortical bone it would be 0.03, with a ratio of osteoclast domain to contact area of about 2.2 (0.03/0.014). Consequently, the proportion of iliac trabecular bone surface normally occupied by active remodeling would be about 0.18, comprising 0.03 for resorption and 0.15 for formation. Since bone turnover in the ilium is higher than in other bones, values more representative of the whole skeleton are 0.072 for total remodeling, comprising 0.012 for resorption and 0.060 for formation (Table 3). If the area of an individual osteoid seam is 0.30 mm² (0.6 mm:0.5 mm), the area of associated active resorption (osteoclast domain) would be 0.06 mm² (0.30:0.30/0.15). Consequently, the total area of an individual trabecular BRU (active resorption and osteoid seam) is about 0.36 mm², and mean values for the whole skeleton are 0.2 BRU/mm² of trabecular bone surface and 4/mm³ of trabecular bone volume (Table 4).

There are five times as many trabecular as cortical BRU, but a fairer comparison is in terms of BRC. Since the mean internal surface of a cortical BRC is about 0.6 times the initial surface of 0.6 mm², a trabecular BRC can be assigned the same surface area as a trabecular BRU. With this convention, the number of BRC is in proportion to the total internal surface. In the average BRU about two thirds of the resorbed cavity has been refilled, because bone formation slows down with time. If the mean depth of the remaining cavity is 14 μm, the volume of bone still missing but potentially replaceable (the remodeling space) is 0.005 mm³/BRU or 2% of active bone volume, amounting to 7.0 cm³ for an average-sized skeleton. The total remodeling space for the whole skeleton is 12.7 cm³, corresponding to 7.6 g of calcium (Table 4). The remodeling space will eventually be converted into bone, so that the combined total of actual bone volume and remodeling space represents the potential bone volume.

b. Bone Resorption — Active, Quiescent, and Arrested

In both trabecular and cortical bone the net effect of osteoclastic resorption is to erode a cavity in a direction perpendicular to the surface, but the three-dimensional path taken by an individual osteoclast is not known in either case. Woods[107] believes that the osteoclasts normally have their ruffled borders facing the exposed ends of the collagen fibers; this may account for the striations visible under higher power light microscopy.[108] If individual osteoclasts erode in a direction parallel to their long axes, they would travel widely across their domains, shaving off thin layers of bone like a carpenter's plane and needing many traverses to reach the prescribed depth. This concept is consistent with the difference between the measured fraction of iliac trabecular surface covered by osteoclasts (0.5 to 1.4%) and the estimated fraction occupied by osteoclast domain (3%) (Figure 16). When resorption surfaces on the cortical-endosteal surface of normal bone are examined by SEM[109] they are of uniform depth, the floor covered by shallow depressions separated by small ridges which are at right angles to the long axes of the collagen fibers (Figure 17). The dimensions of the individual depressions are 10 to 20 μm in width and 2 to 5 μm in depth, much smaller than the dimensions of a single osteoclast, so that the depressions may correspond to separate areas of ruffled border within a single cell.[110] Probably different types of osteoclast movement occur, depending on the spatial and temporal relationship between resorption and formation, but the SEM appearances suggest that an osteoclast moves slowly across a bone surface like an amoeba with multiple areas of ruffled border acting like pseudopodia, each remaining in contact with the bone for long enough to make a small erosion.[109,110] It is premature to conclude that there is a fundamental difference be-

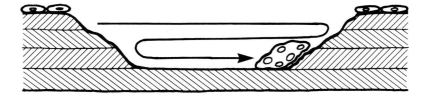

FIGURE 16. One possible mode of osteoclast movement in trabecular bone, with removal of successive lamellae as by a woodworker's plane.

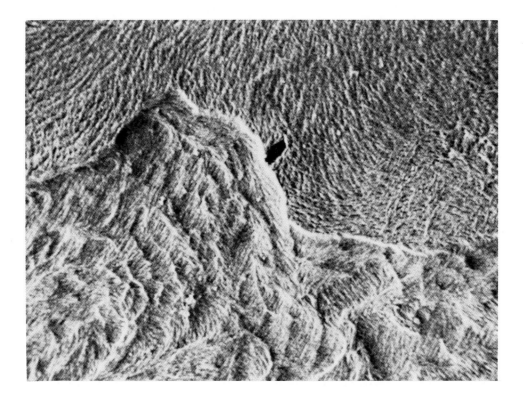

FIGURE 17. Scanning electron micrograph of endosteal resorptive surface of human femur. The shallow pits separated by ridges are believed to correspond to individual osteoclasts or possibly individual osteoclast nuclei. (Micrograph supplied by Krempien[109]).

tween cortical and trabecular bone in the relationship between osteoclast and osteoblast movement (Figure 18).[12]

The relationships between osteoclast domain, quiescent resorption, and arrested resorption are not yet so clear as in cortical bone but the same concepts are applicable. In the ilium the fraction of total trabecular surface occupied by Howship's lacunae is normally about 0.05, whereas the estimated extent of osteoclast domain is 0.03. The difference of 0.02 could represent either quiescent resorption (the reversal phase between completion of resorption and commencement of formation) or arrested resorption (a permanently unrepaired defect on the surface due to interruption of the normal remodeling cycle). In rat alveolar bone the reversal phase is marked by Howship's lacunae containing large mononuclear cells with abundant but poorly staining cytoplasm.[111] Similar cells with a large oval nucleus and dispersed chromatin have been observed in the trabecular bone of the human ilium;[112] they resemble the osteoprogen-

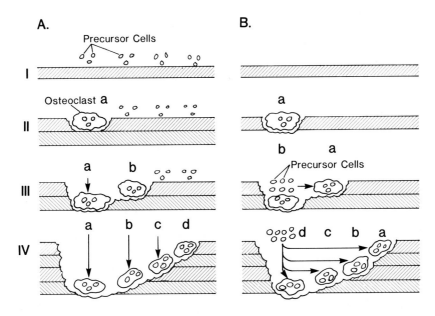

FIGURE 18.   Possible paths of movement of individual osteoclasts in trabecular BRU.
(A) Predominantly vertical movement (perpendicular to the surface); (B) predominantly
longitudinal movement (parallel to the surface). Compare with Figure 12.

itor cells identified by Ham.[108] In the adult dog, cells of identical appearance have the
highest uptake of tritiated thymidine of all the cells on the endosteal surface[113] and so
may be provisionally identified as preosteoblasts. When such cells are present within
a Howship's lacuna they probably identify a reversal phase, but a variety of other cells
can be found which are less well characterized. In older persons (especially those with
osteoporosis) a significant proportion of the total Howship's lacunar surface is covered
by flat lining cells. These resemble those on the inert surface and so identify sites of
arrested resorption which probably contribute significantly to the increase in total
Howship's lacunar surface with age. Unlike the cells adjacent to osteoid seams, the
different types of cell found within Howship's lacunae cannot be placed in temporal
sequence; there was no difference in mean depth from the surface between lacunae
lined by osteoclasts, flat lining cells, or unclassified mononuclear cells.[196] Consequently, osteoclast domain cannot at present be distinguished with certainty from
quiescent resorption by the character of the surface at the site of a grid line intersection, but if there is an osteoclast anywhere within a resorption cavity it seems reasonable to assume that the whole of that cavity is part of a domain, whereas resorption
cavities completely lacking osteoclasts are more likely to be quiescent or arrested.

c. Bone Formation and Osteoid Seam Kinetics

As in cortical bone, the behavior of individual cells is much clearer during bone
formation than during bone resorption. Although the evolution of an individual bone-
forming site cannot be observed, the distance from the cement line provides a marker
for the passage of time. At some sites where bone formation has just begun, cells
previously identified as preosteoblasts may be seen; this probably results from an extremely oblique section plane through a Howship's lacuna adjacent to a nascent seam
(Figure 19), but was responsible for the initial misdescription of these cells as Type I
osteoblasts.[112] With more nearly perpendicular sections the osteoblasts close to the
cement line are columnar or cuboidal in shape, intensely basophilic, and contain abundant pyroninophilic granules indicative of RNA synthesis (Table 10). As distance from

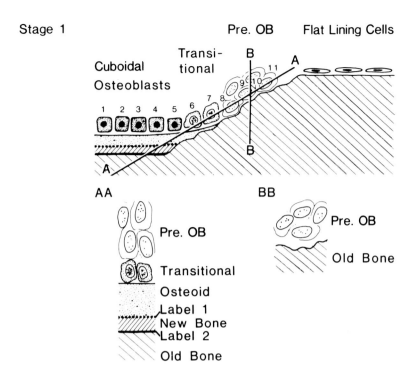

FIGURE 19. Diagram of a Howship's lacuna in which bone formation has just commenced, to illustrate effect of section plane on apparent relationship between preosteoblasts (pre-OB), osteoid seam, and tetracycline label. AA denotes an oblique section, and the lower panel illustrates how this would demonstrate preosteoblasts, osteoid, and tetracycline labels in order. BB denotes a section perpendicular to the trabecular surface, and the lower panel illustrates how this would show preosteoblasts in juxtaposition to the Howship's lacuna.

Table 10
CATEGORIES OF OSTEOBLASTS

	Cuboidal	Intermediate	Terminal	Total
Extent (% OS perimeter)	15	20	65	100
Distance from CL (μm)	20	30	40	35
Width of seam (μm)	15	11	6	8.3
Width of new bone[a] (μm)	5	19	34	26.7
Pyroninophilic granules	+ +	+	+ −	+
MiAR (μm/day)	1.4	1.0	0.6	0.8
f_L (% OS perimeter)	86	80	67	75
MaAR (M_f; μm/day)	1.8	0.8	0.4	0.8
MLT (days)	12.5	13.8	15.0	13.8

Note: Structural and functional characteristics of three kinds of osteoblasts found at three stages of the life history of an osteoid seam. The first three columns are composite values based on measurements by A. R. Villanueva in 6 normal and 19 abnormal cases. Last column gives values of measurements averaged over the entire osteoid perimeter in the usual manner. CL = cement line; MiAR = mineral apposition rate; MaAR = matrix apposition rate; M_f = radial closure rate; F_L = fractional labeling. [a]Distance from cement line less seam width.

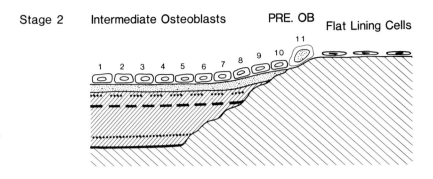

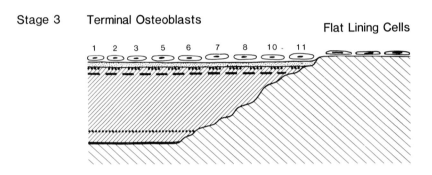

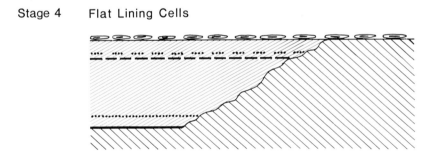

FIGURE 20.   Successive stages of bone formation within Howship's lacuna shown in Figure 19. In stage 2 there is a greater thickness of new bone and the osteoid seam is closer to the surface and is covered by intermediate rather than cuboidal osteoblasts. Cells numbered 8, 9, and 10 could have arisen from preosteoblasts shown in Figure 19 and cells 6 and 7 from transitional cells shown in Figure 19. The continuous solid and dotted lines indicate the original labels shown in Figure 19 in the same location. The interrupted solid and dotted lines show where the labels would have been had the tetracycline been administered shortly before the biopsy. Stage 3 shown osteoid seam almost flush with the surface covered by terminal osteoblasts, and stage 4 shows completion of new BSU covered by flat lining cells.

the cement line increases, the sequence of changes in cell morphology in human trabecular bone is essentially the same as in dog cortical bone (Figure 20). Although the changes in size, shape, and staining characteristics represent a continuous spectrum, it is convenient for descriptive purposes to subdivide the osteoblasts into three categories (Table 10) of mature (cuboidal), intermediate, and terminal (flat). The first two categories correspond to "active" osteoblasts and the last to the "inactive" osteoblasts described by Schenk et al.[114] These terms are convenient but it is often misleading to infer activity from morphologic appearance alone.

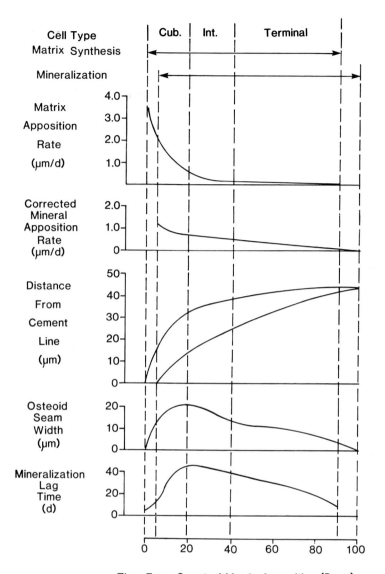

FIGURE 21.   Model of sequence of events at a single site of bone formation, based on measurements (Table 10) of mineral appositional rate and
osteoid seam thickness in relation to distance from the cement line as an
index of the passage of time and endosteal cell morphology (cuboidal, intermediate, and terminal osteoblasts). Horizontal arrows depict durations of
matrix synthesis and mineralization. Changes in matrix apposition rate, corrected mineral apposition rate, distances from the cement line of the matrix
edge (upper line) and the mineralization fronts (lower line), osteoid seam
width, and mineralization lag time are shown as functions of time.

The MaAR cannot be measured directly but it can be inferred by comparing the
directly measured MiAR with the thickness of the immediately adjacent osteoid seam.
By relating these values both to cell morphology and to time, a model can be constructed of the sequence of events at an individual bone-forming site (Figures 19 to
21). The mean osteoid seam thickness adjacent to mature osteoblasts is at a maximum
even though the thickness of each seam is zero at the onset of bone formation. This

suggests that formation starts with a short-lived but very rapid burst of matrix synthesis at a rate which is about three or four times greater than is subsequently found; indeed more than two thirds of the total thickness of new matrix is made in the first 1 to 2 weeks.[197] The osteoid seam attains a thickness of about 15 to 20 μm during the 5 days or so before the onset of mineralization; this has a less rapid initial burst and thereafter follows a time course similar to matrix synthesis but delayed by about 2 weeks, both rates declining with time (Figure 21). Although not directly measured in trabecular bone, the secretory territory of the osteoblast presumably increases with time as it does in cortical bone. It is possible that after the rapid initial phase the rate of matrix synthesis per cell remains constant and the decrease in total rate of matrix synthesis could result from a decrease in the number of cells per unit area of bone-forming surface. This model is consistent with the significant correlations found between total osteoid volume and the fraction of trabecular surface covered by cuboidal osteoblasts,[115] between osteoid seam thickness and the fraction of osteoid surface covered by cuboidal osteoblasts,[114] and (in the dog) between the logarithm of osteoblast height and the thickness of the underlying osteoid seam.[116]

Terminal osteoblasts are found in the latter half of the life history of an individual bone-forming site, when the osteoid seam is getting progressively thinner on the way to its eventual disappearance. As shown by double tetracycline labeling, mineralization is continuing beneath the cells at a rate which, although significantly slower than with mature and intermediate osteoblasts, is inconsistent with complete absence of matrix synthesis. If the corrected MiAR was 0.4 μm/day and MaAR was zero, a seam 6 μm in thickness would disappear in 15 days, and terminal osteoblasts would occupy a much smaller fraction of the bone surface than is found (Table 10). Continuation of matrix synthesis is also suggested by the occurrence of pyroninophilic granules in some terminal osteoblasts. Although the mineralization lag time increases during the life history of the seam, matrix synthesis probably does not normally cease more than 15 days before mineralization (Figure 21).

Bordier has proposed that mineralization is controlled by the osteocytes within the osteoid lying closest to the zone of demarcation ("mineralizing osteoblasts") and that mineralization occurs in short bursts, each corresponding to complete mineralization of a single lamella.[117] In some respects this resembles the cyclical concept of Johnson, but if matrix synthesis and mineralization are controlled by different cells, the amplitude and duration of the cycles could be much smaller than in Johnson's model and mineralization need not be completely interrupted. Such short-term periodicity is consistent with the results of closely spaced multiple labeling in the rabbit which suggested an osteoblast activity cycle of about 48 hr.[90,91] The Bordier model is more easily reconciled with continuity of tetracycline labeling than the Johnson model, but direct evidence in its support is still lacking.

d. Cell Balance and Remodeling Cycle

During completion of a trabecular BSU there is a loss of cells, but of smaller magnitude (Table 6) than in the cortex, the number of osteocytes and lining cells in the complete BRU being about 70% of the number of osteoblasts initially present, rather than only 30% as in cortical remodeling. This reflects important differences in geometry between cortical bone, where the local surface first increases and then decreases about fivefold during a remodeling cycle, and trabecular bone, where the local surface changes very little (Table 6). Another consequence of different geometry is that the availability of bone surface may be a limiting factor in trabecular remodeling even though the S/V ratio is high. Because osteoid is only rarely resorbed, the remodeling sequence must usually begin on the inert surface, which in some diseases may be almost completely replaced by osteoid. By contrast, although the existing surface is quite small in cortical bone, the inert surface is relatively greater and the potential surface is virtually unlimited, because new resorption cavities can arise from Volkmann's canals. Consequently, lack of available surface is never a limiting factor in cortical remodeling.

Despite the uncertainty concerning the three-dimensional organization of a trabecular BRU, the sequence of events at a single location, or remodeling cycle, is essentially the same as in cortical bone, although the absolute (and possibly the relative) duration of the different stages is different. The mean time required to rebuild a trabecular BSU at a single point on the surface ($\sigma_{RC(f)}$) can be calculated from Equation 2 in exactly the same way as the duration of radial closure, except that in the calculation of M_f according to Equation 1 ($= MiAR \cdot f_L$), f_L is the extent of labeled perimeter expressed as a fraction of total osteoid perimeter. Because of the identity in concept, M_f will be referred to as the radial closure rate in trabecular bone also. As in cortical bone, σ_f calculated in this way includes both the initial mineralization lag time and the duration of mineralization and M_f is equivalent to the matrix appositional rate averaged throughout σ_f. By similar reasoning, the time required to excavate to the depth of a BSU ($\sigma_{RC(r)}$) and the duration of the reversal phase between completion of resorption and commencement of formation ($\sigma_{RC(q)}$) can be calculated from Equation 6 using the ratios of either the active or quiescent resorption perimeter to the formation perimeter instead of the ratio of the numbers of active and quiescent resorption spaces to number of seams. The results are equally uncertain because of the difficulty of distinguishing between osteoclast domain, quiescent resorption, and arrested resorption, but an upper limit for $\sigma_{RC(r)}$ can be defined by using the total Howship's lucanar surface and a lower limit by using the osteoclast covered surface.

Because individual BRC cannot be identified, the calculation of their birth rate, or activation frequency, is only possible in terms of an arbitrary unit of osteoid seam area. The area of a trabecular BRC was previously defined as 0.36 mm², the mean internal surface area of a cortical BRC. With this assumption, A_f can be calculated for trabecular bone. For example, if the fraction of trabecular surface covered by osteoid is 0.6 (Table 3) and S/V is 20 mm²/mm³ (Table 1), the area of osteoid per unit volume of bone is 1.2 mm²/mm³ so that A_f is 3.33/mm.³ Since M_f and MWT can be measured by the same methods as in cortical bone, Equation 4 can now be solved (Table 7). Values for A_f and μ_f calculated in this way can also be expressed per unit of surface rather than per unit of volume ($A_{f(s)}$, $\mu_{f(s)}$) simply by dividing by S/V, so that with the previous values $A_{f(s)}$ is 0.167/mm².

The close relationship between volume-based bone turnover and activation frequency shown for cortical bone was critically dependent on the relative constancy of MWT and the same also applies on the trabecular surface. In 36 normal subjects, the largest individual mean value for trabecular MWT was less than twice the smallest and the CV was 17.5%[23] so that individual mean values will virtually never depart from the population mean value by more than 50%.[4] Much of this dispersion is due to varying obliquity of the section plane, a problem which does not usually occur in cortical bone. Consequently, the dependence of bone turnover on activation frequency is not peculiar to cortical bone but is a general property of the skeleton as a whole.

In accordance with the preceding concepts, any element of the trabecular surface can be classified by means of the surface configuration and endosteal cell morphology into one of the seven categories shown in Table 11, each corresponding to a different stage in the remodeling cycle. The further subdivision of the resorption surface which will be needed for a complete understanding must await further improvements in the morphologic and histochemical identification of cells and the more widespread application of electron microscopy to adult human bone. Further subdivision of the formation surface is possible by means of either tetracycline labeling or the histochemical

Table 11
SUBDIVISIONS OF TRABECULAR BONE SURFACE

Type of Surface	%	Days
Total resorption surface	5	25
Osteoclast domain	3	15
Quiescent resorption	1	5
Arrested resorption	1	320
Total formation surface	15	75
Growing seam	2	10
Mature seam	4	20
Terminal seam	9	45
Resting surface	80	400
Total	100	495

Note: Subdivisions of trabecular bone surface in the ilium as % of total surface, with corresponding durations in days calculated from $\sigma_{RC(f)}$ = 80 days as described in text. Duration of arrested resorption is presumed to be the same as the duration of resting surface. Total duration refers to the normal remodeling cycle excluding arrested resorption.

identification of the mineralization front, as will be discussed in more detail in a later section.

D. Noncoupled Bone Turnover During Growth and Repair

The previous account applies only to the process of remodeling in the adult skeleton whereby turnover of bone tissue is accomplished, but both the shape of the bones and the total amount of bone either change very slowly or not at all. Bone turnover during growth, which drastically alters both the size and shape of the bones, will be referred to collectively as modeling.

1. Modeling of Bone During Growth

Bones grow in length by endochondral ossification, a process (the only one) which continually creates new trabeculae until the epiphyses fuse and growth ceases. It is well known that this has no counterpart in the adult skeleton, but it is less widely appreciated that the cellular basis of bone turnover in the shaft of a long bone during growth is also quite different from that occurring in the adult. The general scheme of growth in width is periosteal apposition of new bone without preceding resorption and endosteal resorption of old bone without succeeding formation, so that the osteoclasts and osteoblasts are anatomically segregated to different surfaces. This basic pattern is modified to preserve flaring of the metaphyses and to permit some bones to grow eccentrically rather than concentrically. Slow periosteal growth occurs by simple apposition of circumferential lamellae, whereas rapid periosteal growth is achieved by bars of bone which grow out perpendicular to the surface and fuse to form so-called laminar bone; the longitudinal spaces are then filled centripetally by primary osteons.[6] Less than 10% of total turnover in the long bones of a rapidly growing animal takes place around vascular canals.[118]

Modeling differs from remodeling in six important respects (Table 12). First, resorption and formation occur mainly on different bone surfaces and have no spatial relationship. Second, there is no fixed temporal relation or coupling between them; each activity appears independently of the other to achieve the appropriate changes in size

Table 12
DIFFERENCES BETWEEN MODELING AND
REMODELING

	Remodeling	Modeling
Location	Spatially related	Different surfaces
Coupling	A → R → F	A → F
Timing	Cyclical	Continuous
Extent	Small (<20%)[a]	Large (> 90%)[a]
Apposition rate	Slow (0.3—1.0 μm/day)	Fast (2—10μm/day)
Balance	No change or net loss	Net gain

[a] Of available surface.

and shape and so is subject to independent regulation. Third, modeling involves continuation of the same activity at the same surface without interruption for an extended period of time, whereas in remodeling each surface is subject to repeated cycles of resorption, formation, and quiescence. Fourth, during the growth period almost the entire periosteal and endosteal surface is the site of either formation or resorption at all times, whereas in remodeling only a small fraction of the surface is active at any one time. Fifth, in remodeling the rate of advance of the formation front through tissue space (1 μm/day) is much slower than the rate of advance of the resorption front (5 μm/day), and varies little with age. By contrast, in modeling the rate of periosteal apposition varies with the growth rate; it may be as high as 10 μm/day in 4-week-old rats, falling to about 2.0 μm/day by 16 weeks.[119] In man, it probably averages about 2 to 3 μm/day throughout the growth period. The endosteal resorption front moves at a similar but slightly smaller rate, so that the cortex grows in thickness as well as in external diameter. Finally, modeling produces a net gain in bone volume and mass, whereas remodeling produces either no change or a net loss.

2. Woven and Lamellar Bone[6]

The most characteristic form of woven bone (sometimes known as fetal bone) is formed during early embryonic osteogenesis. In contrast to the regular orientation of collagen fibers in lamellar or parallel fibered bone, in woven bone they run in all directions like a carpet underfelt; this difference is best appreciated by scanning electron microscopy or by examination with polarized light. The trabeculae are randomly arranged without relation to lines of stress and are irregular and variable in thickness. Osteocytes are much more frequent and are scattered at random with no relation to vascular channels. The lacunae are irregular in surface, are considerably larger on the average than in lamellar bone, but are much more variable in size and shape. Because of the increase in both number and size of osteocytes, total fractional lacunar volume is about 7 to 8 times as great as in lamellar bone. Woven bone stains intensely for mucopolysaccharides and shows metachromasia with toluidine blue. Unawareness of these microscopic appearances has misled many observers into overemphasizing the frequency and importance of so-called osteocytic osteolysis.[32]

The differences in structural organization between lamellar and woven bone arise from differences in methods of formation. Lamellar bone is formed only in apposition to an existing surface; each osteoblast is coordinated with its neighbors, and together they make a continuous layer of bone laid down only on one side of the cell. Woven bone is formed directly in condensations of fibrous tissue without the necessity for an adjacent free bone surface, by the rapid, uncoordinated, and nonpolarized action of individual osteoblasts. Each forms an island of new matrix, and adjacent islands coa-

Table 13
CONTRASTING CHARACTERISTICS OF NORMAL AND PATHOLOGIC BONE RESORPTION

	Normal	Pathologic
Purpose	Structural needs of skeleton	"Lebensraum" for abnormal cells
Osteoclast precursors	Mainly local	Mainly blood borne
Bone removed	Fatigued or dead	Indiscriminate
Depth and extent	Controlled	Uncontrolled
Subsequent formation	Coupled	Noncoupled
	Lamellar	Woven

lesce into nodules. Mineralization occurs rapidly and simultaneously in each nodule in the absence of proper seams with diffuse rather than focal uptake of tetracycline and without relation to collagen fibers or blood vessels, the process resembling somewhat the mineralization of cartilage rather than of mature lamellar bone.[11]

Typical lamellar bone and typical woven bone are easy to distinguish, but to classify bone only into these two types is an oversimplification, because all degrees of randomness of collagen fiber orientation can be observed. For example, bone which is less ordered than normal lamellar bone but not so disordered as typical woven bone can be formed in apposition to an existing surface, with both osteoid seams and focal tetracycline uptake. This occurs in Paget's disease, in early osteitis fibrosa, especially in renal osteodystrophy, after fluoride administration, and in the periodontal ligament in response to mechanical manipulation of a tooth. Such bone, although often classified as woven on the basis of its collagen fiber orientation, probably results from a defect in collagen cross linking during the early phase of matrix maturation; it is often associated with (and may be the result of) a faster than normal bone appositional rate. Several other varieties of bone have been described, based on the size as well as the disposition of the collagen fibers.[84] Lamellar or parallel fiber bone may also be further subdivided into primary osteonal bone, secondary osteonal bone, surface bone, and interstitial bone.[120] This is helpful for complete understanding of the complexities of growth, but the details are beyond the scope of the present chapter.

All bone formed in locations where no bone previously existed is initially woven bone. This is true in health and in disease, both inside and outside the skeleton, and at all ages. For example, the two types of embryonic bone formation (intramembranous and endochondral) differ only in the presence or absence of a prior stage of calcified cartilage, and have in common the formation of woven bone followed by its piecemeal resorption and gradual replacement by lamellar bone. Such replacement is the normal fate of woven bone wherever and whenever it is formed.[6] In the adult skeleton typical woven bone is formed in fracture healing, osteogenic sarcoma, and in severe Paget's disease or osteitis fibrosa cystica. The latter two conditions are examples of disruption of the normal coupling between prior resorption and subsequent lamellar bone formation. This break also follows the pathologic bone resorption (around osseous metastases or foci or myeloma) which represents an attempt to find "lebensraum" for the abnormal cells. Pathologic bone resorption occurs without relation to structural requirements, is not subject to the normal spatial and temporal constraints which characterize resorption during normal remodeling, and is not coupled to subsequent lamellar bone formation (Table 13). Increased woven bone formation may occur as a reparative mechanism, but the balance between resorption and formation is much more easily shifted than in normal remodeling so that either osteosclerosis or osteolysis may predominate. Apart from the pathologic significance of woven bone, its formation is relatively refractory to hormonal regulation, although it is subject to both phys-

Table 14
QUANTITATIVE CHARACTERISTICS OF A
TRANSILIAL BIOPSY CORE

	Cortical	Trabecular	Total
Total tissue volume (mm³)	50	150	200
Total bone volume (mm³)	48	30	78
Total surface (mm²)	120	600	720
Number of BSU	700	1200	1900
Number of BRU	40	360	360
Number of BRC	100	360	460
Area of section (mm²)	12.5	37.5	50
Total perimeter (mm)	24	118	142
Remodeling perimeter (mm)	3	21	24
BRC/section	20	42	62
Fraction of total body BRC	1/35,000	1/33,000	1/34,000

Note: Some quantitative characteristics of a transilial biopsy core of 5 mm diameter and 10 mm length. The section is assumed to be through the axis of the core and so of maximum possible area. Number of cortical BRU/mm³ is assumed to be four times higher than in rib. Number of cortical BRU represented in the section is based on the observed mean perimeter of 0.35 mm. Number of trabecular BRU in the section assumes that a random section through a BRU will be 0.5 mm in length. These values apply to ideal conditions in normal bone. A fragmented specimen in an osteoporotic patient may sample less than one fifth of this number of BRU.

ical and chemical influences. For example, corticosteroids may profoundly depress lamellar bone formation but have much less effect on woven bone formation, accounting for the exuberant callus formation sometimes seen in patients with spontaneous or iatrogenic Cushing's syndrome.

III. THE INTERPRETATION OF BONE HISTOMORPHOMETRY

A. Some Limitations of Bone Biopsy as a Diagnostic and Investigative Tool

A large number of different measurements can be made on a bone biopsy specimen, but their reliability is often questioned because the biopsy is such a small sample of the total skeleton. Even with a trephine of 7.5-mm internal diameter a biopsy core from the ilium is often no more than 5 mm in diameter. With a length of 10 mm such a core has a total volume of about 200 mm³ (Table 14) and contains about 1/22,000 of total bone volume and about 1/15,000 of total bone surface. These ratios are no worse than for liver, kidney, or muscle biopsies, and in bone (as in these other tissues) accurate diagnosis can be based on qualitative examination supplemented by a few simple measurements. The problem is more serious when the biopsy is used as an investigative rather than as a diagnostic tool; the larger the number and the greater the complexity of the measurements, the less likely are results in an individual case to be meaningful, so that reliable conclusions can be drawn only from groups of similar cases studied in the same way.

1. The Relationship of Local to Whole-Body Bone Turnover

The volume-based bone formation rate in rib cortical bone determined by the tetracycline labeling technique is about 3.5 to 4%/year in adults, but the distribution of ^{90}Sr in human bone suggests that the turnover of cortical bone in the rib[121] is about

50% more than of cortical bone elsewhere. An overall value of 2.5%/year also agrees with estimates based on the increase in the number of osteons with age[80] and on the amounts of both radioactive strontium[122] and radium[123] which remain in human bone after known periods of time. These latter estimates require two assumptions. The first is that after the radioactive marker has been removed from bone it will be redeposited in accordance with the prevailing plasma-specific activity, and that local recycling within the skeleton can be disregarded. In cortical remodeling the dissolved components of resorbed bone are carried by the Haversian veins away from the sites of bone formation and returned to the systemic circulation so that significant local recycling is unlikely. The second assumption is that turnover is randomly distributed, or that any moiety of bone has the same probability of being turned over in any interval of time. But if some regions turn over more rapidly than others, the observed mean retention will underestimate turnover. For example, if a region of cortical bone turns over randomly at 4%/year, then approximately 67% of the original bone and any strontium or radium it may have contained will remain after 10 years. But if the same mean turnover of 4%/year is made up of 16%/year turnover in 20% of the bone and 1%/year turnover in 80% of the bone then 76% of the original bone and its contained isotopes will remain after 10 years. The annual turnover calculated on the assumption of random distribution would be only 2.7%, so that the indirect estimates could be reconciled with the direct measurements in rib. But it is unlikely that such extreme departures from randomness ever occur, and 3%/year is probably the highest mean value for cortical bone turnover which is consistent with all available data.

The situation in trabecular bone is even more uncertain. If turnover per unit surface was constant throughout the skeleton, the turnover of different types of bone would be in proportion to their S/V ratio. Taking cortical bone turnover as 3%/year, then from the mean S/V ratios previously given (Table 1) mean trabecular bone turnover would be 24%/year and total-body bone turnover 7.2%/year ($3.0 \times 0.8 + 24 \times 0.2$); this value is substantially smaller than estimates of 10 to 15%/year based on various interpretations of radiocalcium kinetic data.[6] The assumption of constant turnover per unit surface is based on limited observations and is subject to many exceptions. In the dog, local values for the accretion of $^{85}Sr/g$ ash suggest that turnover is about six rather than eight times higher in trabecular than in cortical bone, but no S/V measurements were made.[124] In human subjects iliac and lumbar vertebral trabecular bone turnover estimated from the persistence and distribution of ^{90}Sr is only about 10%/year,[125] but the two assumptions underlying this method are less valid than in cortical bone so that the errors are greater. First, because of the network of blood vessels in the marrow there is more opportunity for local recycling of radioactive marker. Second, departures from randomness of turnover are greater than in the cortex. Bone on the surface turns over more readily than bone at the center of a trabecular plate, a difference which becomes greater with increasing thickness of trabeculae. Furthermore, in the vertebral body, the fraction of surface covered by osteoid is much higher for horizontally than for vertically disposed trabeculae.[126]

The most recent published data for surface-based bone formation rate in iliac trabecular bone in normal human subjects[127] correspond to a volume-based bone turnover of 40%/year, assuming a mean value for S_V of 20 mm²/mm³ (Table 1). Turnover in vertebral trabecular bone has not been measured directly, but comparison of values for osteoid surface and volume suggests that turnover is about half that of the ilium, or 20%/year,[126] and comparison of fractions of surface covered by osteoclasts and osteoblasts suggests a value of about one third that of the ilium or 13%/year.[128] If the ilium is taken as representative of trabecular bone throughout the body, the histomorphometric estimate of whole body turnover is 10.4%/year ($3.0 \times 0.8 + 40) \times 0.2$), which is closer to the kinetic estimate, but it seems more likely that the estimated value

Table 15
FACTORS AFFECTING
REMODELING RATES

Demographic — age, sex, race
Systemic — hormonal and metabolic
Regional — biomechanical or disease
Local — trauma or disease
Random — spatial and temporal

Note: Hierarchy of difference mechanisms which can produce changes in bone formation rate and bone turnover in a biopsy specimen.

for the vertebral body is closer to the mean whole-body value, and that turnover of trabecular bone in the ilium is higher than in other bones. The best histomorphometric estimate of whole-body bone turnover is therefore about two thirds of the radiokinetic estimate, the difference probably representing both short-term and medium-term exchange.[6]

In whatever way the difference between these techniques is ultimately resolved, in most circumstances they show the same directional changes. Whole-body turnover can be influenced at three different levels (Table 15). First, there are general factors such as age, sex, and race. Cortical bone turnover in the rib[11] is very high in childhood, falling progressively to a nadir of 2%/year at age 30 to 39, increasing thereafter to 3.5 to 4%/year for the remainder of the adult life; sex differences have not been examined. Trabecular bone turnover in the ilium is about twice as high in males as in females,[127] but age dependence has not been examined. Whole-body turnover determined by radiocalcium kinetics shows similar effects for both age and sex; whether there are racial differences in bone turnover as well as in bone structure and bone mass is not known. Second, there are hormonal and other systemic blood-borne factors such as the level of thyroid or parathyroid hormone, which affect all parts of the skeleton, albeit to different extents. With excess or deficiency of either hormone, both kinetic and histomorphometric methods show the expected increase or decrease in turnover.[34] Third, an increase in regional or local turnover of sufficient magnitude may be reflected in whole-body values. This occurs most strikingly in Paget's disease in which histomorphometric data from the involved ilium correlate well with biochemical and kinetic indices.[100]

The major discrepancy between the histomorphometric and kinetic approaches is in osteomalacia. Bone turnover estimated by histomorphometry may be increased in mild osteomalacia despite the reduction in appositional rate[129] because of secondary hyperparathyroidism, but in severe osteomalacia, histomorphometric bone turnover is usually reduced.[130] On the other hand, kinetically determined bone turnover is usually increased, whatever the severity[1] except in the osteomalacia of renal osteodystrophy.[131] The probable explanation for this discrepancy is that the component of kinetically determined accretion which represents short- and medium-term exchange rather than new bone formation is greatly increased in osteomalacia, because the proportion of incompletely mineralized low-density bone is greatly increased.[6] Such bone is more permeable than normal bone and its mineral is more readily exchangeable; it thus forms a large capacity sump in which labeled calcium ions can be temporarily sequestered.

Table 16
COMPARISON OF DATA ON
CORTICAL REMODELING FROM
RIB AND ILIUM

	Rib	Ilium
Resorbing surface (% total)	0.6	3
Forming surface (% total)	3	12
Inert surface (% total)	96	85
Resorption spaces (/mm²)	0.2	0.4
Osteoid seams (/mm²)	0.4	0.9
Haversian canals (/mm²)	14.4	7.0
Total channels (/mm²)	15.0	8.3
Osteoid volume (% total)	0.1	0.5
F_v (mm³/mm³/y)	0.04	0.10

2. The Ilium as a Site for Biopsy

The ilium is the usual site for biopsies in human subjects because it is readily accessible, but in many ways the bone at this site is atypical. The trabeculae differ significantly in spacing and thickness from other sites and their three-dimensional orientation is less favorable to stereologic analysis than in the spine.[132] The cortical bone in a transilial biopsy is quite unlike typical cortical bone in the rib or other long bones. Although concentric osteons are seen, they run in many different directions so that their cross sections are frequently elliptical rather than circular; their three-dimensional arrangement has never been adequately studied. There are fewer Haversian canals (Table 16) but more of both resorption spaces and osteoid seams. The transitional zone between solid cortex and trabeculae is wider, with many large subendosteal tunnels containing hematopoietic bone marrow which often extends into the Haversian canals. Possibly because of this proximity to the marrow the volume-based bone formation rate in normal subjects is about 8 to 10%/year, or three times higher than the estimated mean value for cortical bone turnover for the whole body, a somewhat greater local increase than for trabecular bone.

When biopsy data are used to illuminate normal bone physiology or the pathogenesis of disease, these differences are important to remember. But the diagnostic value of a measurement does not depend on its meaning, but on how well it enables different disease states to be differentiated from normality and from each other.[10] From this standpoint the unusually high turnover of the ilium is an advantage, since the effects of hormonal or metabolic changes begin sooner and progress more quickly than in other sites. For example, in six patients with metabolic bone disease who had simultaneous rib and iliac biopsies, bone turnover was significantly more depressed in the iliac than in the rib cortex, thus reducing the absolute difference between the two sites.[133] In patients with metabolic bone disease, only rarely would a diagnosis based on ilial biopsy be changed by examination of the whole skeleton.

3. Sampling Problems — Spatial and Temporal

Diagnosis is more concerned with precision than with accuracy; some general aspects of measurement precision are discussed in Chapter 4. Apart from statistical limitations the other major source of imprecision is site-to-site variation within the ilium — it is usually much more difficult to position the biopsy trephine accurately in a living patient than in a cadaver. Deviations of less than 2 cm from the ideal location of the biopsy do not much affect the mean value for surface or osteoid volume measurements[134] but there are significant differences in individual cases, and as the sites become further apart, systematic as well as random differences are observed. For trabecular bone volume, as much as a twofold difference may be found between the largest and the smallest mean value at different sites; the variation is greater for the number of trabeculae than for their mean thickness. In serial biopsies on the same subject, a change of at least 40% in trabecular bone volume is needed for the difference to be significant[135] and a therapeutic response is more accurately judged from changes in mean trabecular thickness.

Because of the sequential evolution of individual BRU and BRC, sampling variation occurs in time as well as in space. If an individual is in a steady state such that the rate of initiation of new BRC (activation frequency) for the whole body has not changed during the preceding 6 months, the skeleton will contain cortical and trabecular BRC which are equally distributed at all stages of their evolution. More rigorously, if the total life span of every BRC is subdivided into the same number of equal intervals of time, the expected number of BRC whose stage of evolution falls within each interval will be the same. In practice, the actual number will vary because the activation frequency will be subject to the random variation which characterizes all biologic quantities, but the departure from theoretical expectation will be minimal provided the number of intervals is small in relation to the total number of BRC in the whole skeleton. If σ_{RC} is taken as 100 days, then even with an interval as short as 1 day, there will be more than 20,000 BRC in each interval (Table 4), so that deviation from equal temporal distribution for the whole body can be disregarded.

But in the sample provided by an average biopsy, this assumption no longer holds. An ideal biopsy of the dimensions previously given will contain about 450 BRC (Table 14) or about 5 for each 1-day interval, and a single section will contain profiles from only 60 BRU, about 6 for every interval of 10 days. In practice, many biopsies are fragmented and incomplete and contain no more than 2 to 3 BRC/10-day interval. With these small numbers there will inevitably be substantial random variation in the number of BRC at different stages of their evolution. Within a region where the factors influencing activation (Table 15) are constant, if activation occurs randomly with equal probability on any element of surface, the numerical data should conform to a Poisson distribution. But in the only study addressed to this point, substantial deviations from a Poisson distribution were found[136] suggesting that an unrecognized nonrandom factor was operating within the samples examined. Whatever the statistical basis, this variation will alter the apparent proportions of the various subdivisions of the bone surface; the error will be greater the smaller the subdivision and the larger the number of subdivisions. Variations in the number of BRC at different stages will also alter the mean value of any measurement which changes during BRC evolution, as explained in a later section.

4. Sampling Variation at the Level of the Tissue and the Cell

In normal subjects the extent of surface involved in remodeling activity depends primarily on the activation frequency or birth rate of remodeling cycles. As well as random fluctuation due to the small size of the sample, this quantity is influenced by regional biomechanical factors such as weight bearing and muscle tension which determine systematic differences between one bone and another, and by the local effect of trauma which may greatly increase turnover in a particular bone or part of a bone (Table 15). Measurements such as appositional rate and sigma which depend on the function of differentiated cells rather than on the rate of creation of new cells, are largely independent of these regional and local factors. In the normal rat and rabbit the appositional rate shows very little variation between different skeletal sites in trabecular bone[137] but on the periosteal surface of the long bones, appositional growth was consistently about 20% slower in the tibia than in the femur.[119] In a group of normal human subjects the mean appositional rate did not differ significantly between the Haversian, cortical-endosteal, and trabecular surfaces within a transilial biopsy, and the intersite variation was less than 20% in any individual subject.[198] In six patients with various metabolic bone diseases subjected to both rib and iliac biopsies at the same time[133] there was a high correlation between the appositional rates in cortical bone at the two sites, but a poorer correlation in various measurements related to bone turnover. In both normal and abnormal subjects the appositional rate is about 20% lower in both cortical and trabecular bone of the ilium than in cortical bone of the rib.[133] The appositional rate is also much less affected by general factors than the bone formation rate; in adult human subjects it shows no sex difference and varies little with age. Consequently it is a more accurate indicator of hormonal or metabolic effects.[137]

B. General Concepts of Interpretation

Some underlying principles implicit in the previous discussion will now be presented explicitly and in greater detail.

1. Remodeling vs. Turnover

In cortical bone each resorption space and each osteoid seam seen in cross section represent different BRU and BRC, but in trabecular bone the results usually obtained are mean values for each BRC represented in the biopsy sample, and the properties of individual BRU and BRC are lost; the measurements are indices of turnover rather than of remodeling.[12]

In any population the mean value of a measurement may be shifted for several reasons. If each member is equally affected, the frequency distribution of the measurement will be displaced but its shape will be normal and the SD unaltered. But the same shift in mean value could result from a large change in a small segment of the population, the remainder being unaffected. In this case the SD will be increased, and the frequency distribution skewed or bimodal (Figure 22). For example, mean osteoid seam width may be increased because each seam is wider than normal, or because a few seams are very wide and the remainder are unaffected. These different causes of the same change in mean value can be distinguished by probit analysis, a technique which has only rarely been applied to histomorphometric data.[138] The mean value may also be shifted because the distribution of the population is abnormal with respect to some property such as age, even if all individual values are normal. For example, because the mineral appositional rate falls progressively during the life history of an individual osteoid seam, the mean value may be altered because the distribution of seams at different stages of their life history is abnormal. One reason for this is random variation in the biopsy sample as already mentioned; if by chance there are too many seams in an early or a late stage, the mean appositional rate will deviate from the correct value for the whole ilium, and *a fortiori,* for the whole skeleton. Another reason is a change in age distribution of the whole skeleton during a transient state, as explained later.

2. Fractions of Space are Equivalent to Fractions of Time

In whatever way the bone surface is subdivided, at any point in time a given segment of surface must be in one of the states specified by the method of subdivision. The fraction of the total surface occupied by a particular subdivision is equal to the average fraction of time for which the corresponding stage of remodeling activity persists, provided remodeling activity is randomly distributed over the surface, that the bone is in

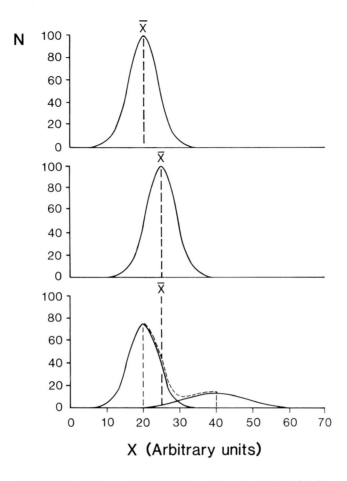

X (Arbitrary units)

FIGURE 22. Diagram to illustrate two different ways in which the mean value of a measurement may be increased (where x is any quantity in arbitrary units). Upper panel shows Gaussian distribution with a mean value of 20 and SD of 5. Middle panel shows an upward shift of the entire population of measurements, with a mean value of 25 and the same SD of 5. Lower panel shows the same mean value with a bimodal distribution in which three quarters of the measurements have the same mean and SD as before, but one quarter have a mean value of 40 and SD of 10. These two situations could be distinguished

a steady state, and that a statistically adequate sample has been examined.[139] For example, if osteoid tissue representing some stage of bone formation is found over 10% of the total trabecular surface, then at every point on the surface osteoid will be present and bone formation occurring for 10% of the time. Another example is the interpretation of the quiescent resorption surface. It is a common misconception that this reflects the difference between the rates of resorption and formation,[140] whereas it is determined entirely by the time interval between these processes, irrespective of how fast they occur.

The same principle applies to the fraction of bone volume occupied by osteoid. For example, if this fraction is 3%, then each moiety of bone is unmineralized for 3% of its total duration and mineralized for 97% of its total duration, with the same constraints concerning steady state and statistically random sampling. As for surface values, these numbers describe the average behavior of bone as a whole, not necessarily of individual moieties. Bone on the surface turns over more frequently than bone at

Table 17
NORMAL REMODELING AND ARRESTED
RESORPTION

Type of Surface	Prevalence	Incidence (/day)	Duration (days)
Normal remodeling	0.99	1/500	495
Arrested resorption	0.01	1/500 · 1/64	320

Note: Calculation of relative incidence of normal remodeling and arrested resorption. For further details see text.

the center of a trabeculum and therefore is unmineralized for a greater proportion of its life history. The same calculation does not apply to the fraction of total marrow space occupied by trabecular bone, because the location of trabeculae changes only very slowly or not at all, so that once a particular anatomical distribution has occurred, the distribution at subsequent times is no longer subject to random variation.

3. Birth Rate vs. Life Span

The concepts of population dynamics applied to individual BRU and BRC can be applied also to each subdivision of the bone surface and bone volume. Rather than considering the birth rate and number of arbitrarily defined individuals as in Equation 3, it is simpler to use the relationship:

$$\text{Prevalence} = \text{Incidence} \cdot \text{Duration} \tag{7}$$

Values for fractional subdivision of a surface or a volume can be treated directly as prevalence data, so that if osteoid occupies 15% of the surface it has a prevalence of 0.15; if the mean seam duration measured by double tetracycline labeling (σ_f) is 75 days we have:

$$0.15 = 1/500 \text{ days} \cdot 75 \text{ days}$$

In other words, from the observed prevalence and duration we calculate that at every point on the surface a new moiety of osteoid appears once every 500 days. This is equivalent to the activation frequency expressed per unit of bone surface instead of per unit of bone volume. The same calculations can be applied to all the subdivisions of the surface shown in Table 11. In a steady state the birth rate or incidence defined in this way will be the same for each segment of the surface which is part of the normal sequence of remodeling, but this does not apply to a segment such as arrested resorption which is not part of the normal cycle. If arrested resorption has a prevalence of 1% and is presumed to have the same duration as the rest of the resting surface, then it can be estimated (Table 17) that approximately 1 out of 64 remodeling cycles will enter an arrested phase.

From Equation 7 it is evident that a change in any fractional subdivision of the surface can result either from a change in its incidence or birth rate, or from a change in its duration, due to a converse change in the work rate of the relevant cell type. A change in activation frequency as the only abnormality will alter the fractional extent of all components of the remodeling cycle in the same proportion excluding arrested resorption. For example, if activation frequency is doubled, then the extent of total resorption and formation surfaces and all their subdivisions will also be doubled. Conversely, a change in the work rate of a cell may produce a change in the fraction of surface corresponding to that cell alone. For example, if the corrected bone apposition rate (M_f) is reduced by half, σ_f will be doubled and the fraction of total formation surface will be doubled, but resorption surface will be unchanged.

Exactly the same kinetic analysis can be applied to the volume of osteoid tissue. The prevalence in this case is defined as the fraction of total bone volume occupied by osteoid, and is equal to the product of osteoid birth rate per unit volume and mean osteoid life span. For example, if osteoid is 2% of total bone volume and the mean osteoid tissue life span (equivalent to the mean mineralization lag time) is 20 days we have:

$$0.02 = 1/1000/\text{day} \cdot 20 \text{ days}$$

In other words, from the observed prevalence and duration of osteoid tissue we calculate that on the average each moiety of bone is replaced by osteoid once every 1000 days, which is equivalent to a bone turnover rate of 36%/year. The osteoid life span in this calculation refers to individual moieties of osteoid tissue and is quite different from (and unrelated to) the life span of the osteoid seam. As with surfaces, the separate contributions of changes in life span and changes in birth rate to the observed change in osteoid volume can only be determined with certainty by the double tetracycline labeling technique, but the osteoid seam thickness gives a good indication as to the likely mechanism. An increase in activation frequency alone will increase the surface extent but not the thickness of osteoid, whereas an increase in osteoid tissue life span alone will increase the thickness but not the surface extent. This analysis will be developed in greater detail in a later section.

4. Transient States and Steady States

The previous interpretations all assume that the bone is in a steady state at the time of the biopsy. This means that for a time period at least as long as σ_{RC} there has been no perturbation of either the activation of new remodeling cycles or the function of individual differentiated cells. If this condition is not met, the bone will be in a transient state, which can only continue for a limited period of time, beyond which the observed measurements cannot be extrapolated.[141] For example, if activation frequency has increased within the last 4 weeks there will be many new sites of bone resorption without a corresponding increase in sites of bone formation, so that fractional resorption surface will be increased but fractional formation surface unchanged. Conversely, if a biopsy is taken within a few months of starting vitamin D treatment for nutritional osteomalacia, the very high bone formation rate is not the result of, and cannot be used to calculate, the activation frequency of new remodeling foci. The time required to reestablish a new steady state depends on the individual value of σ_{RC} for the subject of the biopsy, not the normal value for the population. Furthermore, if the agent which disturbed the steady state also causes a change in σ_{RC}, the duration of the transient will depend on the new rather than the previous value for σ_{RC}. The transient and steady states for resorption and formation surface and formation rate corresponding to various perturbations in cell function are summarized in Table 18. Transient effects are also seen with measurements such as appositional rate which change with time during the life history of an osteoid seam. Following a change in activation frequency there will be a transient change in the age distribution of the osteoid seam population and a consequent change in mean appositional rate. For example, if bone turnover increased between 1 and 2 months before the biopsy, a greater proportion of osteoblasts than normal will be in their stage of maximum vigor, so that the mean appositional rate will be increased. But in each seam, the appositional rate will be appropriate for the osteoblast morphology and distance from the cement line. This is a likely explanation for the reported increase in appositional rate in hyperthy-

Table 18
COMPARISON OF THE TRANSIENT AND STEADY STATES

		Transient State			Steady State		
		R	F	T	R	F	T
Activation frequency (μ_{RC})	↑	↑	NC	NC	↑	↑	↑
	↓	↓	NC	NC	↓	↓	↓
Resorption velocity (M_r)	↑	NC	NC	↑	↓	NC	NC
	↓	NC	NC	↓	↑	NC	NC
Resorption distance	↑	↑	↓	↓	↑	NC or (↑)	NC or (↑)
	↓	↓	↑	↑	↓	NC or (↓)	NC or (↓)
Formation velocity (M_f)	↑	NC	NC	↑	NC	↓	NC
	↓	NC	NC	↓	NC	↑	NC
Formation distance (MWT)	↑	NC	↑	↑	NC	↑	↑
	↓	NC	↓	↓	NC	↓	↓

Note: Comparison of some transient and steady-state effects of changes in function of either precursor cells or differentiated cells. R = fractional resorption surface, F = fractional formation surface, T = bone turnover estimated from bone formation rate, ↑ = increased, ↓ = decreased, NC = no change from values prevailing prior to change in cell function, μ_{RC} = activation frequency of new remodeling cycles, M_r = radial resorption rate, M_f = radial formation rate, MWT = mean wall thickness.

roidism,[142] although it is not certain that the measurements were obtained during a transient rather than a steady state.

5. Other Consequences of a Change in Bone Turnover

As previously explained, a change in activation frequency and bone turnover produces parallel changes in the fraction of bone surface engaged in resorption and formation, the number of resorption spaces and osteoid seams in cortical bone, and the volume and surface, but not the thickness, of osteoid tissue. There will also be a parallel change in the total volume of remodeling space with consequent changes in total bone volume and in external calcium balance. For example, a fivefold increase in bone turnover would produce a fivefold increase in total remodeling space and its calcium equivalent, with a negative calcium balance of 30 g spread over a transient state of approximately 6 months duration, or about 150 mg/day. A change in total-body bone volume of 3% would be difficult to detect, but the proportional change in volume would be much greater in trabecular bone where the remodeling space would increase from 2 to 10% of potential bone volume with a corresponding reduction of actual bone volume; the change will be even greater in the ilium because of its higher than average turnover. Conversely, if bone turnover decreases the remodeling space will contract, with a corresponding positive calcium balance; in either case calcium balance will return towards the initial value when a new steady state is reached. The increased or decreased remodeling space and bone mass will persist for as long as turnover remains at the new level, but will be reversed when the initial level of turnover is restored.[199] As well as this reversible form of bone loss, increased turnover increases the rate of irreversible bone loss with age by increasing the amplitude but not the direction of net loss from the endosteal surface and net gain at the periosteal surface.[143] For a given degree of remodeling imbalance, the rate of loss is directly proportional to the rate of turnover.

Other important consequences of a change in bone turnover relate to mean skeletal age. This is always less than the chronologic age of the organism by an amount which

Table 19
EFFECT OF REMODELING RATE ON BONE AGE

	Cortical (k = 0.03/year)	Trabecular (k = 0.24/year)
Random		
Halflife (years) (0.693/k)	23.1	2.89
MSA (steady state) (1/k)	33.3	4.17
MSA (25 years)	17.6	4.16
Percent with age ⩾ 36 years	34.0	0.02
Selective		
MSA (steady state) (½ k)	16.7	2.08
MSA (25 years)	15.6	2.08
Percent with age ⩾ 36 years	0	0

Note: Relationships between turnover rate (k), halflife, and mean skeletal age (MSA) in cortical and trabecular bone for random (mode 1) and selective (mode 2) remodeling. Values for k are mean values previously estimated. For linear remodeling, halflife and steady-state MSA are identical. Halflife cannot be calculated for selective remodeling.

depends on the rate of bone turnover. The precise relationship between these variables depends on whether the age of a moiety of bone influences the probability that it will be remodeled. Within a region where the S/V ratio and other determining factors are constant, three modes of remodeling can be distinguished. First, if the probability of remodeling is constant and unrelated to the age of the bone, remodeling will be randomly distributed and the bone present at a particular time will disappear in accordance with an exponential model. Second, if the probability of remodeling increases with the age of the bone, old bone will be preferentially removed and the bone will disappear more rapidly and more in accordance with a linear than with an exponential model; this has been called selective remodeling.[7] Finally, if the probability of remodeling decreases with the age of the bone, young bone will be preferentially removed. A small fraction of bone will be constantly renewed but the bulk of the bone will turn over very slowly and its age will be only slightly less than the age of the organism; this has been called redundant remodeling.[7] The same concepts apply to the turnover of red blood cells. This is mainly in accordance with the second mode; most cells are removed only after they have become senescent, but a small number are removed by a random process.[144] In bone the emphasis is reversed; most remodeling is distributed at random, but some is activated in response to biomechanical incompetence due to osteocyte senescence, local trauma, or fatigue microdamange. The third mode of remodeling (redundant) probably occurs only as a manifestation of disease.

The aging of bone tissue can be expressed in various ways. The halflife ($T_{1/2}$) is the time required for any moiety of bone to decline to one half of its initial amount. A quantity of more biologic significance for bone is the mean skeletal age (MSA), which is the weighted mean age of all the moieties of bone present at one time. For random remodeling (first mode), if k is the turnover rate expressed as a rate constant, $T_{1/2}$ is given by 0.693/k, where 0.693 is the natural logarithm of 2. In a steady state MSA is given by 1/k, but at low rates of remodeling the steady state may never be achieved within a normal life span and the mean age at time t is given by $1/k \, (1 - e^{-kt})$.[145] For selective remodeling (second mode) the halflife and the mean age are the same and are given by $-\frac{1}{2}k$ or $\frac{1}{2} \cdot 1/k$. Various indices of aging of cortical and trabecular bone calculated in accordance with these expressions from the previously derived values for bone turnover rate are given in Table 19.

As bone ages its density increases because of the progression of secondary mineralization. As a result it becomes less permeable and the mineral becomes less accessible to exchange with the extracellular fluid. The mineral density of perilacunar bone increases still further when the osteocyte dies — the phenomenon of micropetrosis which is associated with increased brittleness.[54] Osteocytes in cortical bone have a halflife of about 25 years[11] which corresponds to a mean life span of 36 years. With random remodeling of cortical bone at a rate of 3%/year, 34% of the bone would be more than 36 years old (Table 19), but in trabecular bone with a turnover rate of 24%/year the corresponding proportion is only 0.02%. Conversely, the proportion of young bone increases with the rate of turnover. For example with a turnover rate of 50%/year, 11.8% of the bone will be less than 3 months old and 4.1% less than 1 month old; if turnover increases to 150%/year these proportions increase to 31.3 and 11.8%. Consequently, the proportion of immature collagen and incompletely mineralized bone will increase, as is found in experimental renal osteodystrophy in the rat.[146]

Because of its high turnover it is clear that aging of trabecular bone is of little consequence. Even if the turnover rate was reduced by half, the proportion of bone old enough to be susceptible to osteocyte death would only increase to 1.3%, but a similar proportional reduction in turnover in cortical bone would increase the proportion from 34 to 58%. The ilium is reasonably representative of the whole skeleton with respect to the directional effect of disease; consequently a low rate of turnover in an iliac biopsy, although not important in itself, suggests that the mean age of the whole skeleton is increased. This means that cortical bone throughout the body has more dead osteocytes, is less able to repair fatigue microdamage, and is more brittle than normal.

C. The Significance of Some Particular Measurements

1. Amount and Distribution of Bone

Both cortical and trabecular bone are lost with age. The former is indicated by reduction in the thickness of the cortex and the latter by reduction in the fraction of marrow space occupied by bone, or trabecular bone volume (TBV). It is also possible to calculate the density, expressed as amount per unit volume of bone within the periosteum, of cortical bone, trabecular bone, and total bone, but these quantities discriminate less well between normal and abnormal subjects.[147] Trabecular bone density defined in this way shows little change with age because loss of trabecular bone is partly offset by the addition of new trabeculae from endosteal tunneling of the inner third of the original cortex.[148] The distinction between actual bone and potential bone must be recalled. The former is the bone present at the time of the biopsy, the latter includes all the additional bone which will be added when the present remodeling space has been converted to completed bone structural units. The magnitude of the remodeling space cannot be measured directly in trabecular bone but it can be estimated from the extent of surface remodeling activity. In the trabecular bone of the ilium the remodeling space is about 6% of the potential bone volume. Consequently, a reduction of 12% in TBV from 0.2 to 0.175 mm³/mm³ could be a temporary consequence of a threefold increase in turnover with the potential for complete replacement when turnover returns to normal. The same increase in surface remodeling activity and consequently in remodeling space will be produced by a threefold prolongation of sigma, but this is less commonly reversible than an increase in turnover.

As with any other measurement of amount of bone, the result can be compared with three different standards; ideally, these comparisons should be made both for the actual and the potential bone volume. Comparison with normal values at skeletal maturity for a person of the same sex and race is an index of the absolute deficit of bone, whether due to defective accumulation or excessive loss.[149] Comparison with standards based on age as well as sex and race indicates whether or not the observed deficit is more than can be attributed to age; the comparison is more accurate if year specific values are calculated by interpolation from the usual decade specific values.[148] Finally, comparison with an absolute threshold which is independent of sex, race, and age indicates the risk of structural failure. In the ilium this threshold is 11%, a value which separates the majority of patients who have sustained vertebral compression fractures from those who have not.[147] But TBV correlates less well than the ash weight with the compressive strength both of the ilium itself and of the vertebral body, indicating that biomechanical competence depends on the quality of bone as well as its quantity.[150]

Because of the large sampling error in TBV, a biopsy is more useful in studying the pathogenesis of osteoporosis than in establishing a diagnosis. The recognition of an increase in remodeling space as an explanation for a reduction in actual TBV has already been mentioned. All loss of potential TBV occurs because of terminal remodeling imbalance — new BSU are smaller than the resorption spaces in which they are built. The cavities may be made too large, refilling of normal size cavities may be incomplete, or there may be some combination of these abnormalities. Whatever the explanation, the loss of bone is irreversible because when construction of the new BSU has been completed, the biologic capital created by activation has been spent. In cortical bone the size of the cavities is indicated by the diameter of the cement line and the degree of refilling by the diameter of the Haversian canal and the MWT. In trabecular bone, the original location of the surface from which the cavity was eroded cannot be determined with any accuracy, but the MWT is an index of the size of new BSU. This measurement declines significantly with age,[23] indicating that loss of trabecular bone is at least partly due to a reduction in the amount of bone formed in each bone remodeling cycle. Whether there is in addition an increase in the amount of bone resorbed in each remodeling cycle is not known, although good evidence for this was found at the cortical-endosteal surface of the rib.[151] Other evidence for the contribution of depressed bone formation to bone loss is an increase in the surface extent of arrested resorption and a decrease in both matrix and mineral apposition rates.

As well as the amount of trabecular bone, it is important to consider its distribution. Conversion of trabecular plates to bars or rods can be recognized by the presence of circular or elliptical profiles in the biopsy rather than the elongated profiles produced by sections through plates.[200] In osteoporosis there is also an increase in the number of profiles which are isolated rather than being connected. At times abnormally thickened trabeculae are observed, probably representing a compensatory hypertrophy occurring in response to an earlier loss of trabeculae. Focal thickening of trabeculae may also represent callus formation around a trabecular microfracture.

2. Relationships of Surface, Volume, and Porosity

The S/V increases progressively as bone becomes more porous and is always higher in trabecular than in cortical bone. However, the surface area per unit volume of tissue (S_v) is highest for incompletely trabeculized cortical bone with a porosity of about 0.35 and is lower for both typical cortical and typical trabecular bone.[7] As a result, an increase in porosity leads to an increase in surface area in cortical bone but a decrease in surface area in trabecular bone (Figure 23). If remodeling activity per unit of surface and terminal remodeling imbalance in each BRU are the same in the two types of bone, these geometric relationships will tend to increase the rate of loss of cortical bone and to decrease the rate of loss of trabecular bone with time. Comparing individual values for S_v and porosity in this way could be of prognostic importance, and could clarify the interpretation of other data.

The relationship of volume to surface can also be used to calculate mean trabecular thickness (MTT), which gives an indication of the distribution of trabecular bone. If TBV and MTT are reduced in the same proportion, then the number of trabeculae has

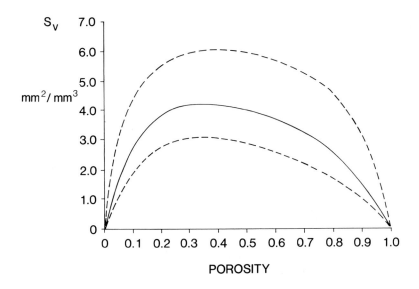

FIGURE 23.   Relationship between specific surface (S_v) and porosity of bone. The solid line indicates theoretical relationship based on analysis of the three-dimensional geometry of bone. Dotted lines enclose the range of measurements found on actual samples of human bone. Note that S_v is the total surface per unit volume of bone tissue and must be distinguished from S/V, which is the total surface per unit volume of bone. (Redrawn from Martin, R. B., *Crit. Rev. Bioeng.*, in preparation, 1982.)

not changed. However, if the same reduction in TBV is associated with no change in MTT, then the number of trabeculae must be reduced. This can be formalized by the calculation of mean trabecular density (MTD) from the following relationship:

$$MTD(/mm) = \frac{TBV_f \cdot 1000}{MTT(\mu m)}$$

Since new lamellar bone can only be made in apposition to existing bone, reduction in thickness but not density has a greater potential for regeneration in response to treatment. However, it may also have a poorer prognostic significance, since uniform thinning of trabeculae probably indicates failure of the compensatory thickening which can be observed in normal subjects.

3. Resorption Surface Measurements

If each subdivision of the surface has changed in the same direction and by about the same proportional extent, a likely explanation is a change in activation frequency and bone turnover with no change in differentiated cell function. However, significant departures from proportionality can arise for a variety of reasons even when a change in activation frequency is the main cause of a change in surface measurements. For example, an increase in surface extent of "active" resorption or osteoclast domain without a corresponding increase in surface extent of bone formation could be a transient effect due to a recent increase in activation frequency or a result of random variation due to the sampling problems discussed earlier. More commonly, an increase in surface extent of bone formation is found without a corresponding increase in surface extent of bone resorption. This could be due to inability to recognize mononuclear osteoclasts or other types of resorbing cell, to the transient effect of a recent fall in activation frequency, to the presence of Type I BRU in which the resorptive phase is so rapid that its traces are only rarely seen, or simply to random variation. If these possibilities can all be ruled out, bone formation must have appeared on quiescent bone surfaces, without prior resorption, as in the early stages of sodium fluoride treatment in osteoporosis.[152]

The interpretation of resorption values is hampered by the difficulty of reliably distinguishing between the three subdivisions of the total resorption surface. As a rough guide the extent of osteoclast domain can be taken as five times the surface extent of multinucleated osteoclasts, or twice the surface extent if mononuclear osteoclasts can be identified histochemically. Like all subdivisions of the surface, an increase in osteoclast domain may be due to either an increase in birth rate (activation frequency), or an increase in life span ($\sigma_{RC(r)}$) due to depression of osteoclast vigor; both mechanisms operate in most patients with hyperparathyroidism, primary or secondary. Unlike osteoblasts, the functional state of the osteoclast cannot be determined from its morphology at the light microscopic level; indirect methods of estimating osteoclast function by the double tetracycline method are described in Chapter 7. A disproportionate increase in reversal phase indicates partial uncoupling of resorption and formation due to prolongation of the quiescent interval between them, and an increase in arrested resorption surface represents the accumulation of permanently unrepaired defects as previously explained.

Because osteoclasts usually avoid unmineralized bone surfaces, it is useful to express the resorption surface data as fractions of the nonosteoid surface as well as fractions of the total surface. For example, in renal osteodystrophy where the osteoid surface frequently exceeds 80% of the total surface, the osteoclast surface per unit of total surface is often normal, but per unit of nonosteoid surface is almost invariably increased and correlates better with serum immunoreactive parathyroid hormone.[36]

There are important qualitative changes in resorption space morphology in some disorders. As is evident from the usual dimensions of a BSU, a normal resorption space does not extend more than 40 to 60 μm from the surface. In severe hyperparathyroidism, primary or secondary with osteitis fibrosa cystica, the link between resorption and formation is broken. The resorption front is no longer subject to the normal geometrical constraints and travels for a greater distance through the bone, even though its rate of advance may be slower than normal. This leads to the well-known histologic appearances of dissecting resorption, characterized by large intratrabecular cavities filled with fibrous tissue and sometimes containing islands of woven bone.[143] This is a qualitative departure from normal which is never caused by an increase in turnover alone (as in hyperthyroidism), however great the increase in surface resorption. By contrast, intracortical porosity will increase because of increased longitudinal tunneling with increased turnover from any cause.

A more elusive change in resorption space morphology probably occurs in rapidly progressive osteoporosis from any cause, with increased depth of resorption cavities progressing to complete perforation of trabecular or subendosteal plates rather than dissecting along the trabeculae as in osteitis fibrosa. This process occurs quickly, so that its prevalence is low, but it probably initiates the loss of whole trabeculae by conversion from plates to bars, which is an important component of trabecular bone loss.[201]

4. Tetracycline Fluorescence Measurements

The linear extent of fluorescence is usually expressed as a fraction of total osteoid perimeter. This is reasonable provided that the total length of osteoid can be accurately defined, but the recognition of osteoid is more affected by section preparation and staining in trabecular than in cortical bone. The thin seams at the terminal stage of bone formation when the new BSU is almost completed must be distinguished from

Table 20
FACTORS AFFECTING TETRACYCLINE
FLUORESCENCE IN BONE SECTIONS

Thickness of diffusion barrier due to new mineral deposition
Kinetics of binding to different forms of bone mineral
Variation in local bone blood flow
Rate of fall in plasma tetracycline level due to renal clearance
Time interval between label administration and biopsy
Chemical nature of fixative and other details of specimen handling

the lamina limitans, the thin border of unmineralized tissue which covers the whole of the inert surface. Although there are subtle differences in texture and staining color, the thin flat cells on the two types of surface may be indistinguishable and interobserver error is reduced by the adoption of some arbitrary lower limit for the thickness of an osteoid seam, such as 1.5 μm.[153] With this proviso, the fraction of osteoid perimeter which is labeled is usually equated with the extent of osteoid perimeter at which mineralization was active, but this simple interpretation must be reexamined in light of the dual mechanism of tetracycline fixation — selective binding to the mineral which is on the bone surface at the time of the peak blood level of tetracycline and trapping by new mineral which is deposited subsequent to the peak blood level. Selective binding occurs mainly to recently deposited amorphous or immature mineral but may also occur to some component of the cement line, a point which is of particular importance in the interpretation of single labels.

The persistence of sufficient tetracycline in bone to cause microscopically detectable fluorescence at sites of bone formation depends on at least six factors (Table 20). The rate of new mineral deposition determines the thickness of the barrier to outward diffusion as a function of time. This thickness and the kinetics of binding to mineral together determine the ease with which tetracycline can escape from the bone surface and return to the circulation. Although initial binding is stronger to amorphous than to crystalline mineral, it is not known whether tetracycline already bound to amorphous mineral becomes less firmly bound with transformation to the crystalline state. The rate at which tetracycline diffuses away from the osteoid surface also depends on the local bone blood flow and the rate at which the plasma tetracycline level falls again to zero, a process which is governed mainly by renal clearance of the drug. The time for which outward diffusion can occur in vivo clearly depends on the time interval between label administration and biopsy. This interval also affects the postbiopsy loss of tetracycline which continues in vitro, depending on the chemical characteristics of the fluid in which the biopsy specimen is immersed and the details of subsequent handling.

When iliac biopsies were taken at varying periods of time after administration of tetracycline for 1 day and placed in formaldehyde solution buffered to pH 7.0, the fraction of trabecular osteoid perimeter showing fluorescence increased rapidly to 50% during the first 24 hr and thereafter increased more slowly to a peak of about 60% at 3 days.[117] Since the linear extent of tetracycline fixation could not increase in vivo while the blood level of tetracycline was falling, the results indicate that the label became progressively less elutable in vitro, presumably because of continued mineral apposition in vivo. Further support for this interpretation is that in patients with osteomalacia the fraction of labeled osteoid perimeter increased more slowly but continued to increase for 5 days, as would be expected with a slower mineral apposition rate. Similar experiments have not been performed with other fixatives such as 70% ethanol, but with fresh untreated bone the label is readily elutable if given within 24 hr of examination, but cannot be removed without demineralization if it has existed in vivo for more than 7 days.[154]

Clearly the interval between label administration and biopsy must be standardized. If the interval is too short, some in vitro loss of label will occur, but some fluorescence will persist at quiescent bone surfaces and either of these opposing processes may predominate. If the interval is too long, some of the osteoid which was labeled will have completed its mineralization by the time of the biopsy, its place taken by new osteoid which has had no chance to become labeled. Consequently, the fraction of the current osteoid perimeter which is labeled will have diminished. However, if the patient is in a steady state the total linear extent of fluorescence, whether or not there is overlying osteoid, can be related to the total osteoid perimeter with little loss of accuracy.

Although retention of tetracycline is usually a consequence of the mineralization process, detectable fluorescence is sometimes found in the absence of mineralization. Occasionally, resting seams as defined in Section II.C.2.a. may take a single label, but the mechanism of tetracycline fixation is unknown. Such paradoxically labeled resting seams may occur both in osteoporosis and in osteomalacia, but the existence of this phenomenon can only be established in patients who received a double tetracycline label, in whom the expected proportion of single- and double-labeled surfaces can be estimated (Chapter 8). Another mechanism of single-label uptake in the absence of mineralization occurs in an occasional patient with severe osteomalacia where the osteoid-bone interface remains at the cement line, indicating that none of the osteoid formed at that location has been mineralized. At such sites there may be a thin line of fluorescence at the cement line even though the zone of demarcation is narrow or absent and there is no mineralization front. Some characteristic of the cement line appears to lead to preferential binding of tetracycline, an interpretation which is supported by the occasional finding of a similar thin fluorescent band at Howship's lacunae where bone resorption has recently ceased.

The foregoing analysis applies mainly to the administration of a single label; giving a double label introduces an additional complication to determining the extent of the mineralizing perimeter. Clearly all of the double-labeled perimeter must be included but different investigators include none, all, or half of the single-labeled perimeter. The latter method, which is equivalent to taking the mean of the individual label perimeters, is most accurate provided that labeled cement lines and labeled resting seams can be excluded and that the two labels are distinguishable. If two labels of the same color are superimposed, taking half the apparent single label will seriously underestimate the extent of mineralizing perimeter. The matter is discussed at length in Chapter 8. The problem is to distinguish between the inevitable lack of one of the labels at the beginning or end of the life history of the osteoid seam from temporary interruption of tetracycline fixation at some time in between and from labeled resting seams. The mechanism and significance of this interruption, whether for single or double labels, must now be examined.

Of the six factors which determine the extent of fluorescence, the label-to-biopsy interval and the details of specimen processing can be standardized and variation in renal function and in bone blood flow are probably of minor importance in most cases, but the relative importance of the other two factors is uncertain both in general and in specific cases. If at a particular location mineralization were to abruptly cease after label administration, but a large amount of amorphous mineral was present, some fluorescence would probably still be detectable a few days later. Conversely, the combination of a reduced amount of amorphous mineral and a low mineral appositional rate might permit all the tetracycline to escape before the biopsy was taken. Consequently, the absence of label does not necessarily mean that mineralization has stopped altogether, only that its rate has been markedly reduced.

The extent of osteoid which is labeled gives similar but not necessarily identical information to the extent of osteoid showing toluidine blue stainable granules at the zone of demarcation, the so-called mineralization front. One advantage of this method is that if the patient's condition is not rapidly changing, the results are not dependent on precise timing of the biopsy. The values for the two methods correlate well in most instances[36] but detailed side-by-side comparison in specific locations in the bone has not been reported, and they may represent different phenomena. If the absence of one or more bands of tetracycline fluorescence is correlated with cell morphology and distance from the cement line, periodic interruption or slowing down of mineralization can be demonstrated in normal bone, not only at the beginning and end of the life history of an osteoid seam but at some stage in between.[112] These short-term interruptions could occur between the mineralization of successive lamellae as suggested by Bordier et al.[117] Absence of both fluorescent bands is normally found at the very beginning of the seam, but otherwise is indicative of longer interruption in mineralization unless fortuitously the two labels were given in phase with a short-term periodicity in mineralization.

Similar correlations have not been studied for the mineralization front, although it has been stated that this is invariably present beneath so-called "active" osteoblasts. This suggests that the granules persist unchanged during short-term interruptions of mineralization. Identification of resting seams is more difficult in trabecular than in cortical bone because fuchsin permeability is a less reliable indicator of the duration of mineralization arrest in thin plastic-embedded sections, than in thick hand-ground sections. If absence of the mineralization front indicates a more prolonged interruption of mineralization than absence of tetracycline fluorescence alone, it could be a more reliable indicator of a resting seam and of lack of vitamin D action on bone. However, the length of the tetracycline-labeled perimeter is normally longer rather than shorter than the length of the mineralization front perimeter.

A decrease in the labeled fraction of osteoid perimeter, however determined, can therefore result from a number of different mechanisms (Table 21). First, it can reflect a prolongation of the time interval between onset of matrix synthesis and onset of mineralization (the initial mineralization lag time), in which case the unlabeled osteoid will be close to the cement line covered by cuboidal osteoblasts and presumably lacking a mineralization front. Second, it may result from an increase in the frequency or duration of the normal pauses in mineralization. In this case the unlabeled osteoid will be further from the cement line and more likely covered by intermediate or terminal osteoblasts. The presence or absence of the mineralization front probably depends on the duration of the pause; if this is increased, it may indicate that a sick osteoblast is able to engage in either matrix synthesis or mineralization but not both at the same time. Finally, there may be premature cessation of mineralization to form a true resting seam; in this case the unlabeled osteoid will be even further from the cement line but may otherwise be indistinguishable from the previous category. In the most severe form of reduction in fractional labeling in osteomalacia, all three mechanisms are operative; the entire osteoid surface may be covered by flat cells and lack a mineralization front, so that the previous distinctions are no longer applicable. Depending on the mechanism, a reduction in fractional labeling may or may not be accompanied by a reduction in the extent of osteoid covered by so-called "active" osteoblasts, so that these indices, although related in general, may nevertheless vary independently.

As well as being of interest in itself, the labeled fraction of osteoid perimeter is used to convert the mineral appositional rate to the radial closure rate (M_f). These quantities embody the most important information provided by the double tetracycline labeling technique but they are not, as is commonly believed, specific indices of mineralization alone. Rather they reflect different aspects of osteoblast function, and low values most

Table 21
MECHANISMS FOR REDUCTION IN FRACTION OF LABELED OSTEOID

	Distance from cement line	Endosteal cells	Mineralization front
Delayed onset	Close	Cuboidal	Absent
Longer or more frequent pauses	Intermediate	Intermediate	Depends on duration
Premature cessation	Far	Flat	Absent

Note: Characteristics of three different mechanisms for reduction in fraction of osteoid perimeter showing tetracycline labeling. These distinctions are not applicable in severe osteomalacia, in which all three mechanisms are operative. Characteristics for mineralization front are inferential, not observational.

commonly result from impaired matrix synthesis. It is clear that mineralization can only proceed faster than matrix synthesis during the latter part of the life history of a seam when it is getting thinner. Furthermore, in a completed BSU the total volume of matrix and the total volume of mineralized bone are necessarily the same, so that the matrix apposition rate and mineral apposition rate averaged throughout the life history of a bone-forming site must be equal. Consequently, a primary reduction in the rate of matrix synthesis must inevitably lead to a secondary reduction in the mean rate of mineralization, either in appositional rate or fractional labeling or both. The mineral appositional rate is therefore, in most circumstances, a measure of the maximum rate of matrix apposition at a particular surface, and the radial closure rate is a measure of the average rate of matrix apposition throughout the life history of the seam. This average rate can be reduced either because the maximum rate is reduced (low appositional rate) or because the pauses between successive bursts of bone formation are increased (reduction in fractional labeling). Frequently, these different abnormalities of osteoblast function go together but they can be dissociated. For example, in chronic renal failure there may be a marked reduction of fractional labeling combined with a normal appositional rate.[155]

What has just been said concerning the significance of the mineral apposition rate applies to measurements averaged over the entire double-labeled surface. The apposition rate is fastest at the start of bone formation and declines progressively until the new BSU is completed (Table 10), so that measurements restricted to one segment of the osteoid seam life history may show effects of disease or treatment which are obscured if the values are averaged for the entire duration of the seam. For example, when measurements are classified according to cell morphology and distance from the cement line, the appositional rate in some patients with osteoporosis is reduced during the initial period of maximum osteoblast vigor, even though the mean value determined in the usual way is normal. Another technique is to give multiple closely spaced labels and to measure the appositional rate only at the most active bone-forming sites which show all labels, disregarding other sites where one or more labels are missing. When measured in this way, the appositional rate is increased in primary hyperparathyroidism[156] and by PTH administration,[157] even though the mean value averaged over the entire double-labeled surface in the usual way is reduced in hyperparathyroidism.[158,159] This suggests that the osteoblasts are stimulated by PTH when they are young and vigorous, but are inhibited by PTH, or by some other consequence of the hyperparathyroid state, when they get older and less active.

5. Mineralization Lag Time, Osteoid Maturation Time, and Osteoid Seam Thickness

In order to clarify the relationships between the rates of matrix synthesis and mineralization, and the thickness of osteoid seams, the concept of mineralization lag time

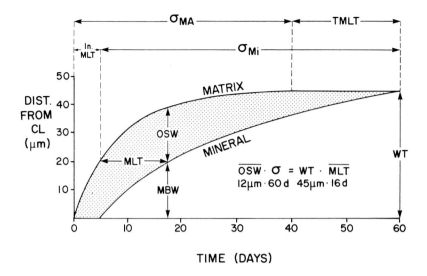

FIGURE 24.   More detailed version of middle panel of Figure 21. Movement of the edge of the bone matrix and of the osteoid:bone or mineral interface away from the cement line (CL) is shown as a function of time from the onset of matrix apposition. The horizontal distance between these lines at any distance is the mineralization lag time (MLT), and the vertical distance between these lines at any time is the osteoid seam width (OSW). σ_{Ma} is total duration of matrix apposition. σ_{Mi} is total duration of mineral apposition. TMLT = terminal mineralization lag time. In·MLT = initial mineralization lag time or osteoid maturation. WT = wall thickness of completed BSU. σ_f = the total duration of bone formation, = σ_{Ma} + TMLT = σ_{Mi} + In·MLT. For further details see text.

(MLT) first formulated by Baylink[44] must be further analyzed. An osteoid seam represents the advancing edge of a new BSU, so that during its life span it travels from the cement line to the new bone surface. In Figure 24 the separate time courses of matrix apposition and mineral apposition are represented by plotting the locations of the outer border of the osteoid seam and of the osteoid-bone interface as functions of time. On such a plot the vertical distance between the lines represents the thickness of the osteoid seam at any time and the horizontal distance between the lines represents the interval between apposition of matrix and apposition of mineral (or MLT) at any distance from the cement line. Clearly the thickness of the osteoid seam is determined by the rate of matrix apposition and the time that it continues before mineralization begins at a particular location, or in symbols

$$OSW = MLT \cdot MaAR$$

which, on rearrangement gives:

$$MLT = \frac{OSW}{MaAR} \tag{8}$$

OSW and MLT can also be shown as functions of time and distance (Figure 24). These individual values must be distinguished from the mean values averaged over the whole bone-forming surface and throughout the life history of the seams.

The time from the onset of matrix synthesis to the completion of mineralization is σ_f, and the distance from the cement line to the bone surface is the wall thickness (WT) of the new BSU. If, in Figure 24, the area enclosed by the lines depicting the time courses of matrix synthesis and mineralization is denoted by A and using a horizontal line above the symbol to denote a mean value rather than an individual value, then:

Therefore: $\quad A = WT \cdot \overline{MLT}$; also, $A = \sigma_f \cdot \overline{OSW}$

Therefore: $\quad$ mean $\overline{MLT} \cdot WT = \sigma_f \cdot \overline{OSW}$

or: $\quad \overline{MLT} = \dfrac{\sigma_f \cdot \overline{OSW}}{WT} = \dfrac{\overline{OSW}}{M_f}$

$$\overline{OSW} = \overline{MLT} \cdot M_f$$

This proof is quite general and does not depend on any particular mathematical form for the time functions of matrix synthesis or mineralization. It also demonstrates once more the identity between M_f and the mean matrix apposition rate averaged throughout the life span of the seam.

MLT is the time during which newly formed osteoid remains unmineralized, and both collagen and ground substance undergo a variety of biochemical changes. These changes, collectively known as maturation, precede and are presumably necessary precursors for mineralization, and the time which they require will be referred to as the osteoid maturation time (OMT).[160] OMT cannot at present be measured, but since mineralization can begin after but not before the completion of osteoid maturation, OMT cannot be greater than the lowest value for MLT. If osteoid maturation is the main determinant of the onset of mineralization, OMT will be equal to MLT, but if other aspects of the mineralization process become rate limiting so that mineralization fails to occur even when osteoid maturation is complete, MLT will be longer than OMT.

The total duration of bone formation (σ_f) can be partitioned into the initial MLT (the time between initiation of matrix synthesis and initiation of mineralization) and the time required to complete mineralization (σ_{Mi}); σ_f can also be partitioned into the time required to complete matrix synthesis (σ_{Ma}) and the terminal MLT (the time between termination of matrix synthesis and termination of mineralization) or in symbols:

$$\sigma_f = \sigma_{Mi} + IMLT = \sigma_{Ma} + TMLT \qquad (9)$$

It is usually tacitly assumed that σ_{Ma} and σ_{Mi} must be the same, but this is evidently not the case unless the initial and terminal MLT values are equal (Figure 24). There are limited data on this point, but it seems likely that MLT increases during the life history of the seam. In normal subjects without resting seams, complete absence of both labels is found only at the beginning of the seam. In this case the initial MLT can be estimated from σ_f calculated in the usual way, multiplied by the fraction of osteoid which is completely unlabeled, or in symbols:

$$IMLT = \sigma_f \cdot f_{UL} \qquad (10)$$

Application of this calculation to normal subjects gives values in the range of 4 to 6 days, substantially shorter than the mean MLT of about 15 days. An approximation to OMT is also given by dividing the mean seam width by the appositional rate rather than by the corrected appositional rate, or in symbols: OMT = MOSW/M, since M is a closer estimate than M_f of the maximum rate of matrix apposition at the onset of seam formation; OMT estimated in this way is normally about 10 days. Which ever method is used, TMLT is longer than IMLT and σ_{Ma} is shorter than σ_{Mi}. It was stated earlier that the MaAR and MiAR averaged throughout the total period of bone formation must be equal, but if they are averaged only during their own time periods, they are not equal since:

$$\sigma_{Ma} = \text{WT} \, \overline{\text{MaAR}} \therefore \overline{\text{MaAR}} = \text{WT}/\sigma_{Ma} \quad \text{and}$$

$$\sigma_{Mi} = \text{WT} \cdot \overline{\text{MiAR}} \therefore \overline{\text{MiAR}} = \text{WT}/\sigma_{Mi}$$

Consequently, since σ_{Mi} is longer than σ_{Ma}, $\overline{\text{MaAR}}$ is faster than $\overline{\text{MiAR}}$.

6. Fluorescent Label Width as an Index of Mineral Accumulation

In vivo labeling with tetracycline enables measurement, at least in principle, of mineral accumulation as well as mineral apposition or "vertical" as well as "horizontal" mineralization in Figure 2. In the rat, recently deposited bone is permeable to tetracycline until its mineral density exceeds 20% of maximum.[44] The time required to attain this critical value can be determined from the mean label width and the mineral appositional rate.[44] For example, if the duration of tetracycline administration is 3 days, mean label width 8.0 μm, and mineral apposition rate 0.8 μm/day, the mean width of the zone of instantaneous labeling is 5.6 μm [8.0 to (3 × 0.8)]. This width corresponds to a time of 7 days (5.6/0.8), which is the time needed for mineral density to increase from zero to the critical value of 20% of maximum. Consequently, the mean rate of mineral accumulation averaged over the first 7 days is 2.56% of maximum per day (20/7). These calculations can be summarized in the following equation:

$$\text{RIMA} = \frac{20}{\text{LW/MAR} - \text{LD}}$$

where RIMA = rate of initial mineral accumulation in percent per day, LW = label width in microns, MAR = mineral apposition rate in microns per day, and LD = label duration in days.

In the rat, RIMA correlates very closely with the osteoid maturation rate (OMR) under a variety of different experimental conditions.[161] On the assumption that osteoid maturation is zero when initially formed and 100% at the onset of mineralization, at least in the rat osteoid maturation and initial mineral accumulation are different expressions of a more fundamental process, possibly related to the function of the mineralizing osteoblast or osteoid osteocyte.

Unfortunately, in human trabecular bone the mineral density needed to prevent tetracycline penetration is unknown, and it is difficult to measure the label width accurately in oblique sections. Nevertheless, label width is increased in vitamin D refractory osteomalacia[162] and in renal osteodystrophy[62] and decreases towards normal in both conditions with vitamin D, either with[162] or without[62] phosphate. These changes in label width are associated with differences in the ratio of amorphous to crystalline mineral and it was initially proposed that the change in mineral maturation and composition was the cause of the change in label width because of preferential binding of tetracycline to amorphous mineral. However, it seems more likely that both the change in mineral composition and in label width are dependent in different ways on the rate of initial mineral accumulation.

7. Osteoid Volume, Surface, and Thickness

The total volume of osteoid is the product of the total surface and the mean thickness of osteoid, which can vary independently. As explained in Chapter 3, thickness is the three-dimensional counterpart of the two-dimensional measurement of width in a histologic section. From Equation 7 the osteoid surface is determined by the surface-based activation frequency (birth rate) and σ_f (seam life span), and from Equation 8, OSW is the product of M_f and $\overline{MLT}$ (osteoid tissue life span). But if wall thickness is treated as constant, σ_f varies inversely with M_f, and M_f does not affect osteoid volume but simply alters the partition between thickness and surface. For example, if M_f is reduced by half, σ_f and the total osteoid surface will be doubled, but osteoid volume will remain the same because osteoid thickness will be halved.

It was shown earlier in terms of a simple population analogy that the volume of osteoid is determined by the volume-based bone formation rate (F_v) and the mean osteoid life span (which is the same as $\overline{MLT}$), both of which are independent of M_f. This independence can be shown more rigorously in terms of the primary measurements (Chapter 3). If M_f is expressed in millimeters per year, $\overline{OSW}$ in millimeters, $\overline{MLT}$ in years, d (the grid constant) in millimeters, and i_{os} is the number of intersections on osteoid, h_{TB} is the number of hits on bone (mineralized or unmineralized), and h_{os} is the number of hits on osteoid then:

$$F_v \ (/y) \ = \ \frac{M_f \cdot i_{os}}{h_{TB} \cdot d}$$

The indirect estimate of osteoid width from volume and perimeter is given by:

$$\overline{OSW} \ = \ \frac{h_{os} \cdot d}{i_{os}}$$

As previously shown:

$$\overline{MLT} \ = \ \frac{\overline{OSW}}{M_f} \ = \ \frac{h_{os} \cdot d}{i_{os}} \cdot \frac{1}{M_f}$$

$$\therefore F_v \cdot \overline{MLT} \ = \ \frac{M_f \cdot i_{os}}{h_{TB} \cdot d} \cdot \frac{h_{os} \cdot d}{i_{os}} \cdot \frac{1}{M_f} \ = \ \frac{h_{os}}{h_{TB}} \ = \ ROV$$

Note that M_f is a component of both F_v and $\overline{MLT}$ but is cancelled out when these expressions are multiplied together, the product giving relative osteoid volume (ROV) expressed as a decimal fraction. Whether MLT is determined by osteoid maturation or by the process of mineralization, its effect on osteoid volume is the same.

In summary (Figure 25), the birth rate of new osteoid (μ_f) affects the surface and volume but not the thickness, the mean osteoid life span $(\overline{MLT})$ affects the thickness and volume but not the surface, and the mean corrected mineral apposition rate (M_f) affects the surface and thickness in opposite directions but does not affect the volume. Conversely, the osteoid surface is determined by μ_f and M_f but not by $\overline{MLT}$, the osteoid thickness is determined by $\overline{MLT}$ and M_f but not by μ_f, and osteoid volume is determined by μ_f and $\overline{MLT}$ but not by M_f. These relationships apply only to the new steady state established after a change. For example, a fall in M_f without a change in $\overline{MLT}$ will lead immediately to a fall in osteoid thickness but only after a delay to an increase in osteoid surface; consequently, there will be a transient decrease in osteoid volume.

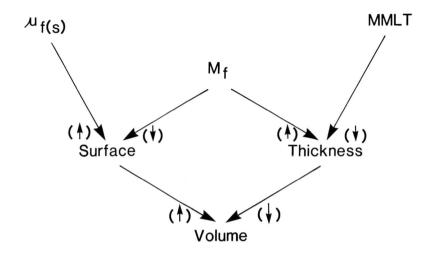

FIGURE 25.   Relationships of osteoid surface, thickness, and volume to surface-based birth rate of osteoid ($\mu_{f(s)}$), mean mineralization lag time (MMLT), and mean corrected mineral apposition rate (M_f). Note that M_f influences surface and thickness in opposite directions but has no effect on volume provided that a new steady state is reached. For further details see text.

Rigorous interpretation of the various osteoid measurements in relation to their primary determinants is clearly impossible without double tetracycline labeling, but some clues may be given by knowledge of the constraints to which the primary determinants are subject. The birth rate (μ_f) may be either increased or decreased by about fivefold or more. M_f may be increased by about 50 to 100% when there is appositional woven bone formation as in Paget's disease,[163] after fluoride administration, and in renal osteodystrophy.[164] In all these conditions seam thickness can be increased in the absence of a mineralization defect and osteoid surface will be increased either as a result of an increase in turnover or because bone formation occurs on quiescent bone surfaces without relation to previous resorption. M_f may also be increased by 10 to 40% in hyperthyroidism and as a transient consequence of a recent increase in bone turnover, but with these exceptions, all easy to recognize, M_f can only be reduced, sometimes to as little as 1/20 of normal or even less. Similarly, $\overline{MLT}$ can be slightly shortened in hyperthyroidism[142] but otherwise can only be prolonged, sometimes to as much as 20-fold or even more. It is also helpful to relate the measurements to endosteal cell morphology. The osteoid surface can be increased because of an increase either in the birth rate or the life span of new osteoid seams of arbitrary area. In the absence of tetracycline labeling these different mechanisms can be distinguished to some extent by the morphology of the cells on the surface; a reduction in M_f and consequent prolongation of σ_f for any reason is usually accompanied by a reduction in the fraction of osteoid covered by "active" osteoblasts and a corresponding accumulation of terminal osteoblasts, whereas an increase in birth rate will increase the number of "active" osteoblasts and the surface extent of seam in the same proportion, so that the "active" fraction will be unchanged.

8. The Recognition of Osteomalacia

The histomorphometric findings in different disease states are covered in Chapter 14, but some general points concerning osteomalacia arise out of the preceding analysis. Traditionally, an increase in both osteoid volume and osteoid seam width have been regarded as the most characteristic morphologic findings in osteomalacia. As

Table 22
THREE CONDITIONS WITH RESTING SEAMS[a]

	Osteomalacia	Osteoporosis	Serious Illness
Duration of abnormal OB function	Years	Years	Months
Relation to BRC: Depression of MaAR	Throughout	Throughout	Random[b]
Prolongation of MLT	Throughout	Frequently	Random[b]
Osteoid seam thickness — absolute	N or ↑	N or ↓	N
Related to cell morphology	Always ↑	N or ↓	N
Frequency distribution	Skewed to R[c]	Skewed to L[d]	N

Note: OB = osteoblast; BRC = bone remodeling cycle; MaAR = matrix appositional rate; MLT = mineralization lag time.

[a]   Comparison of three conditions in which resting seams may occur.
[b]   Affects all BRU present during a defined time interval, at whatever stage.
[c]   In direction of increase.
[d]   In direction of decrease.

mentioned previously, osteoid volume may be increased because of an increase either in F_v or in $\overline{MLT}$, but very large increases (more than tenfold) are seen only in osteomalacia with marked prolongation of $\overline{MLT}$. The most characteristic radiographic feature of osteomalacia is the Looser zone, which results from an unhealed stress fracture.[165] Looser zones are rarely found unless the relative osteoid volume exceeds 20%,[166] first because there may be a corresponding reduction in mineralized bone volume and second because the degree of increase in osteoid is a good indication of the severity and duration of the osteomalacic process.

More recently, several additional or alternative criteria for osteomalacia have been proposed. These include a reduction in the fraction of osteoid seam showing a tetracycline label or a toluidine blue stainable mineralization front,[117] a reduction in the MiAR,[167] and absence of low-density bone adjacent to the osteoid seam.[168] But these are all nonspecific consequences of impaired osteoblast function which do not specifically relate to a defect in mineralization but identify an increase in the proportion of resting seams. As such, they are all found in some patients with osteoporosis, and occasionally in seriously ill patients without metabolic bone disease as well as in osteomalacia. The characteristics of these three different situations in which resting seams occur are contrasted in Table 22. In osteoporosis, the several consequences of impaired osteoblast function tend to be most evident in the latter part of the life history of the osteoid seam, whereas in osteomalacia they are present from the outset. Consequently, in some cases of osteoporosis there is an accumulation of thin terminal seams at which appositional rate, fractional labeling, and fractional mineralization front are very low, and mineralization lag time markedly prolonged. With some methods of section preparation and staining these mini-seams are not detected, in which case the values for fractional labeling and fractional mineralization front are likely to be normal. Recognition of mini-seams identifies an important defect in osteoblast function in osteoporosis, but nevertheless may decrease the apparent diagnostic value of the bone biopsy because the numerical differences between osteomalacia and osteoporosis are reduced. Consequently, for routine clinical use, mini-seams can be considered separately and not included in the calculated values. For the same reason, the estimation of OMT from the formula $\overline{OSW}/M$ gives better discrimination between osteoporosis and osteomalacia than the calculation of mean MLT from $\overline{OSW}/M_f$.

Some of these difficulties can be overcome by paying closer attention to the traditional criterion of osteoid seam width in the diagnosis of osteomalacia. This is determined as the product of M_f (as an estimate of matrix appositional rate) and MLT, but these quantities usually change in opposite directions so that both may be abnormal even if osteoid width is normal. If the osteoid seam width is decreased, M_f is reduced by more than MLT is increased (the usual situation in osteoporosis); the prolonged MLT most likely reflects a primary disorder of matrix maturation (prolonged OMT). Conversely, if the osteoid seam width is increased, MLT is increased by more than M_f is reduced, the usual situation in osteomalacia: the prolonged MLT most likely reflects a primary defect in mineralization in addition to any defect in osteoid maturation (MLT > OMT). This interpretation is strengthened by the significant correlation between OSW and M_f in normal subjects and in involutional osteoporosis.[200] In mild cases of osteomalacia seam width may be normal in absolute value, but increased in relation to the reduced M_f and to the increased surface extent of terminal osteoblasts, which in osteoporosis are both usually associated with thin seams. Furthermore, the frequency distribution of osteoid seam width will be skewed, due to the presence of some seams which are much wider than are ever normally encountered.[130] Since the osteoid seam width is crucial to the interpretation of increased osteoid volume (or hyperosteoidosis), it should be measured directly rather than indirectly calculated from perimeter and area measurements.

Little is known about the kinetics of evolution of individual osteoid seams in osteomalacia. In most cases MLT is prolonged both because of an increase in OMT and because of an independent defect in mineralization. If MLT is prolonged by more than M_f is depressed from the onset of matrix apposition but does not increase further, the osteoid seam will grow slowly to an abnormal thickness before mineralization begins and wide seams would be present even if all osteoid were ultimately to become mineralized. However, if MLT increases progressively, mineralization might never catch up with matrix synthesis and the seam width would continue to grow (although very slowly) in the absence of treatment until matrix synthesis stopped completely. This is presumably the mechanism for the increase in total trabecular width and total trabecular bone volume which occurs in some patients with severe osteomalacia.[130] Observations on patients who for various reasons have received tetracycline continuously for long periods of time indicate that at least some bone-forming sites in some osteomalacic patients do eventually complete mineralization. In some cases, this occurs only close to the surface, so that the deeper layers of osteoid are buried beneath mineralized bone. For any seam which remains permanently unmineralized the loop in Figure 24 does not close. The relationships previously deduced between matrix apposition, mineral apposition, and mineralization lag time are therefore no longer valid, and the identity between MaAR and MiAR no longer holds. However, the rate of increase in osteoid seam width can be estimated by comparing values for M_f and mean osteoid seam width in the same biopsy. Such calculations indicate that matrix apposition is reduced in all forms of osteomalacia (with the possible exception of that due to diphosphonate administration) and that the degree of uncoupling between the matrix and mineral apposition rates almost never exceeds 10% of the mean values.

9. Variations in Mineral Density

The subdivision of total bone into mineralized bone and osteoid is oversimplified; variation in mineral density occurs within each category. Normal osteoid is completely unmineralized, but osteomalacic osteoid may paradoxically be partly mineralized with irregularly distributed small clusters of mineral detectable by electron microscopy[169] and in toluidine blue stained sections.[151] Variations in the completeness of mineralization of bone have been subject more to qualitative than quantitative description; such variations can be studied in several ways, for which analytic precision and morphologic detail tend to vary inversely. The simplest methods are also destructive, such as the weight of ash per unit volume[170] or the phosphorus to hydroxyproline ratio.[171] Both of these correlate with fractional osteoid volume but cannot distinguish an increase in osteoid from an increase in the proportion of incompletely mineralized bone. More complex measurements related to mineral structure and composition are beyond the scope of this text. Nondestructive chemical analysis can be performed by electron probe[172] and by neutron activation.[173] The latter method combined with microscopic determination of bone and osteoid volume (and an assumed value for the mineral density of osteoid) can give values for the mineral density of mineralized bone; this can vary independently from the proportion of osteoid.[202]

The distribution of mineral density is of greater morphologic interest than a mean value for the whole skeleton. It has been most studied by microradiography, in which X-rays are taken of ground sections.[56] With care to minimize variations in section thickness and appropriate calibration, the mineral density of individual regions of bone can be estimated within about 10 to 20%, precise enough to allow the accumulation of mineral by newly formed bone to be studied as a function of time.[54] More commonly, the technique is used to give qualitative information such as the presence of low-density bone on the surface as a sign of recent mineralization and the persistence of low-density bone beneath the surface in various pathologic conditions.[56] The fraction of bone surface with adjacent low-density bone determined by microradiography is similar to the fraction covered by "active" osteoblasts[174] and therefore corresponds to about one third of the total osteoid perimeter. The combination of microradiography with automated methods of analyzing visual information such as the quantimet should greatly increase our understanding of such phenomena, but so far has been little used.

Mineral density can also be studied indirectly by means of differential permeability to stains of different ionic radius. As judged by comparison with microradiography, the penetration of bone by organic molecules such as basic fuchsin indicates a mineral density of less than about 80% of maximum, whereas penetration by smaller inorganic molecules such as the permanganate ion indicates a density of less than about 90% of maximum.[55] These criteria have only been validated for fresh hand-ground sections of cortical bone after adequate removal of surface stain and may not apply to plastic-embedded sections of trabecular bone. Several kinds of abnormality can be detected with these methods; those related to the osteocyte and perilacunar bone are described separately in the next section.

Areas of fuchsin-permeable low-density bone may be found both in biopsies of patients with osteomalacia and hyperparathyroidism and in autopsy and operative specimens from persons with no evidence of metabolic bone disease. They are of spotty distribution, usually involving only part of a BSU, e.g., one quadrant of an osteon. Although amenable to measurement by the grid-point method, no systematic quantitative study has been undertaken. The low-density areas are not a transitional state between young and old bone because they are not concentrated near the bone surface or beneath osteoid seams. They are not a reflection of decreased mean density due to decreased skeletal age because neither their frequency nor extent are related to bone turnover, nor do they represent confluence of low-density perilacunar bone.[55] In patients with metabolic bone disease the low-density areas are detectable by microradiography as well as fuchsin permeability and are often associated with apparent enlargement of osteocyte lacunae;[36,174] they must not be confused with buried osteoid in partially treated osteomalacia. In persons without disordered mineral metabolism or metabolic bone disease, low-density bone (referred to in this context as "feathered bone")[55] has been detected only by fuchsin permeability. The two techniques have rarely been used concurrently in the same specimen, but fuchsin permeability is probably more sensitive to modest decreases in density than is microradiography since it is

Table 23
PHENOMENA CONFUSED WITH "OSTEOCYTIC OSTEOLYSIS"

Normal
    The resorptive phase of osteocytic mini-modeling
    The resorptive phase of osteocytic mini-remodeling
    Permanent lacunar enlargement due to osteocyte death
    Temporary removal of mineral from perilacunar bone[a]
Abnormal

Without lacunar enlargement[b]

Primary defect in mineralization or perilacunar bone[c,d]
Defective mineralization during mini-remodeling[c,d]
Partial demineralization with dissolution when exposed to nitric acid[c,e]
Presence of fuchsin-permeable perilacunar zone in XLH
Extreme perilacunar demineralization in proximity to pathologic bone resorption

With lacunar enlargement[b]

Primary enlargement in woven bone
Transient increase in mean value after increase in turnover[f]
Apparent enlargement due to section obliquity
Delay or failure of periosteocytic bone formation during mini-remodeling
Pathologic enlargement of lacunae in response to an abnormal resorptive stimulus[g]

Note: Different phenomena which have been or could be confused with "osteocytic osteolysis".[6,32] For
        further explanation and citation see text.

[a]   A component of plasma calcium homeostasis which probably affects pericanalicular bone also. The
        amount of calcium removed is usually too small to be detectable by histologic methods.
[b]   Distinction requires careful attention to methodologic detail.
[c]   With enlargement of permanganate-permeable perilacunar zone.
[d]   In rickets and osteomalacia.
[e]   In response to acute or chronic PTH excess, possibly also metabolic acidosis; may be an extreme form
        of a normal phenomenon.
[f]   An increase in the number of recently initiated bone-forming surfaces will increase the proportion of
        large cuboidal osteoblasts, which will give rise to larger osteocytes and, therefore, to larger lacunae.
[g]   Only this should be referred to as osteolytic osteolysis; it is not known whether it is always irreversible
        or sometimes reversible.

present beneath the entire osteoid perimeter except at resting seams. There is no relationship between feathered bone and abnormal osteocytes, either in affected areas or
in the skeleton as a whole.

The mechanism by which low-density bone is produced and its pathologic and clinical significance are uncertain. In osteomalacia it probably reflects a delay in secondary
as well as in primary mineralization. In hypercalcemic hyperparathyroidism the presence of low-density bone has been ascribed to a previous episode of vitamin D deficiency which may have initiated parathyroid stimulation,[175] a reasonable but so far
unproven suggestion. In vivo demineralization or halisteresis is generally assumed not
to occur, but there is no *a priori* reason why secondary mineralization should not be
reversed, given the appropriate change in bone ECF composition. This mechanism has
been suggested for both primary and secondary hyperparathyroidism.[36] In chronic
renal failure it has been proposed that metabolic acidosis selectively dissolves carbonate from bone,[176] but the observed deficit in carbonate could result equally well from
deposition of mineral of abnormal composition *de novo*.[35] Low-density bone in per-

sons without metabolic bone disease results from a profound delay in secondary mineralization, possibly because of a localized disturbance in blood supply, such that affected bone remains below the 80% threshold for fuchsin permeability for several years rather than only for a few months.[177]

10. Osteocytes, Osteocyte Lacunae, and Perilacunar Bone

Several methods are used for determining the frequency and extent of so-called osteocytic osteolysis. This term has unfortunately been applied indiscriminately to several different processes which must be clearly distinguished (Table 23). To unravel the confusion it is first necessary to describe in more detail the special characteristics of perilacunar bone — the 1- to 2-μm thick layer of bone adjacent to the wall of the osteocyte lacuna which differs from interlacunar bone in several respects.[6,32] In perilacunar bone the mineral is less densely packed, more soluble, more amorphous and less crystalline, and more reactive with silver nitrate. Most of the mineral is more easily removed by acid or EDTA but a small fraction of the mineral is more tightly bound than in bone generally.[178] The matrix of perilacunar bone is also different — the collagen fibers are fewer and most loosely packed and there is preferential staining with methylene blue after demineralization.[6,178] As already mentioned, the lower mineral density of perilacunar bone is associated with permeability to permanganate but not to basic fuchsin. Maintenance of the differences between perilacunar and interlacunar bone is dependent in some way on the presence of a living cell — osteocyte death leads to increased perilacunar mineral density and loss of permanganate permeability.[54]

In young and rapidly growing animals, osteocytes undergo a series of changes between laying down of the bone on their periosteal surface and resorption of the same bone from the endosteal surface. Recently entombed osteocytes retain some of the ultrastructural appearances of osteoblasts and continue for a limited period to form a new bone around themselves. This process is then reversed as the endosteal resorbing surface gets closer, with disintegration of intralacunar matrix and resorption of perilacunar bone, ending in degeneration and death of the osteocyte with irreversible lacunar enlargement shortly before the bone is resorbed.[179] Resorbing osteocytes stain for acid phosphatase like the osteoclasts on the surface. The whole process may be referred to as osteocytic mini-modeling, and has been observed in several species including the rat, the chicken, and man.[180] Like modeling in general, it has no exact counterpart in the normal nongrowing adult skeleton, although in adult man recently formed bone near the surface contains bone-forming osteocytes[181] and irreversible lacunar enlargement sometimes precedes pathologic bone resorption from the surface.

In adult man, osteocytic lacunae are flattened ellipsoids, with a mean long axis of about 20 μm, a mean breadth of 10 μm, and a mean thickness of 4 μm (Figure 26). The major longitudinal plane faces the bone surface, so that histologic sections perpendicular to the surface will show profiles of the minor longitudinal plane, with a maximum length of 20 μm and maximum width of 4 μm, and mean values of 15 and 3 μm, respectively.[182,183] In cortical bone, lacunae have been classified by electron microscopy into four types:[184] first, a small lacuna with a smooth border containing an inactive cell; second, an enlarged lacuna with a rough, irregular border containing an osteocyte with abundant lysosomes ("osteolytic"); third, an enlarged lacuna containing an osteocyte with abundant endoplasmic reticulum ("osteoplastic") surrounded by recently formed intralacunar matrix in which the collagen fibers are arranged radially; and finally, an enlarged but empty lacuna containing only cellular debris.

Lacunar enlargement involves mainly an increase in thickness, not in length or breadth (Figure 26). An apparent increase in thickness is produced by obliquity of the section plane, which is much easier to control in cortical bone than in trabecular bone. Classification then depends only on irregularity of the border, a much more subjective

93. **Kelin, M. and Frost, H. M.,** Evidence of periodic changes in the rate of formation of individual osteons in human bone, *Henry Ford Hosp. Med. Bull.,* 12, 565, 1964.

94. **Boyde, A.,** Scanning electron microscope studies of bone, in *The Biochemistry and Physiology of Bone,* Vol. 1, Bourne, B. H., Ed., Academic Press, New York, 1972, 259.

95. **Frost, H. M.,** Measurement of osteocytes per unit volume and volume components of osteocytes and canaliculae in man, *Henry Ford Hosp. Med. Bull.,* 8, 208, 1960.

96. **Frost, H. M.,** Observations on osteoid seams: the existence of a resting state, *Henry Ford Hosp. Med. Bull.,* 8, 220, 1960.

97. **Frost, H. M.,** Human osteoid seams, *J. Clin. Endocrinol. Metab.,* 22, 631, 1962.

98. **Frost, H. M.,** A new bone affection: feathering, *J. Bone Jt. Surg.,* 42A, 447, 1960.

99. **Landeros, O. and Frost, H. M.,** The cross section size of the osteon, *Henry Ford Hosp. Med. Bull.,* 12, 517, 1964.

100. **Lauffenburger, Th., Olah, A. J., Dambacher, M. A., Guncaga, J., Lentner, Ch., and Haas, H. G.,** Bone remodeling and calcium metabolism: a correlated histomorphometric, calcium kinetic, and biochemical study in patients with osteoporosis and Paget's disease, *Metabolism,* 26, 589, 1977.

101. **Kelin, M. and Frost, H. M.,** The numbers of bone resorption and formation foci in rib, *Henry Ford Hosp. Med. Bull.,* 12, 527, 1964.

102. **Lacroix, P.,** The internal remodeling of bones, in *The Biochemistry and Physiology of Bone,* Vol. 3, 2nd ed., Bourne, G. H., Ed., Academic Press, New York, 1974.

103. **Kolbel, R.,** Spontane angiogene Knochenneubildung in spongiosem Knochen, *Z. Orthopaed.,* 116, 682, 1978.

104. **Hattner, R., Epker, B.N., and Frost, H. M.,** Suggested sequential mode of control of changes in cell behaviour in adult bone remodeling, *Nature (London),* 206, 489, 1965.

105. **Schenk, R. K., Merz, W. A., and Muller, J.,** A quantitative histological study on bone resoprtion in human cancellous bone, *Acta Anat.,* 74, 44, 1969.

106. **Evans, R. A., Dunstan, C. R., and Baylink, D. J.,** Histochemical identification of osteoclasts in undecalcified sections of human bone, *Min. Elect. Metab.,* 2, 179, 1979.

107. **Woods, C. G.,** *Diagnostic Orthopaedic Pathology,* Blackwell Scientific, Oxford, 1972.

108. **Ham, A. W.,** Some histophysiological problems peculiar to calcified tissues, *J. Bone Jt. Surg.,* 34A, 701, 1952.

109. **Krempien, B.,** Bone modeling processes at the endosteal surface of human femora: Scanning electron microscopical studies in normal bone and in renal osteodystrophy, *Virchows Arch. Pathol. Anat. Histol.,* 382, 73, 1979.

110. **Boyde, A. and Jones, S. J.,** Estimation of the size of resorption lacunae in mammalian calcified tissues using SEM stereophotogrammetry. Report on Scanning Electron Microscopy/1979/II SEM Inc., AMF O'Hare, Ill., 393.

111. **Baron, R.,** Remaniement de l'os alveolarie et des fibres desmondontales au cours de la migration physiologique, *J. Biol. Buccale,* 1, 151, 1973.

112. **Parfitt, A. M., Villanueva, A. R., Crouch, M. M., Mathews, C. H. E., and Duncan, H.,** Classification of osteoid seams by combined use of cell morphology and tetracycline labeling. Evidence for intermittency of mineralization, in *Bone Histomorphometry,* 2nd Int. Workshop, Meunier, P. J., Ed., Armour Montague, Paris, 1977.

113. **Kimmel, D.,** Personal communication.

114. **Schenk, R. K., Olah, A. J., and Merz, W. A.,** Bone cell counts, in *Clinical Aspects of Metabolic Bone Disease,* Frame, B., Parfitt, A. M., and Duncan, H., Eds., Excerpta Medica, Amsterdam, 1973.

115. **Olah, A. J.,** Quantitative relations between osteoblasts and osteoid in primary hyperparathyroidism, intestinal malabsorption and renal osteodystrophy, *Virchows Arch. Pathol. Anat. Histol.,* 358, 301, 1973.

116. **Zallone, A. Z.,** Relationships between shape and size of the osteoblasts and the accretion rate of trabecular bone surfaces, *Anat. Embryol.,* 152, 65, 1977.

117. **Bordier, P. J., Marie, P., Miravet, L., Ryckewaert, A., and Rasmussen, H.,** Morphological and morphometrical characteristics of the mineralization front. A vitamin D regulated sequence of the bone remodeling, in *Bone Histomorphometry,* 2nd Int. Workshop, Meunier, P. J., Ed., Armour Montague, Paris, 1977.

118. **Baylink, D., Morey, E., and Rich, C.,** Effect of calcitonin on the rates of bone formation and resorption in the rat, *Endocrinology,* 84, 261, 1969.

119. **Raman, A.,** Appositional growth rate in rat bones using the tetracycline labeling method, *Acta Orthopaed. Scand.,* 40, 193, 1969.

120. **Smith, J. W.,** Collagen fibre patterns in mammalian bone, *J. Anat.,* 94, 329, 1960.

121. **Kulp, J. L. and Schulert, A. R.,** Strontium 90 in man. V., *Science,* 136, 619, 1962.

122. **Carr, T. E. F., Harrison, G. E., Loutit, J. F., and Sutton, A.,** Movement of strontium in the human body, *Br. Med. J.,* 2, 773, 1962.

123. **Rowland, R. E.**, Resorption and bone physiology, in *Bone Biodynamics*, Frost, H. M., Ed., Little, Brown, Boston, 1964.

124. **Gong, J. K., Burgess, E., and Bacalao, P.**, Accretion and exchange of Strontium-85 in trabecular and cortical bones, *Radiat. Res.*, 28, 753, 1966.

125. **Rivera, J.**, Strontium-90 in human vertebrae, 1962—1963, *Radiol. Health Data*, 5, 511, 1964.

126. **Arnold, J. and Wei, C. T.**, Quantitative morphology of vertebral trabecular bone, in *Radiobiology of Plutonium*, Stover, B. J. and Jee, W. S. S., Eds., J.W. Press, Salt Lake City, Utah, 1972.

127. **Melsen, F. and Mosekilde, L.**, Tetracycline double-labeling of iliac trabecular bone in 41 normal adults, *Calcif. Tissue Res.*, 26, 99, 1978.

128. **Krempien, B., Lemminger, F. M., Rita, E., and Weber, E.**, The reaction of different skeletal sites to metabolic bone disease — a micromorphometric study, *Klin. Wochenschr.*, 56, 755, 1978.

129. **Mosekilde, L. and Melsen, F.**, Anticonvulsant osteomalacia determined by quantitative analysis of bone changes: population study and possible risk factors, *Acta Med. Scand.*, 199, 349, 1976.

130. **Frame, B. and Parfitt, A. M.**, Osteomalacia: current concepts, *Ann. Int. Med.*, 89, 966, 1978.

131. **Gonick, H. C., Drinkard, J. P., Hertoghe, J., and Rubin, M. E.**, A calcium kinetic approach to the problem of renal osteodystrophy, *Clin. Orthoped.*, 100, 315, 1974.

132. **Whitehouse, W. J.**, Cancellous bone in the anterior part of the iliac crest, *Calcif. Tisue Res.*, 23, 67, 1977.

133. **Villanueva, A. R., Parfitt, A. M., and Duncan, H.**, Comparison of haversian bone dynamics between 11th rib and iliac trephine biopsies, in *Bone Histomorphometry*, 2nd Int. Workshop, Meunier, P. J., Ed., Armour Montague, Paris, 1977.

134. **Melsen, F., Melsen, B., and Mosekilde, L.**, An evaluation of the quantitative parameters applied in bone histology, *Acta Pathol. Microbiol. Scand.*, 86, 63, 1978.

135. **Visser, W. J., Niermans, H. J., Roelofs, J. M. M., Raymakers, J. A., and Duursma, S. A.**, Comparative morphometry of bone biopsies obtained by two different methods from the right and the left iliac crest, in *Bone Histomorphometry*, 2nd Int. Workshop, Meunier, P. J., Ed., Armour Montague, Paris, 1977.

136. **Ritz, E., Krempien, B., Bommer, J., and Jesdinski, H. J.**, A critical analysis of micromorphometry in metabolic osteopathy, *Verh. Dtsch. Ges. Pathol.*, 58, 363, 1974.

137. **Tam, C. S., Harrison, J. E., Reed, R., and Cruickshank, B.**, Bone apposition rate as an index of bone metabolism, *Metabolism*, 27, 143, 1978.

138. **Morgan, D. G.**, *Osteomalacia, Renal Osteodystrophy and Osteoporosis*, Charles C Thomas, Springfield, Ill., 1973.

139. **Parfitt, A. M.**, The quantitative approach to bone morphology: a critique of current methods and their interpretation, in *Clinical Aspects of Metabolic Bone Disease*, Frame, B., Parfitt, A. M., and Duncan, H., Eds., Excerpta Medica, Amsterdam, 1973, 86.

140. **Ritz, E., Malluche, H. H., Krempien, B., and Mehls, O.**, Bone histology in renal insufficiency, in *Calcium Metabolism in Renal Failure and Nephrolithiasis*, David, D. S., Ed., John Wiley & Sons, New York, 1977, 197.

141. **Frost, H. M.**, The origin and nature of transients in human bone remodeling dynamics, in *Clinical Aspects of Metabolic Bone Disease*, Frame, B., Parfitt, A.M., and Duncan, H., Eds., Excerpta Medica, Amsterdam, 1973, 124.

142. **Meunier, P. J., Bianchi, G. G. S., Edouard, C. M., Bernard, J. C., Courpron, P., and Vignon, G. E.**, Bony manifestations of thyrotoxicosis. Symposium on Metabolic Bone Disease, *Orthoped. Clin. North Am.*, 3, 501, 1972.

143. **Parfitt, A. M.**, The actions of parathyroid hormone on bone. Relation to bone remodeling and turnover, calcium homeostasis and metabolic bone disease. III. PTH and osteoblasts, the relationship between bone turnover and bone loss, and the state of the bones in primary hyperparathyroidism, *Metabolism*, 25, 1033, 1976.

144. **Berlin, N. I. and Berk, P. D.**, The biological life of the red cell, in *The Red Blood Cell*, 2nd ed., Mack, D., Ed., Academic Press, New York, 1975.

145. **Halvorsen, H. and Parfitt, A. M.**, unpublished data.

146. **Russell, J. E. and Avioli, L. V.**, Effect of progressive end stage renal insufficiency on bone mineral-collagen maturation, *Kidney Int.*, 7, S97, 1975.

147. **Courpron, P., Meunier, P., Bressot, C., and Giroux, J. M.**, Amount of bone in iliac crest biopsy. Significance of the trabecular bone volume. Its values in normal and in pathological conditions, in *Bone Histomorphometry*, 2nd Int. Workshop, Meunier, P. J., Ed., Armour Montague, Paris, 1977.

148. **Parfitt, A. M.**, Some problems in measuring the amount of bone by histologic methods, in *Bone Histomorphometry*, 2nd Int. Workshop, Meunier, P. J., Ed., Armour Montague, Paris, 1977.

149. **Parfitt, A. M. and Duncan, H.**, Metabolic bone disease affecting the spine, in *The Spine*, Rothman, R. and Simeone, F., Eds., W. B. Saunders, Philadelphia, 1975, 599.

150. Melsen, F., Viidik, A., Melsen, B., and Mosekilde, L., Some relations between bone strength, ash weight and histomorphometry, in *Bone Histomorphometry,* 2nd Int. Workshop, Meunier, P. J., Ed., Armour Montague, Paris, 1977.

151. Wu, K. and Frost, H. M., Bone resorption rates in physiological, senile and postmenopausal osteoporosis, *J. Lab. Clin. Med.,* 69, 810, 1967.

152. Kleerekoper, M., Crouch, M., Frame, B., Mathews, C., Matkovic, V., Oliver, I., Rao, D., and Parfitt, A. M., Effect of sodium fluoride alone on iliac bone remodeling dynamics in osteoporosis, *Calcif. Tissue Int.,* in press.

153. Duncan, H., Osteoblasts and osteoid, a hard look, in *Bone Histomorphometry,* 2nd Int. Workshop, Meunier, P. J., Ed., Armour Montague, Paris, 1977.

154. Frost, H. M., Tetracycline labeling of bone and the zone of demarcation of osteoid seams, *Can. J. Biochem. Physiol.,* 40, 485, 1962.

155. Frost, H. M., Jee, W. S. S., Kimmel, D., et al: Histomorphometric analysis of trabecular bone in renal dialysis patients treated with 25-hydroxyvitamin D_3: preliminary report, in *Vitamin D: Biochemical, Chemical and Clinical Aspects Related to Calcium Metabolism,* Norman, A. W., et al (Eds.). Walter deGruyter, Berlin, 1977, 885.

156. Tam, C. S., Bayley, T. A., Harrison, J. E., Murray, T. M., Birkin, B. L., and Thomson, D., Bone biopsy in the diagnosis of primary hyperparathyroidism, in *Endocrinology of Calcium Metabolism,* Copp, D. H. and Talmage, R. V., Eds., Excerpta Medica, Amsterdam, 1978.

157. Tam, C. S., Cruickshank, B., Swinson, D. R., Anderson, W., and Little, H. A., The response of bone apposition rate to some nonphysiologic conditions, *Metabolism,* 28, 751, 1979.

158. Meunier, P. J., Bressot, C., and Edouard, C., Dynamics of bone remodeling in primary hyperparathyroidism. Histomorphometric data, in *Endocrinology of Calcium Metabolism,* Copp, D. H. and Talmage, R. V., Eds., Excerpta Medica, Amsterdam, 1978.

159. Parfitt, A. M., Rao, D. S., Crouch, M., Frame, B., Matkovic, V., Villanueva, A. R., Walczak, N., and Kleerekoper, M., Iliac histomorphometry and radial photon absorptiometry in primary hyperparathyroidism, *Calcif. Tissue Int.,* 28, 170, 1979.

160. Melsen, F. and Mosekilde, L., Dynamic studies of trabecular bone formation and osteoid maturation in normal and certain pathological conditions, *Metab. Bone Dis. Rel. Res.,* 1, 45, 1978.

161. Stauffer, M., Baylink, D., Wergedal, J., and Rich, C., Decreased bone formation, mineralization, and enhanced resorption in calcium-deficient rats, *Am. J. Physiol.,* 225, 269, 1975.

162. Teitelbaum, S. L., Rosenberg, E. M., Bates, M., and Avioli, L. V., The effects of phosphate and vitamin D therapy on osteopenic hypophosphatemic osteomalacia of childhood: a morphometric study, *Clin. Orthoped.,* 116, 38, 1976.

163. Meunier, P., La maladie osseuse de Paget. Histologie quantitative, histopathogenie et perspectives therapeutiques, *Lyon Med.,* 233, 839, 1975.

164. Sherrard, D. J., Baylink, D. J., Wergedal, J., and Maloney, N. A., Quantitative histological studies on the pathogenesis of uremic bone disease, *J. Clin. Endocrinol.,* 39, 119, 1974.

165. Parfitt, A. M., The clinical and radiographic manifestations of renal osteodystrophy, in *Perspectives in Hypertension and Nephrology: Calcium Metabolism In Renal Failure and Nephrolithiasis,* David, D. J., Ed., John Wiley & Son, New York, 1977.

166. Meunier, P., Edouard, C., Richard, D., and Laurent, J., Histomorphometry of osteoid tissue: the hyperosteoidoses, in *Bone Histomorphometry,* 2nd Int. Workshop, Meunier, P. J., Ed., Armour Montague, Paris, 1977.

167. Ramser, J. R., Frost, H. M., Frame, B., Arnstein, A. R., and Smith, R., Tetracycline-based studies of bone dynamics in rib of 6 cases of osteomalacia, *Clin. Orthoped.,* 46, 219, 1966.

168. Kelly, P. J., Jowsey, J., and Riggs, B. L., A comparison of different morphologic methods of determining bone formation, *Clin. Orthoped.,* 41, 7, 1965.

169. Bonucci, E., Matrayt, H., Tunchot, S., and Hioco, D. J., Bone structure in osteomalacia with special reference to ultrastructure, *J. Bone Jt. Surg.,* 51B, 511, 1969.

170. Morgan, D. B., Stanley, J., and Fourman, P., The mineral deficit in osteomalacic bone, *Clin. Sci.,* 35, 377, 1968.

171. Tougaard, L., Melsen, F., and Mosekilde, L., Bone phosphorus/hydroxyproline ratio and bone histomorphometry in normals, *Scand. J. Clin. Lab. Invest.,* 38, 89, 1978.

172. Hefferen, J. J., Kim, D., Lenke, J., DelGreco, F., Timmerman, M., and Levin, N. W., Distribution of fluoride in uremic bone, in *Clinical Aspects of Metabolic Bone Disease,* Frame, B., Parfitt, A. M., and Duncan, H., Eds., Excerpta Medica, Amsterdam, 1973.

173. Batra, G. J. and Bewley, D. K,. Analysis of small biopsy samples by neutron activation analysis, *J. Radioanal. Chem.,* 16(1), 275, 1973.

174. Jowsey, J., *The Bone Biopsy,* Plenum Press, New York, 1977.

175. Jowsey, J., Bone histology and hyperparathyroidism, *Clin. Endocrinol. Metab.,* 3, 267, 1974.

176. Pellegrino, E. D. and Biltz, R. M., The composition of human bone in uremia. Observations on the reservoir functions of bone and demonstration of a labile fraction of bone carbonate, *Medicine*, 44, 397, 1965.

177. Frost, H. M., Feathering: a theory of genesis, *Henry Ford Hosp. Med. Bull.*, 9, 103, 1961.

178. Villanueva, A. R., Frost, H. M., and Roth, H., Halo volume — Part III. Existence of a pattern in the matrix, *Henry Ford Hosp. Med. Bull.*, 9, 133, 1961.

179. Jande, S. S. and Belanger, L. F., The life cycle of the osteocyte, *Clin. Orthoped.*, 94, 281, 1973.

180. Whalen, J. P., Winchester, P., Krook, I., O'Donohue, N., Dische, R., and Nunez, E. A., Neonatal transplacental rubella syndrome. The effect on normal maturation of the diaphysis, *Am. J. Roentgenol. Radium Ther. Nucl. Med.*, 121, 166, 1974.

181. Bonucci, E. and Gherardi, G., Electron microscope investigations of osteocytes in renal osteodystrophy, in *Bone Histomorphometry*, 2nd Int. Workshop, Meunier, P. J., Ed., Armour Montague, Paris, 1977.

182. Baud, C., Histophysiology of the osteocyte: an introduction to the morphometry of peri-osteocytic lacunae, in *Proceedings of the First Workshop on Bone Morphometry*, Jaworski, Z. F. G., Ed., University of Ottawa Press, Ottawa, Canada, 1976.

183. Meunier, P. and Bernard, J., Morphometric analysis of periosteocytic osteolysis, in *Proceedings of the First Workshop on Bone Morphometry*, Jaworski, Z. F. G., Ed., University of Ottawa Press, Ottawa, Canada, 1976.

184. Baud, C. A. and Auil, E., Osteocyte differential count in normal human alveolar bone, *Acta Anat.*, 78, 321, 1971.

185. Krempien, B., Geiger, G., Ritz, E., and Buttner, S., Osteocytes in chronic uremia. Differential count of osteocytes in human femoral bone, *Virchows Arch. A.*, 360, 1, 1973.

186. Krempien, B., Ritz, E., and Geiger, G., Behaviour of osteocytes in various ages and chronic uremia. Morphological studies in human cortical bone, in *Proceedings of the First Workshop on Bone Morphometry*, Jaworski, Z. F. G., Ed., University of Ottawa Press, Ottawa, Canada, 1976.

187. Urist, M. R., Gurvey, M. S., and Fareed, D. O., Long-term observations on aged women with pathologic osteoporosis, in *Osteoporosis*, Barzel, U. S., Ed., Grune & Stratton, New York, 1970, 3.

188. Frost, H. M., A unique histological feature of vitamin D resistant rickets observed in four cases, *Acta Orthopaed. Scand.*, 33, 220, 1963.

189. Frost, H. M., Bosworth, D. M., Halliburton, R. H., and Sezgin, A., A report of some examples of abnormal halo volume, *Henry Ford Hosp. Med. Bull.*, 9, 404, 1961.

190. Steendijk, R., Vanden Hoofff, A., Nielsen, H. K. L., and Jowsey, J., Lesion of the bone matrix in vitamin D resistant rickets, *Nature (London)*, 207, 426, 1965.

191. Steendijk, R. and Boyde, A., Scanning electron microscopic observations on bone from patients with hypophosphatemic (vitamin D resistant) rickets, *Calcif. Tissue Res.*, 11, 242, 1973.

192. Jaffe, H. L., *Metabolic, Degenerative, and Inflammatory Diseases of Bones and Joints*, Lea & Febiger, Philadelphia, 1972.

193. Villanueva, A. R. and Frost, H. M., Aging change in the kinetics of lamellar bone resorption in normal human rib, *J. Gerontol.*, 19, 462, 1964.

194. Oliver, I. and Crouch, M., unpublished data.

195. Crouch, M., Oliver, I., and Parfitt, A. M., unpublished data.

196. Crouch, M., unpublished data.

197. Parfitt, A. M., Villanueva, A. R., Mathews, C. H. E., and Aswani, J. A., Kinetics of matrix and mineral apposition in osteoporosis and renal osteodystrophy: relationship to rate of turnover and to cell morphology, in *Bone Histomorphometry*, 3rd Int. Workshop, Jee, W. S. S. and Parfitt, A. M., Eds., Armour Montague, Paris, 1981, 213.

198. Villanueva, A. R., unpublished data.

199. Parfitt, A. M., The morphologic basis of bone mineral measurements: transient and steady state effects of treatment in osteoporosis, *Miner. Electrolyte Metab.*, 4, 273, 1980.

200. Parfitt, A. M., Mathews, C., Rao, D., Frame, B., Kleerekoper M., and Villanueva, A. R., Impaired osteoblast function in metabolic bone disease, in *Osteoporosis: Recent Advances in Pathogenesis and Treatment*, DeLuca, H. F., Frost, H., Jee, W., Johnston, C., and Parfitt, A. M., Eds., University Park Press, Baltimore, 1981, 321.

201. Arnold, J. S., Trabecular patterns and shapes in aging and osteoporosis, in *Bone Histomorphometry*, 3rd Int. Workshop, Jee, W. S. S. Parfitt, A. M., Eds., Armour Montague, Paris, 1981, 297.

202. Preuss, L. and Parfitt, A. M., unpublished data.

Chapter 10

CELL KINETICS UNDERLYING SKELETAL GROWTH AND BONE TISSUE TURNOVER

Z. F. G. Jaworski, Donald B. Kimmel, and Webster S. S. Jee

TABLE OF CONTENTS

I. INTRODUCTION

The skeleton's competence as an organ of physical support depends not only upon the quantity, but also upon the microscopic and gross organization of the bone tissue. The majority of the increase in bone volume during growth results from the production of new bone tissue in subchondral and subperiosteal locations. In contrast, both the microscopic structure of bone tissue and fine adjustment of bone volume depend upon the bone turnover process, not only during growth but also after maturity. The turnover process is sustained by the self-renewing populations of osteoclasts and osteoblasts. Any increase in volume during turnover is a consequence of net excess of bone formed by osteoblasts over that resorbed by osteoclasts.

Parallel changes occur over the skeleton's life span in the microscopic and gross organization of the bone tissue and in the numerical and spatial relationship between the osteoblast and osteoclast cell populations. These changes imply that their relative numbers and rate of function mediate the adjustment of bone volume and structure to meet the changing mechanical demands of the skeleton. Structural changes that account for the failure of the skeleton in its supportive function and during various bone diseases are probably the consequence of direct or indirect effects of a given pathogenic factor on the kinetics or function of one or both of these cell populations. Clarification of the normal and pathogenic factors operating at both levels is essential for both understanding the pathogenesis and planning proper preventive and therapeutic strategies of any metabolic bone diseases.

Knowledge of the bone cell kinetics underlying the bone remodeling process both under normal conditions and during disease is still largely incomplete. This is partially because of the inapplicability of such specific methods as tritiated thymidine-based autoradiography (^{3}HTdR) in clinical medicine. However, it is also because there is a conceptual gap, in that alterations in the structure and volume of bone tissue during disease are mainly considered as consequences of altered bone cell function, probably because cell function is more easily studied.

This review will summarize our current knowledge about bone cell kinetics as it bears on clinical problems and will construct a frame of reference for additional research. We will first review the processes by which the skeleton grows, second we will discuss cell population kinetics generally and its application in the skeleton, third we will list methods and terms used in the study of bone cell population kinetics, and last we will discuss the results of their application to various natural systems and to experimental models.

II. GROWTH AND THE MATURE SKELETON

During the growth period, the skeleton simultaneously undergoes two processes: (1) the production of new bone tissue at endochondral and periosteal sites which accounts for longitudinal and transverse bone growth and (2) the turnover of bone tissue, the resorption of preexisting bone by osteoclasts, and the simultaneous deposition of new bone by osteoblasts. Only turnover persists into adult life and is of concern because of its role in producing metabolic bone disease.

Within cortical bone, turnover occurs in a form that persists through the life of the skeleton; the osteoclasts, which appear first, form a cutting cone which is subsequently refilled by the osteoblasts, which form a closing cone. In this manner, new osteons are formed. This sequence of cellular events can be represented schematically as $A_{cl} \rightarrow R \rightarrow A_{bl} \rightarrow F$, the recruitment of osteoclasts (A_{cl}) which resorb bone (R), which is followed by the recruitment of osteoblasts (A_{bl}) which form bone (F). In contrast, at the periosteal and endosteal surfaces, bone resorption and formation occur separately, though bone turnover is well coordinated. The sequence of cellular events is better represented as A_{ct}-R in some areas and A_{bt}-F in other areas.[1,2]

During growth, bone turnover in larger animals serves a dual purpose: (1) to substitute mature lamellar bone for the original woven bone and (2) to develop and preserve the gross and microanatomic structure of the growing bones. The funneling of the metaphysis, the continuous enlargement of the bone marrow cavity within the long bones,[3] and the drift of alveolar bone[4] are examples of this process.

When growth ceases, the resorption and formation processes on the periosteal and endosteal surfaces become locally coupled, bone formation following resorption *in situ*. Consequently, although the Haversian or intracortical turnover process remains more or less the same throughout life, the turnover process on the periosteal and endosteal surfaces changes in spatial organization. At this point, the function of bone turnover on the periosteal and endosteal surfaces changes to the renewal, restructuring, and autorepair of bone tissue and is accomplished in discrete units known as bone structural units (BSU). The same factors probably regulate these processes during the growing and adult periods, with their intensity and spatial organization differing because of the rapid increase in the size of the skeleton during the growth period.[5]

During growth, bone resorption and formation occur on bone surfaces in areas distant from each other, so bone balance can be considered only from the point of view of the whole bone. After maturity, all changes in bone volume may be considered in terms of the bone turnover process of individual envelopes or even in terms of the individual BSU of each envelope. For example, a cumulative negative bone balance, achieved in individual BSU on the endosteal bone surface over a full life span accounts for the enlargement of the bone marrow cavity during life and coincidentally, the age-related loss of bone after maturity.

III. CELL KINETICS

A. Types of Cell Populations

During the life of an organism, cell populations pass through various stages.[6] During embryogenesis, rapidly proliferating cells of the fertilized ovum first differentiate into three basic layers: the endoderm, the mesoderm, and the ectoderm. Subsequently, the cells of each layer, as they divide, differentiate further and provide the functional cells that constitute the tissues, organs, and systems of the organism. During growth, the progenitor cells for each tissue are recruited from the undifferentiated but variously committed stem cells of each layer which were left behind at early stages of development. The progenitor cells retain both the capacity to replicate themselves and to give rise to functional cells, during both the growth and adult periods. The functionally differentiated cells, on the other hand, lose most of their potential to proliferate.

During growth all cell populations expand, but some populations continue to renew themselves throughout life (e.g., the epithelial cells of the gut, bone marrow cells, and bone cells) while other cell populations become static (e.g., neurons of the central nervous system). After growth ceases, the self-renewing populations become stable, meaning that the recruitment and attrition of cells become balanced. Some cell populations retain the potential to expand under special circumstances (e.g., the endocrine glands under proper stimulation, connective tissue after injury, bone marrow at high altitude). The regeneration of some tissues (e.g., liver, skin, epithelium, bone) represents a special case of a potential for expanding of a cell population.

B. Factors that Determine the Size and Life Span of a Cell Population

The size of a mature functional cell population at any given time depends on the recruitment rate of new cells. This is a function of the number of stem cells differentiating into functional cells and their mean life span. This relation is expressed in the general formula:[7]

$$A = \mu \times s \tag{1}$$

in which A equals the number of cells in the population, u the recruitment rate, and s a cell's mean life span.

It follows that the size and the life span of a static cell population, like the cells of the nervous system proper after its formation, depend only on the life span of the neuron. Since the life span of a neuron approaches that of the organism itself, the nervous system persists indefinitely. However, the size of a stable self-renewing functional cell population is basically determined by the duration of cell recruitment, i.e., the period that the local precursor cells proliferate. If proliferation ceases in due course, the cell population will disappear. The time of its disappearance will depend on the mean life span of the functional cells.

C. Status of Cells Within the Functional and Progenitor Cell Populations

All cell populations are constituted and the self-renewing kinds are maintained by mitosis. A cell may divide either to reproduce its own kind (unicompartmental cell population) or it may divide to supply more fully differentiated functional cells (multicompartmental cell population). In multicompartmental self-renewing cell populations (such as osteoblasts, osteoclasts, and osteoprogenitor cells) the recruitment of functional cells implies a parallel renewal of the progenitor cell population. If this did not happen, because the life span of functional cells is limited and they usually lack the capacity to divide further, the functional cell population would become extinct. The cycles of cells that reproduce themselves, therefore, must be distinguished from the cycles of functional cells.

A proliferating cell population contains three types of cells: those within the proliferating cell cycle, known as the population's growth fraction; G_0 cells, which under appropriate stimuli respond by proliferating; and resting cells, which have no capacity for proliferation even under stimulation. Cells in the growth fraction of any homogenous population are in one of four cycle phases: S — the phase during which new DNA is made and [3]HTdR is incorporated; G_2 — the phase during which the cells complete the synthesis of structural protein necessary for cell division; M — the mitotic period composed of prophase, metaphase, anaphase, and telophase; or G_1 — the phase during which many of the enzymes and structural proteins responsible for normal cell function are synthesized and the phase during which cell differentiation is thought to occur. The proliferative cells of a population are in a constant progression: G_1-S- G_2-M. The phase durations are referred to as T_s, T_{G2}, T_m, T_{G1}, and T_c. Typically T_s lasts 5 to 10 hr, T_{Gs} 1 to 4 hr, and T_m about 0.5 to 1 hr. T_{G1} varies from a few hours to indeterminate lengths.[8]

D. Bone Cell Populations

During growth, the overall osteoblast and osteoclast populations within the skeleton (or within individual bones) increase in size as long as the net bone mass continues to increase. This factor is reflected by the increase in bone resorption and formation surfaces, both within the cortex and at the periosteal and endosteal envelopes. This overall increase may be due either to the increase in the size of local cell populations and/or in their number which occurs despite the fact that these local populations are transient. The areas of bone resorption and of compaction of the spongiosa that persists into the funneled region of the metaphysis of growing long bones are examples of this activity.

When longitudinal bone growth ceases, the overall size of these two cell populations stabilizes and an important change takes place in their spatial relationship: the osteoblasts on the periosteal and endosteal bone surface locally and sequentially replace the osteoclasts. From that time on, because bone forms in areas where resorption has taken place, a new mechanism by which bone balance is achieved and regulated becomes operative. Bone balance is determined, on one hand, by the difference between the volume of bone resorbed and formed by local osteoclast and osteoblast cell populations and, on the other hand, by the number of local sites of bone turnover operating on the bone surfaces. It thus becomes apparent that the generation, life span, and performance of local cell populations both on the bone surfaces and within the cortex continue to be a part of the system that operates throughout the whole skeleton during its full lifetime. In addition, the adult organism retains a population of precursor cells with the potential of generating functional cells; these precursor cells are also involved in primitive bone formation and turnover needed for fracture repair.

Although it has been proven that osteoclast cell populations are derived from the mononuclear phagocytic system and osteoblasts from locally resident stroma or connective tissue cells,[9,10] their respective precursor cells are difficult to distinguish from each other and are intermingled in areas of resorption or formation both in growing and adult systems. The proliferating precursor cell population is the osteoprogenitor cell population. Only ultrastructural studies permit direct distinguishing of the preosteoclasts and preosteoblasts within the osteoprogenitor cell population.[11-13]

Establishing the separate lineages of osteoclast and osteoblast populations was done by a number of alternate approaches: by the use of the newt-regenerative limb model,[14] by bone marrow and splenic transplants in osteopetrotic mice,[15-17] and by two-way limb transplants between animals with different cell karyotypes.[18-20] The derivation of osteoblasts from the stroma cells was established by using the techniques of local X-irradiation, bone marrow regeneration after depletion in the bone marrow cavity,[21] transplantion of marrow to ectopic sites,[22-25] fibroblast tissue culture,[26] and induction of heterotopic bone formation.[27]

As far as function is concerned, the resorptive capacity of an osteoclast per nucleus is significantly greater than is the bone-forming capacity of an individual osteoblast. Consequently when net bone mass is not changing, fewer osteoclast nuclei are required to remove a given amount of bone than the number of osteoblasts to form it initially. This implies that recruitment of a given number of new osteoclast nuclei demands the recruitment of proportionately more osteoblasts, meaning the change in the preosteoclast proliferation rate relative to preosteoblasts.

IV. CELL KINETIC METHODS

A. Tritiated Thymidine Labeling (Rationale)

Although certain aspects of bone cell kinetics can be inferred from cell and mitosis counts, from histomorphometric analysis of undecalcified bone sections sampled by biopsy in vivo, or from autopsy material after in vivo labeling with tetracycline, the study of bone cell kinetics proper requires special techniques. The advent of such cell markers as ³HTdR has revolutionized these studies. ³HTdR has the advantage of being able to identify about ten times as many proliferative cells as studies of mitotic indexes. Because it is metabolized within 1 hr of injection, it is incorporated into S-phase cells only briefly and leaves the cells labeled with a marker which is diluted only by cell division. In autoradiographs of cells harvested 1 hr after the injection of ³HTdR, only cells in S phase at the time of injection are found labeled. At later times, at least two kinds of labeled cells can be identified, those that differentiate after division into functional cells and those that continue to replicate. The former have a high probability of retaining the label for the rest of their life span and through all their transformations (in bone, in the cell nuclei of osteoclast and osteoblast), and the latter lose their label within several divisions because of dilution. In this manner, labeled functional cells can be observed as they undergo various morphologic changes allowing one to draw some inferences about the origin, maturation, migration, and fate of such cells. The proportion and location of labeled cells within the population at various times after labeling may also give some indication of the rate of recruitment of various functional cell types.[8]

B. Parameters Used in Studies of Proliferating Cell Populations

Parameters commonly used are the growth fraction, the labeling index, and the multiple labeling index.

Growth fraction (GF) — The fraction of cells in a cell population that is in some phase of the proliferative cycle.

Labeling index (LI) — The percentage of a homogenoeus cell population that is labeled with tritiated thymidine 1 hr after a single injection. This percentage reflects the ratio of T_s to T_c.

$$LI = T_s/T_c \times GF \times 100 \qquad (2)$$

After 1 hr, cells proceed to mitosis and give a nonrepresentative value for the ratio of T_s to T_c.

Multiple labeling index (MLI) — The percentage of cells of a population labeled with ^{3}HTdR thymidine 1 hr after the last of a series of injections that have been spaced at intervals less than T_s. If the injections are extended for a period longer than the estimated T_c, the MLI will assume a value equal to GF. This is particularly useful in studying cell populations showing very limited proliferation.

V. APPLICATIONS

A. Bone Cell Kinetics in Natural Systems

1. Bone Turnover in the Metaphysis During Endochondral Growth

In the primary and secondary spongiosa of the proximal tibia of 6-week-old rats, the principal DNA-synthesizing bone cells, the osteoprogenitor cells, are located within 1 mm of the growth cartilage-metaphyseal junction in the primary spongiosa.[28] They are found mainly between the blood vessels and the calcified cartilage septae that are covered with bone. They have spindle-to-ovoid-shaped nuclei with little cytoplasm. There are also a few proliferating osteoblasts in the same region. Data derived from observation periods up to 28 days after injection of ^{3}HTdR showed that the labeling index of osteoblasts was quite low and the labeling index of osteoclast nuclei was zero. Each rose to achieve the same fractional labeling as that initially found in the osteoprogenitor cell population. From this it was concluded that this osteoprogenitor cell population gives rise to both osteoblasts and osteoclasts.[29] Ultrastructural autoradiographic studies of these cells found that the osteoprogenitor cell population is actually composed of two cell types of strikingly different cytoplasmic appearance. One, the preosteoclast, is a cell with many mitochondria and free ribosomes, similar to a monocyte, whose features are also found in osteoclasts. The other, the preosteoblast, is a cell with a rudimentary Golgi body and much rough endoplasmic reticulum, features also found in osteoblasts.[10] Confirmatory studies noted that the former cells had a much greater ability to ingest ferritin and thorium dioxide than did the latter.[11] Another study located the same cells along endosteal surfaces of young rabbits.[12] A detailed cell kinetic study of these osteoprogenitor cells found that there were two kinetic populations within the osteoprogenitor cell populations. One, the preosteoblast, has members which divide and give one progeny to migrate into new bone tissue and the other to stay behind and differentiate into an osteoblast. The other, the preosteoclast, has members which may or may not divide, but are all included into osteoclasts within 3 to 5 days of the time they arrive in the metaphysis. New members of this population arrive through the blood vessels.[30]

There have been several studies of cell cycle time in the osteoprogenitor cell population, but because the population is inhomogenous it is not realistic to ascribe great significance to them. One such study gave a value of 33 ± 34 hr,[28] while another more recent value was 39 ± 17 hr.[30] For each study, T_s is 6 to 8 hr, T_{G2} is 2 to 4 hr, T_m is 0.5 to 1 hr, and T_{G1} is 20 to 30 hr.

2. Periosteal Growth

A forming site at the periosteum of a growing bone represents a clear picture of the kinetics and fate of preosteoblasts and osteoblasts. The newly formed bone can be readily delimited with an injection of ^{3}H-glycine, given prior to ^{3}HTdR label. In these studies, preosteoblasts were often seen labeled along with a few osteoblasts. The flow of the labeled cells was traced from preosteoblasts to osteoblasts to osteocytes over a period of several days.[31]

3. Adult Bone Remodeling

a. Background

Although the origin and lineage of bone cells during bone growth and during bone remodeling probably do not differ, bone turnover in larger adult animals is so specifically organized that the cell kinetics at the remodeling sites warrants special study. The studies of the kinetics of adult bone remodeling are discussed in some detail as they bear directly on the problems of metabolic bone disease in the adult.

The dimensions of remodeling sites within the cortex and on the periosteal or endosteal surfaces in larger mammals vary from 100 to 800 μm in length. Their architecture is best studied under low magnification in relatively thick (50 to 100 μm), undecalcified sections. The resorption sites are recognized by the presence of Howship's lacunae and formation sites by the presence of the osteoid border. The resulting units of bone tissue (BSU) are clearly recognized by the presence of cement lines delineating their boundary with other units of bone. The study of cell kinetics via ^{3}HTdR requires the use of 3-μm thick sections and much greater magnification. Consequently, proper reconstruction of the three-dimensional cytoarchitecture from the two-dimensional aspects of such sites and correct interpretation of the location of various cells and their interrelationships demands the study of serial sections.

b. Cortical Bone (Haversian Remodeling)

The recruitment and migration of cells within the evolving secondary Haversian system in young adult beagle dogs has been studied. Autoradiographs of serial longitudinal sections of rib biopsy specimens, taken from 1 hr to 15 days after the injection of ^{3}HTdR, were used. The bone sections were subjected to semiquantitative analysis for the time of appearance, number, location, and transformation of the various cells seen labeled initially.

At 1 hr after injection, the following types of labeled cells were seen:

1. Cells with large oval nuclei with pale cytoplasm located in the vicinity of the osteoclastic fronts
2. Spindle cells on the surface of the eroded cavity just behind the osteoclast and in front of the osteoblasts

3.   A few osteoblasts
4.   Elongated cells with similarly oblong nuclei, grouped in two or three loose layers, located in front of the vascular axis and between the vascular axis and the osteoblasts
5.   Endothelial cell nuclei in the tip of the vascular bud located in the cutting cone behind the osteoclastic front

Some of these labeled cells could still be seen in a similar location 24 hr after ^{3}HTdR injection.

The labeled osteoblasts increased rapidly in number from 14 to 24 hr being found mainly in the most proximal part of the closing cone and then, progressively with the passage of time, in its more distant parts. Labeled osteocytes were first seen at 9 days in the zone with the labeled osteoblasts in the distal part of the cutting cone.

During the time when the number of labeled osteoblasts increased rapidly, labeled nuclei were only rarely found in osteoclasts. Only after 3 days, when the osteoclastic front had moved about 150 μm from the position at 1 hr, could osteoclasts with two or three labeled nuclei be seen, always located quite near the advancing resorption front. Subsequently, a decline in the number of labeled osteoclastic nuclei could be observed.

The following conclusions are made.

1.   The osteoclasts appear to form by fusion of the mononuclear cells. The turnover of osteoclast nuclei is about 8%/day and the mean life span of the osteoclast nuclei is about 11.5 days. The life span of osteoclasts must have been much longer than the 15-day observation period.
2.   In contrast, both the proliferation of osteoblast precursors and osteoblast differentiation are rapid. The life span of an osteoblast, on the other hand, appears indeterminate, ranging from a few days to perhaps several weeks.
3.   It is difficult to speculate on the derivation of the local precursors of the various functional cells on the basis of this semiquantitative data. It would appear, however, that the osteoclast nuclei and the osteoblasts in the Haversian systems are not kinetically related to allow flow of nuclei from resorbing to forming cells, because ^{3}HTdR labeling appears first in the osteoblasts and at the same time or later in the osteoclast nuclei.[32]

c. Trabecular Bone

Remodeling sites in trabecular bone occupy 200 to 300 μm of bone surface in all dimensions. They usually contain both osteoclasts and osteoblasts. When such areas are small they contain mainly osteoclasts, representing newly initiated sites, or osteoblasts, representing old remodeling units approaching the end of their cycle. In most areas cells similar in appearance to osteoprogenitor cells of the growing long bone metaphysis are seen. The nuclei of the cells are predominantly elliptical in shape, whereas the nuclei of the other osteoprogenitor cell populations are a continuum of spindle-to-elliptical-shaped members. Other ultrastructural studies seem to confirm that the population consists of both preosteoclasts and preosteoblasts, as in the growing long bone metaphysis.[33] It seems that remodeling of trabecular bone, as in the metaphysis and the Haversian systems, demands the presence of both preosteoblasts and preosteoclasts to accomplish bone resorption and formation in close proximity.

At 1 hr after ^{3}HTdR injection, the labeling index of osteoprogenitor cells of the thoracic vertebral body of the young adult beagle, in which the turnover rate is about 180%/year, is 1.5%. Labeled osteoclast nuclei are seen as early as 16 hr after injection, and increase in number up to 4 days postinjection.[34]

4. Fracture Healing

After the fracture of the femur of a 5-week-old mouse, the osteoprogenitor cells of the periosteum respond by proliferating. The peak labeling index is about 25% at 32 hr after fracture. Although during the first 2 days after fracture the whole periosteum responds, later only the preosteoblasts in the immediate area around the fracture proliferate significantly above normal.[35]

5. Effect of Age

The labeling index of periosteal osteoprogenitor cells of the distal femur of mice is 8.5% at 1 week of age but only 0.07% at 8 weeks of age and thereafter. One would expect this because large numbers of osteoblasts and osteoclast nuclei are required during the growth period. In testing the effect of age upon the proliferative capacity of G_0 periosteal osteoprogenitor cells, the response to fracture in an 18-month-old mouse was somewhat slower and of smaller magnitude than that seen in a 5-week-old mouse.[36-37]

6. Periodontal Ligament

The time course of the proliferative response of periodontal ligament fibroblasts to orthodontic stimulation is about the same as that for the fracture site. The osteoblasts which eventually form bone after tooth repositioning are clearly derived from the periodontal ligament fibroblast populations, either with or without fibroblast division, by direct differentiation or release from G_2 block.[38-39] This system is quite interesting because bone resorption and bone formation are predictably activated by the movement of the tooth in the socket, something which represents a direct way to study the effect of mechanical factors on the skeleton.

7. Egg Laying Cycle of Birds

During the period of ovulation and for a few hours thereafter, there is an increasing accumulation of woven bone in the marrow cavity of long bones of hens. As the egg enters the shell gland where its calcification begins, there is an increase in resorption of the accumulated woven bone, such that it is completely removed. It was concluded from the original work that the occurrence of daily alternating resorption and formation of medullary bone required individual bone cells to completely switch their function from resorption to formation daily.[40]

At one time it was held that this model system represented an excellent area in which to pursue studies of bone cell differentiation and proliferation. It has subsequently been shown that regardless of the phase of ovulation, the same number of osteoclasts is present. It was noted that it is a cell function problem, the osteoclasts establishing ruffled borders and increasing their attachment to bone surfaces as calcium for eggshell calcification is required. As calcium is to be sequestered, they dissociate themselves from bone surfaces and begin to store membrane which may have once been a ruffled border.[41] However, this model retains great potential because changes in medullary bone volume are linked with the levels of estrogens in the body fluid. This system may permit the study of the poorly understood effects of estrogen on bone.

B. Bone Cell Kinetics in Experimental Models
1. Background

Changes in bone volume and structure that occur during the life span imply the existence of agents or factors that under normal circumstances regulate bone cell kinetics. On the other hand, structural changes that account for the skeleton's mechanical failure are associated with an excess or a deficiency of a normal factor or presence of an abnormal one. Study of direct effects of some agents may demand the study of surgically modified animals or ideally, if adequate model systems exist, the use of organ culture. The study of bone cell kinetics in disease by means of ³HTdR requires the use of animal models because, unlike tetracycline, ³HTdR cannot be used in clinical medicine.

2. Hormones

a. Corticosteroids

Osteopenia occurs both in Cushing's disease and after the administration of corticosteroids. In adults, the initial effect of corticosteroids seems to be to increase bone resorption and to decrease bone formation while a later effect is to depress the rate of bone remodeling.

It has been shown that administration of cortisone acetate to rats reduces the labeling of metaphyseal osteoblasts after 1 day and that of osteoprogenitor cells after 3 days. This is followed by a decrease in the total number of cells in the metaphysis and a retarded rate of bone elongation, consistent with the early depression of cell proliferation. After 2 or 3 days a transient drop in the number of osteoclasts occurs, which does not persist at 19 days.[42,43] The depression in bone remodeling activity in adults[44] probably reflects the reduced cell proliferation and number of osteoblasts observed during the cell kinetic study.

b. Estrogens

In adults, the fall of estrogen levels at menopause is associated with an increase in bone remodeling rates, an accelerated rate of loss of bone mass, and an increasingly negative calcium balance.[45] However, cell kinetic findings during experimental studies do not appear to shed any light on this problem.

Single injections of estrogen into mice are known to stimulate bone formation[46] and osteoblast proliferation for 1 to 2 weeks[47] without notable effect upon osteoprogenitor cells. In an organ culture system of fetal rat fibulae treated with estrogen, higher proliferation than normal is seen.[48] The labeling index of fibroblasts in the stimulated periodontal ligament and the primary spongiosa from female rats is highest during the peak concentration of circulating estrogens and lowest during their lowest phase.[49] Cumulative bone elongation is least during the days of highest estrogen levels, suggesting the possibility that chondrocytes and bone cell kinetic responses may differ.[50]

The cell kinetic findings do not seem to bear at all on the role of estrogen in postmenopausal osteoporosis because they lead one to the conclusion that estrogens stimulate cell proliferation, particularly of osteoblasts, while evidence from human studies leads one to an opposite conclusion. This signifies the obvious need for more studies of the basic effects of estrogen.

c. Parathyroid Hormone

An excess of circulating parathyroid hormone (PTH) during adulthood produces osteitis fibrosa associated with an elevated level of bone remodeling. This suggests that the hormone stimulates cell proliferation, first of osteoclasts then later of osteoblasts.[51]

In the metaphysis of a growing rat, a single injection of PTH produces a transient increase in the number of osteoclasts followed by a similar transient increase in the number of osteoblasts. If ³HTdR is injected simultaneously with PTH, most of the labeled nuclei that appear during the next 24 hr are within osteoclasts. If ³HTdR is injected 24 hr after PTH, most label appears in osteoblasts.[52]

Simulating spontaneously occurring hyperparathyroidism of humans requires a continuous infusion of hormone at low concentrations into experimental animals to mimic the constant secretory activity of the gland. In 5-month-old beagles that had been infused with 50 pg/kg/day of PTH for 15 weeks, there was a great increase in the fraction of bone surfaces undergoing resorption and formation in all regions of trabecular bone. Quantitative studies indicated that each bone remodeling unit was functioning normally, but that the number of remodeling units was at least twice normal. The labeling index of osteoprogenitor cells of the remodeling system in PTH-infused dogs was about 5% when compared to the control 2.5%.[53]

The cell kinetic studies seem to explain the observation in humans, because the increased bone cellularity in bones of hyperparathyroid patients is probably a consequence of stimulated proliferation within the osteoprogenitor cell population.

d. Vitamin D Metabolites

It is securely established that D metabolites are required to allow PTH responsiveness of the osteocyte bone lining cell pump system which is responsible for momentary control of serum calcium.[51] Many bone effects seen during D deficiency in adults are indirect, principally due to the hypersecretion of PTH which occurs as hyporesponsive osteocytes are called upon to exercise momentary control over serum calcium. Osteitis fibrosa, as a consequence of secondary hyperparathyroidism, is the first level of D deficiency. A more severe hard tissue consequence of D deficiency is mineralization defects indicating that there may be some direct effect of the D metabolites on the osteoblasts' role in mineralization.[54]

One study of the effect of $1,25(OH)_2D_3$ showed an increased number of osteoclasts when compared to their control.[55] Other studies have shown that acute administration of $1,25(OH_2)D_3$ stimulates osteoclastic activity and enhances their size in both organ culture and intact animals.[56] It remains to be established whether the D metabolites have direct effects on the cell kinetics of osteoprogenitor cells.

3. Drugs

Various drugs have been employed as treatments for Paget's disease and osteoporosis. In Paget's disease, the objective is to reduce the high rate of bone turnover which is associated with the disease by using agents such as calcitonin, mithramycin, and diphosphonates. In osteoporosis, the objective is to reduce or reverse the loss of bone mass by using agents which shift the bone balance to a less negative or even positive state.

The main drugs which have been studied by cell kinetic methods are the diphosphonates. In the rat long bone metaphysis, a low dose of EHDP produces an increase in the number of osteoclasts, which is credited to an increased rate of osteoclast differentiation. At higher doses, a rapid rate of osteoclast evolution persists, but overall osteoprogenitor cell proliferation is decreased and osteoblast numbers are decreased, indicating probable decreased preosteoblast proliferation in the presence of stimulated osteoclast differentiation.[57] In another experiment done in alveolar bone of the periodontal ligament, it was seen that EHDP at a high dose (10 mg/kg/day) inhibited both the proliferation of the periodontal ligament preosteoblasts and their rate of differentiation into osteoblasts. This is an example of a study in which the inability to identify the two subtypes of the osteoprogenitor cell population leads to unclear results.[58]

Still, these findings are quite consistent with observations made on patients with Paget's disease of bone who were treated with diphosphonates, where an increase in the number and size of osteoclasts is associated with a depression in the rate of bone turnover. There are few experiments using diphosphonates as a direct treatment for osteoporosis.

4. Diseases
a. Disuse Osteoporosis

There is no ^{3}HTdR labeling data concerning disuse osteoporosis, but indirect data suggest that the rapid bone loss which occurs following immobilization is mediated by the altered relationship between the osteoclast and osteoblast populations. Permanent bone loss appears to occur mainly as a consequence of prolonged increase in the rate of osteoclast recruitment and the inhibited recruitment of osteoblasts. Only after several months of continuous immobilization does the rate of osteoclast recruitment decline and do osteoblasts reappear. Normal bone turnover seems finally reestablished, although bone volume remains substantially and presumably permanently reduced.[59]

b. Uremia

Renal osteodystrophy is a complex disorder in which the impaired metabolism of vitamin D and subsequent secondary hyperparathyroidism play the most prominent pathogenic role. Although the course of this condition in humans has been studied extensively, only a few studies of cell kinetics during experimental uremia which might allow deeper understanding of the condition have been completed. A preliminary study of growing uremic rats sacrificed from 1 to 144 hr after [3]HTdR injection noted an increase in both the total number of labeled and unlabeled osteoclast nuclei and of labeled and unlabeled osteoblasts in the metaphysis of the growing long bone. An increase in the rate of accumulation of labeled cells of both types was also noted in the uremic animals. The increased cellularity was attributed to secondary hyperparathyroidism induced by renal insufficiency, a similar finding to that seen in adult humans. There was no alteration in the cell kinetics of the growth cartilage in uremic animals when compared to controls. This suggests that the modeling of the metaphysis and bone elongation are under different controls and respond differently to hyperparathyroidism.[60]

VI. SUMMARY AND CONCLUSIONS

Turnover is a feature of both the growing and adult skeleton. However, turnover during growth results in shape changes due to linked, but distantly placed resorption and formation, while turnover during adulthood results mainly in local tissue renewal with only minor shape changes. The structure of the units is best understood in cortical bone where it does not vary with age but is also reasonably well studied in trabecular bone.

Bone remodeling in the adult is accomplished by a continuous series of individual units going through predictably coordinated phases of resorption of preexisting bone followed by *in situ* formation of new bone. The proper completion of this process is dictated by at least two levels of control, cell kinetics (recruitment) and cell function. In the past, the primary emphasis has been on cell function, because it is relatively straightforward to study during both disease and health of both human subjects and experimental animals by the use of in vivo tetracycline labeling. The study of cell kinetics has been limited largely because the methods by which it can be directly studied are applicable only to laboratory animals, but the specific disease conditions are found only in humans. Knowledge about cell kinetics of remodeling is particularly important when considering that the overall rate of remodeling is principally a function of activation of new sites of turnover, which is dictated by cell kinetics and its controlling factors.

There is little reason to doubt that the osteoclasts and osteoblasts of the adult have the same origins as in growing animals, that is, hematogenous for osteoclasts and local connective tissue for osteoblasts. A confounding problem of past studies has been that the proliferating bone cells, the osteoprogenitor cells, look alike under the light microscope but are actually of two kinetically and morphologically different types, preosteoclasts and preosteoblasts, which are readily identified with low power electron microscopic examination. It is desirable to know the effects of various agents on the kinetics of each of these cell types, but a complete set of studies is not currently available. Bone cells are an example of a multicompartmental cell population, preosteoclasts and preosteoblasts, respectively, being the proliferative precursors of osteoclast nuclei and osteoblasts.

The most common cell kinetic tool in the experimental laboratory is ^{3}HTdR. It can be used as a single injection to identify proliferating cells by studying animals killed 1 hr after injection to find the labeling index, as a tracer of cell fates for several weeks after injection, and as a multiple label to assess the total number of proliferating cells.

Understanding metabolic bone disease and how to best treat it calls for understanding the level of dysfunction. In some cases, for example corticosteroid therapy, the problem is at least partially one of cell function. In others, like the hyperparathyroidisms and Paget's disease, it is clearly a problem of accelerated cell recruitment. In still others, particularly the osteoporoses, neither a marked cell functional nor kinetic problem has been identified. The effects of some conditions and agents on bone cell proliferation have been investigated, and in some instances, where reasonable data are available, the cell kinetic findings seem to bear on clinical problems. In other cases, particularly for the estrogens and D metabolites, a clear relationship is yet to be established.

REFERENCES

1. Rasmussen, H. and Bordier, P. J., *Physiological and Cellular Basis of Metabolic Bone Disease,* Williams & Wilkins, Baltimore, 1974.
2. Johnson, L. C., *Morphologic Analysis in Pathology in Bone Biodynamics,* Little, Brown, Boston, 1964.
3. LaCroix, P., *The Internal Remodeling of Bones in the Biochemistry and Physiology of Bone,* Vol. 3, Bourne, G. H., Ed., Academic Press, New York, 1971.
4. Baron, R. and Saffar, J. L., A quantitative study of bone remodeling during experimental periodontal disease in the golden hamster, *J. Periodontal Res.,* 13, 309, 1978.
5. Frost, H. M., *Bone Remodeling and Its Relationship to Metabolic Bone Diseases,* Charles C Thomas, Springfield, Ill., 1973.
6. Cheng, H. and Lablond, C. P., Origin, differentiations and renewal of the four main epithelial cell types in the mouse small intestine. I—V, *Am. J. Anat.,* 141, 461, 1974.
7. Frost, H. M., Relation between bone tissue and cell population dynamics, histology, and tetracycline labeling, *Clin. Orthoped. Rel. Res.,* 49, 65, 1966.
8. Cleaver, J. L., *Thymidine Metabolism and Cell Kinetics,* John Wiley & Sons, New York, 1967.
9. Jee, W. S. S. and Kimmel, D. B., *Bone Cell Origin at the Endosteal Surface in Bone Histomorphometry,* Meunier, P., Ed., Armour Montagu, Paris, 113, 1976.
10. Owen, M., Histogenesis of bone cells, *Calcif. Tissue Res.,* 25, 205, 1978.
11. Scott, B. L., Thymidine-^{3}H electron microscope radioautography of the osteogenic cells of the fetal rat, *J. Cell Biol.,* 35, 115, 1967.
12. Luk, S. C., Nopajaroonsri, C., and Simon, G. T., The ultrastructure of the endosteum: a topographic study in young adult rabbits, *J. Ultrastruct. Res.,* 46, 165, 1974.
13. Thyberg, J., Electron microscopic studies on the uptake of exogenous marker particles by different cell types in the guinea pig metaphysis, *Cell Tissue Res.,* 156, 301, 1975.
14. Fischman, D. A. and Hay, E. D., Origin of osteoclasts from mononuclear leucocytes in regenerating newt limbs, *Anat. Rec.,* 143, 329, 1962.
15. Walker, D. G., Bone resorption restored in osteopetrotic mice by transplants of normal bone marrow and spleen cells, *Science,* 190, 784, 1975a.
16. Walker, D. G., Spleen cells transmit osteopetrosis in mice, *Science,* 190, 785, 1975b.
17. Walker, D. G., Control of bone resorption by hematopoietic tissue. The induction and reversal of congenital osteopetrosis in mice through use of bone marrow and splenic transplants, *J. Exp. Med.,* 142, 651, 1975.

18. Jotereau, F. A. and LeDouarin, N. M., The developmental relationship between osteocytes and osteoclasts: a study using the quail-chick nuclear marker in endochondral ossification, *Dev. Biol.*, 63, 253, 1978.

19. Kahn, A. J. and Simmons, D. J., Investigation of cell lineage in bone using a chiera of chick and quail embryonic tissue, *Nature (London)*, 258, 325, 1975.

20. Simmons, D. J. and Kahn, A. J., Cell lineage in fracture healing in chimeric bone grafts, *Calcif. Tissue Int.*, 27, 247, 1979.

21. Patt, H. M. and Maloney, M. A., Bone marrow regeneration after local injury: a review, *Exp. Hematol.*, 3, 135, 1975.

22. Friedenstein, A. J., Piatetzy-Shaprio, I. I., and Petrakova, K. V., Osteogenesis in transplants of bone marrow cells, *J. Embryol. Exp. Morphol.*, 16, 381, 1966.

23. Friedenstein, A. J., Petrakova, K. V., Kurolesova, A. I., and Frolova, G. P., Heterotopic transplants of bone marrow, *Transplantation*, 6, 230, 1968.

24. Friedenstein, A. J. and Kuralesova, A. I., Osteogenic precursor cells of bone marrow in radiation chimeras, *Transplantation*, 12, 99, 1971.

25. Friedenstein, A. J., *Determined and Inducible Osteogenic Precursor Cells in Hard Tissue Growth, Repair, and Mineralization,* Ciba Found. Symp., Blackwell Scientific, Oxford, 169, 1973.

26. Wilson, F. D., Pool, R. R., Stitzel, K., and Momeni, M. H., The marrow interface (endosteum) potential relationship of microenvironments in the regulation of response to internal emitters, in *Health Effects of Plutonium and Radium*, Jee, W. S. S., Ed., J. W. Press, Salt Lake City, 617, 1976.

27. Buring, K., On the origin of cells in heterotopic bone formation, *Clin. Orthoped. Rel. Res.*, 110, 293, 1975.

28. Kember, N. F., Cell division in endochondral ossification, *J. Bone Jt. Surg.*, 42B, 824, 1960.

29. Young, R. W., Cell proliferation and specialization during endochrondral osteogenesis in young rats, *J. Cell Biol.*, 14, 357, 1962.

30. Kimmel, D. B. and Jee, W. S. S., Bone cell kinetics during bone elongation in the rat, *Calcif. Tissue Int.*, 32, 123, 1980.

31. Owen, M., Cell population kinetics of an osteogenic tissue, *J. Cell Biol.*, 19, 19, 1962.

32. Jaworski, Z. F. G. and Hooper, C., Study of cell kinetics within evolving secondary haversian systems, *J. Anat.*, 131, 91, 1980.

33. Baron, R., The significance of lacunar erosion without osteoclasts: studies on the reversal phase of the remodeling sequence, Jee, W. S. S. and Parfitt, A. M., Eds., *S.N.P.M.D.*, Paris, 1981, 35.

34. Kimmel, D., A light microscopic description of osteoprogenitor cells of remodeling bone in the adult. in *Bone Histomorphometry*, Jee, W. S. S. and Parfitt, A. M., Eds., S.N.P.M.D., Paris, 1981, 181.

35. Tonna, E. A. and Cronkite, E. P., Cellular response to fracture studied with tritiated thymidine, *J. Bone Jt. Surg.*, 43A, 352, 1961.

36. Tonna, E. A., The cellular complement of the skeletal system studied autoradiographically during growth and aging, *J. Biophys. Biochem. Cytol.*, 9, 813, 1961.

37. Tonna, E. A. and Cronkite, E. P., Changes in the skeletal cell proliferative response to trauma concomitant with aging, *J. Bone Jt. Surg.*, 44A, 1557, 1962.

38. Roberts, W. E. and Jee, W. S. S., Cell kinetics of orthodontically-stimulated and non-stimulated periodontal ligament in the rat, *Arch. Oral Biol.*, 19, 17, 1974.

39. Yee, J. A., Kimmel, D. B., and Jee, W. S. S., Cell kinetics of periodontal ligament fibroblasts after orthodontic stimulation, *Cell Tissue Kinet.*, 9, 293, 1976.

40. Bloom, W., Bloom, M. A., and McLean, F. C., Calcification and ossification, medullary bone changes in the reproductive cycle of female pigeons, *Anat. Rec.*, 81, 443, 1941.

41. Miller, S. C., Osteoclast cell-surface changes during the egg-laying cycle in japanese quail, *J. Cell Biol.*, 75, 104, 1977.

42. Simmons, D. J. and Kunin, A. S., Autoradiographic and biochemical investigations of the effect of cortisone on the bones of the rat, *Clin. Orthoped. Rel. Res.*, 55, 201, 1967.

43. Young, R. H. and Crane, W. A. J., Effect of hydrocortisone on the utilization of tritiated thymidine for skeletal growth in the rat, *Ann. Rheum. Dis.*, 23, 163, 1964.

44. Jett, S., Wu, K., Duncan, H., and Frost, H. M., Adrenalcorticosteroid and salicylate actions on human and canine haversian bone formation and resorption, *Clin. Orthoped. Rel. Res.*, 68, 301, 1970.

45. Nordin, B. E. C., Peacock, M., Crilly, R. G., Francis, R. M., Speed, R., and Barkworth, S., Summation of risk factors, in *Osteoporosis: Recent Advances in Pathogenesis and Treatment,* Deluca, H. F., Frost, H. M., Jee, W. S. S., Johnston, C. C., Jr., and Parfitt, A. M., Eds., University Park Press, Baltimore, 1981, 359.

46. Urist, M. R., Budy, A. M., and McLean, F. C., Endosteal bone formation in estrogen-treated mice, *J. Bone Jt. Surg.*, 32A, 143, 1950.

47. Simmons, D. J., Cellular changes in the bones of mice as studied with tritiated thymidine and the effects of estrogen, *Clin. Orthoped. Rel. Res.*, 26, 176, 1963.

48. Liskova-Kiar, M., Effect of estradiol benzoate on the proliferation of osteogenic cells in fetal rat fibulae cultures in vitro, *Rev. Can. Biol.*, 37, 35, 1978.

49. Dawson, L. R., Kimmel, D. B., Miller, S. C., and Jee, W. S. S., Estrogen effects on the periodontal ligament, *J. Dent. Res.*, 52, 136, 1973.

50. Whitson, S. W., Dawson, L. R., and Jee, W. S. S., A tetracycline study of cyclic longitudinal bone growth in the female rate, *Endocrinology*, 103, 2006, 1978.

51. Parfitt, A. M., Actions of parathyroid hormone on bone: relation to bone remodeling and turnover, calcium homeostasis and metabolic bone disease. II. PTH and bone cells: bone turnover and plasma calcium regulation, *Metabolism*, 25, 909, 1976.

52. Young, R. W., The specialization of bone cells, in *Bone Biodynamics*, Frost, H. M., Ed., Little, Brown, Boston, 1964, 117.

53. Malluche, H. H., Sherman, D., Meyer, W., Ritz, E., Norman, A. W., and Massry, S. G., Effects of long term infusion of physiologic doses of 1-34 PTH on bone, *Am. J. Physiol.*, 242, 197, 1982.

54. Parfitt, A. M., Villaneuva, A. R., Mathews, C. H. E., and Aswani, S. A., Kinetics of matrix and mineral apposition in osteoporosis and renal osteodystrophy. Relationship of rate of turnover and cell morphology, in *Bone Histomorphometry*, Jee, W. S. S. and Parfitt, A. M., Eds., S.N.P.M.D., Paris, 1981, 213.

55. Holtrop, M. E., Effects of 1,25 dihydroxy vitamin D_3 osteoclasts: studies by TEM and LM morphometry, in *Bone Histomorphometry*, Jee, W. S. S. and Parfitt, A. M., Eds., S.N.P.M.D., Paris, 1981, 213.

56. Liskova-Kiar, M. and Proschek, L., Influence of partially purified extracts of solanum malacoxylon on bone resorption in organ culture, *Calcif. Tissue Res.*, 26, 39, 1978.

57. Miller, S. C., Jee, W. S. S., Kimmel, D. B., and Woodbury, L. A., Ethane-1-hydroxy-1,1-diphosphate (EHDP) effects on incorporation and accumulation of osteoclast nuclei, *Calcif. Tissue Res.*, 22, 243, 1977.

58. Yee, J. A. and Jee, W. S. S., The effects of EHDP on the proliferation and differentiation of stimulated periodontal ligaments fibroblasts, *J. Metab. Bone Dis. Rel. Res.*, 3, 55, 1981.

59. Uthoff, H. K. and Jaworski, A. F. G., Bone loss in response to long-term immobilization, *J. Bone Jt. Surg.*, 60B, 420, 1978.

60. Kimmel, D. B., Ritz, E., Krempien, B., and Mehls, O., Bone cell kinetics and function during experimental uremia, *J. Metab. Bone Dis. Rel. Res.*, 3, 191, 1981.

Chapter 11

HISTOMORPHOMETRIC CHARACTERISTICS OF METABOLIC BONE DISEASE

Z. F. G. Jaworski

TABLE OF CONTENTS

I. INTRODUCTION

A. Morphometry and Histomorphometry

In general terms morphometry deals with the quantitative aspects of structures, i.e., applied to the skeleton it becomes a method to count and to measure its elemental structures. Two counts or measurements at different times provide the rate of change in their number or configuration and dimensions. Labeling of bone with tetracycline in vivo for the purpose of measuring the rate of new bone formation in biopsy or autopsy material is an application of that general principle.

Considering that the competence of the skeleton as an organ of physical support depends on the structure and volume of bone tissue (Table 1) the significance of bone morphometry becomes apparent.

In adults bone tissue undergoes a continuous rearrangement by means of constantly arising discrete, microscopic sites of lamellar bone tissue turnover having a standard dimension, configuration, and orientation. Morphometry applied to this remodeling process becomes histomorphometry.

A variety of factors may affect this process and consequently produce more or less characteristic changes in the quality and quantity of bone tissue. These changes in the final analysis are responsible for the symptoms of bone disease. Bone histomorphometry in this context becomes both an extension of microscopic pathology and a method of analysis. First, it allows us to recognize and better classify various metabolic bone diseases on morphological grounds, hence its diagnostic applications; second, as a method of analysis, it allows us to describe the morphologic changes in terms of altered bone tissue turnover and balance. This will lead to the identification of basic disturbances in bone cell kinetics and function underlying these diseases. Ultimately, rational therapeutic strategies depend on such information.

In this text we propose to review critically the data provided by histomorphometric analysis of both rib and iliac bone biopsy material as well as their contribution to the present understanding of major metabolic bone diseases. This requires some familiarity with the normal processes with which the abnormal can be compared. Thus, first we will review the concept of the bone remodeling system and then the type and the quality of data obtained by histomorphometry as well as the way they can be manipulated to obtain the information sought.

B. The Remodeling System

In fact, the concept of a bone remodeling system advanced by Frost in the 1960s and the realization that metabolic bone diseases in adults result from derangements of its function[2,3] were themselves an outcome of histomorphometry. Thus, the older data collected by anatomists and pathologists could be reinterpreted in the light of new data yielded by new techniques, i.e., biopsy of the rib, rapid preparation of undecalcified bone sections, in vivo tetracycline labeling, special staining techniques, and methods for counting and measuring.[2]

This system operates by means of constantly arising microscopic sites of bone turnover, basic multicellular units of bone remodeling (BMU),[2,3] in which the osteoclasts first erode a cavity of standard size, shape, and orientation on each of the bone envelopes (periosteal, intracortical or Haversian, and endosteal). It is refilled subsequently, more or less completely, by new bone produced by osteoblasts. They leave behind new basic structural units of bone tissue (BSU),[4] of which the Haversian systems in the cortex are the best example. Bone apposition occurs in three stages: osteoid deposition, its initial mineralization after a certain lag time accounting for the presence of osteoid borders, and accretion, i.e., the subsequent growth of the appatite crystals. Thus, the

Table 1
STRUCTURAL CAUSES OF MECHANICAL FAILURE OF THE SKELETON

Decreased total bone volume (irreversible osteopenia; osteoporosis)
Decreased mineralized bone volume due to expanded remodeling space (reversible osteopenia)
Increased mean age of bone tissue
Disorganized structure

mineral content may vary even under normal circumstances from one region of bone to another according to the prevailing local remodeling rate.[5] This discrepancy between the mineral content and the bone volume may be greatly exaggerated in disease.

The consequences of this system dysfunction derive from the fact that in larger adult animals, including man, its main role is to maintain the competence of the skeleton as an organ of physical support and locomotion.[3] Consequently, the remodeling sites, individually and collectively, can be viewed as instruments for: (1) a piecemeal renewal and repair of the preexisting bone tissue, (2) its reconstruction, i.e., the spatial realignment of bone tissue structures within the compacta and the realignment of the trabecular pattern in the spongious bone, and (3) the regulation of bone tissue balance; all of which may evolve differently on various bone envelopes, in various regions of the same bone envelope, and in different parts of the skeleton.[5] For instance, in adults bone is lost normally from the endosteal trabecular envelope and accumulates slowly on the periosteal envelope.[3,6,7]

Bone tissue turnover during the growth period, as well as after maturity, appears regulated primarily by mechanical forces. For its proper function, however, the system requires a supply of permissive or modulatory factors such as nutrients, hormones, etc. Ultimately it is at the level of the bone cell that the skeleton's competence or failure as an organ of support is determined, and to be understood (as shown in Figure 1) the pathogenesis of metabolic bone disease has to be studied at this level. It follows that a variety of pathogenic factors may derange the bone remodeling system and produce more or less specific structural changes responsible for the mechanical failure of the skeleton and the symptoms of bone disease.

C. Control Sites of Bone Remodeling

The identification of sites of interaction between an agent (Levels 2, 3 in Figure 1) and the remodeling system, i.e., whether it affects kinetics or function of one or both cell populations involved (Level 3), is crucial for understanding the pathogenesis of bone disease as well as for the elaboration of therapeutic strategies.

The organization and operation of the remodeling system appear to be regulated at the following sites:

1. The activation of new remodeling sites. Local proliferation of osteoclast precursors located on, or close to, the bone surfaces starts locally the sequence $A_{c1} \rightarrow R \rightarrow A_{b1} \rightarrow F$. Activation of osteoclast precursors (A_{c1}) results in the appearance of osteoclasts and bone resorption (R) followed by the activation of osteoblast precursors (A_{b1}), appearance of osteoblasts, and bone formation (F). It determines the number and location of remodeling sites.
2. The number of osteoclasts recruited within the remodeling site during the life span of the resorption phase determines the size of the resorption cavity.
3. The spatial organization of osteoclasts. The configuration of osteoclastic fronts and the distance of their advance determines the configuration and orientation of the new BSU; the role of osteoblasts in this regard appears to be passive, i.e., they refill the cavity more or less completely with new bone. Thus, the restructur-

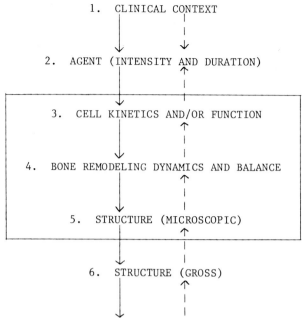

FIGURE 1.   Pathogenesis of metabolic bone disease.

ing of the bones is dependent on factors which control these aspects of osteoclast activity.

4.   The velocity of advance of the osteoclast front or linear resorption rate depending on the efficiency of the osteoclast per nucleus and its domain (resorption surface area per osteoclast).

5.   The activation of the bone formation phase. The emergence of osteoblast precursors located on or close to the bone surface causes conversion of cavities from bone resorbing into bone-forming sites. This activation may be delayed or sometimes fail altogether.

6.   The number of osteoblasts recruited within the remodeling site during the life span of the bone-forming phase balanced against the number of osteoclasts recruited during resorption determines the BSU net balance and volume.

7.   The osteoblast function in terms of both the rate of matrix synthesis and its mineralization.

The tissue level bone turnover rate depends mostly on the new site activation frequency (Item 1) while the bone balance achieved at the remodeling site level depends mostly on the Items 2 and 6, namely the respective numbers of osteoclasts and osteoblasts recruited within the individual sites. The tissue level balance (distinguished from BSU level balance) will be also influenced by the new site activation frequency (Item 1).

D. Remodeling System and Metabolic Bone Disease

Major metabolic bone diseases were first described and classified by great pathologists of the late 19th and early 20th century on the basis of autopsy material. With the advent of radiology, some of these diseases could be recognized also in vivo. Even today, the X-ray examination of the patient constitutes the essential step toward diagnosis.

The advances in biochemistry and endocrinology of Albright's era which followed revealed that metabolic bone diseases often constituted part of a larger disease entity, i.e., osteitis fibrosa cystica of hyperparathyroidism, osteomalacia of vitamin D deficiency, etc.[8] Furthermore, gross and microscopic pathology of some metabolic bone diseases and the action of some hormones could be connected, such as parathyroid hormone and osteoclastic bone resorption, estrogens and osteoblast function, vitamin D and bone mineralization. Thus one tended to explain the gross appearance of the skeleton in terms of bone balance or the difference between the bone resorption and formation, each related to a more or less well-identified etiological factor; bone resorption controlled by parathyroid hormone and bone formation by estrogens, etc.[1] With the available technology, bone balance could be estimated only by means of metabolic balance studies and rates of bone resorption and formation by radiokinetics,[9] which yield information concerning the whole skeleton. However, these methods do not allow one to distinguish between various mechanisms affecting bone balance, and therefore the effects of therapeutic agents on bone volume were often erroneously interpreted.

Toward the later 1950s it became apparent that a stalemate in treatment of some metabolic bone diseases was reached and that new approaches to their study were needed. The introduction of iliac crest biopsy by Beck and Nordin,[10] transilial biopsy by Bordier,[11] and rib biopsy by Frost[12] in the early 1960s offered the possibility of examining the bone tissue directly in vivo, first for diagnostic purposes and then to study the pathogenesis of metabolic bone diseases or to assess the effect of treatment.

The way histomorphometry contributed to the understanding of the pathogenesis of metabolic bone disease is shown in Figure 1.

In the sequence of events leading to bone disease, the pathogenic factor (Level 2 in Figure 1) affects first the cell kinetics or function of one or both cell populations involved (Level 3) which will alter the bone remodeling dynamics and bone tissue balance (Level 4) producing in turn with time microscopic alterations (Level 5) and ultimately gross structural changes (Level 6). Sooner or later, depending on the intensity and duration of the pathogenic factor, these changes will affect the mechanical competence of the skeleton (Level 7) producing symptoms such as pain, fractures, and deformities.

Thus, the Levels 3, 4, and 5 encased in the rectangle in Figure 1, until the introduction of bone biopsy, constituted a black box studied in vivo only from the outside by means of whole skeletal turnover and balance. The study of Levels 3, 4, and 5 required access to the skeleton in vivo. This was provided by introduction of iliac and rib biopsies with tetracycline-based histomorphometry.

II. HISTOMORPHOMETRY

A. Access to Bone Tissue In Vivo

In patients, practical considerations dictated the selection of the sampling site and the sample size. Beck and Nordin, who introduced bone biopsy in 1960, for the purpose of diagnosis selected the iliac crest;[10] the transilial biopsy adopted subsequently by Bordier[11] has now become the preferred method and site of bone sampling in vivo. The eleventh rib biopsy introduced by Frost in the early 1960s[12] was combined with tetracycline labeling in vivo from the beginning; it offered, therefore, the possibility to measure the bone volume in addition to static and dynamic parameters of bone remodeling and turnover. Until the early 1970s Frost and his school accumulated a large body of data based on the rib biopsy.[13] On the other hand the iliac bone biopsy, which samples mostly trabecular bone, presented specific problems which delayed the use of tetracycline labeling; consequently, at first, only bone volume and static parameters of bone remodeling such as resorption and formation surface perimeter length, osteoid volume, and osteoid seam thickness were measured. Since the early 1970s when prebiopsy tetracycline labeling became a routine procedure, the gap between our knowledge of remodeling dynamics in the compacta and in the trabecular bone began to close,[14] and the iliac bone became the preferred site of biopsy. Its wide use led to a rapid accumulation of data concerning the structural and dynamic features of a variety of metabolic bone diseases.[15] Therefore, the advantages and limitations of rib and iliac bone biopsy as well as the specificity of data which they respectively provide will be reviewed next.

Preparation of sections from rib biopsy is easy and rapid[2] providing cross sections of the cortex without major geometric distortion because the Haversian systems are sectioned very close to the perpendicular. Its major drawback was the magnitude of the surgical procedure. Presently, rib biopsies are seldom performed apart from experimental studies in animals.

Iliac bone biopsy, on the other hand, is easily obtained; its shortcoming is the unavoidable tangential cut of structures in bone sections due to the geometry of the trabeculae. This requires the use of correction factors which render the measurements and computations of data more laborious and less precise. Furthermore, some parameters such as new site activation frequency cannot be directly determined in the spongy bone.

Rib biopsy samples a fragment of tubular bone and hence offers the possibility of studying bone remodeling on three surfaces: periosteal, cortical-endosteal, and with the cortex (Haversian turnover). It does not provide a representative sample of trabecular bone which may be more important in the pathogenesis of mechanical failure.

Certain aspects of bone remodeling are envelope-specific; for example, age-related bone loss occurs mostly on the endosteal bone envelope. This applies particularly to bone cell kinetics and the control of bone balance within the remodeling sites. To a certain extent new site activation frequency is also envelope specific, but the function of cells both under normal circumstances and in disease appears to be equally affected on all bone envelopes.

From the point of view of the pathogenesis of most metabolic bone diseases and the study of chronic effects of endocrine, nutritional, circulatory, and biochemical factors on bone and the skeleton, the iliac bone biopsy data, particularly combined with the double tetracycline labeling, did more than just confirm the information which the rib biopsy data had provided. This is particularly true in regard to our present understanding of conditions such as renal osteodystrophy, Paget's disease of bone, and some others. Its most useful application is in the assessment of effects of therapy such as medical treatment of Paget's disease of bone, osteomalacia, and osteoporosis. Furthermore, while measurement of cortical bone volume (cortical thickness) can be obtained reliably by means of radiomorphometry, the trabecular bone volume, which is crucial in the pathogenesis of irreversible osteoporosis, can be only measured precisely in iliac bone biopsy material (Table 2).

B. Static Remodeling Parameters and the Concept of a Bone Remodeling Space (RSp)

The Bone Remodeling Space (RSp) can be defined as the space within the total bone volume occupied by ongoing remodeling. It consists of all bone resorption and formation sites and includes that volume resorbed at each resorption site as well as that volume remaining to be completed at each formation site at the time of biopsy.

The concept is important in two ways. First, the RSp is related to structural causes of skeletal mechanical failure in various metabolic bone diseases. For example, in some cases of renal osteodystrophy, the greatly expanded RSp at the expense of mineralized bone volume accounts for symptoms (bone pain, fracture). Second, the volume of the

Table 2
BONE MEASUREMENTS

Method	Procedure[a]	Approach[b]	Source[c]	Parameters								
							Turnover			Remodeling		Bone cell morphology
				Amount of bone	Balance	Mean age	Remodel. space	Static	Dynamic	Static	Dynamic	
Radiology	NI	D	WS	X ——————	X[d]							
Microradioscopy	NI	D	S							X		
Densitometry												
radio-	NI	ID	S	X ——	X							
Photon abstr.	NI	ID	S	X ——	X							
Neutron activ.	NI	ID	WS	X ——	X							
Microradiography	B	D	S	X ——	X	X		X		X		
Morphometry												
radio-	NI	D	S	X ——	X							
Histo-												
Decalcified	B	D	S	X ——	X							
Nondecalcified	B	D	S	X ——	X		X	X		X		X
Tetracycline	B	D	S			X			X		X	
Radiokinetics	NI	ID	WS		X				X			
Biochemistry												
metab. balance	NI	ID	WS		X							
urinary hy-pro	NI	ID	WS						X			
alk. phosphatase	NI	ID	WS						X			

[a] Procedure: invasive — biopsy (B), Noninvasive (NI).
[b] Approach: direct (D), indirect (ID).
[c] Source: sampling (S), whole skeleton (WS).
[d] X —— X: Serial measurements provide rates.

RSp is needed to properly evaluate alterations in total bone volume. For example, total bone loss may occur by two different mechanisms, one reversible (expansion of RSp) and the other practically irreversible (negative balance at the BSU). These cannot be distinguished by indirect methods of measuring bone volume relying on bone mineral content such as densitometry or external calcium balance.

Under normal circumstances within the total skeleton, the volume occupied by ongoing remodeling is insignificant and does not affect skeletal strength; in pathological situations, however, it may expand considerably and reduce the amount of mineralized, functional, bone tissue to the level of the fracture threshold.

The relation between the static remodeling parameters and the RSp and their relation to bone remodeling dynamics within the cortex is expressed in the following equations:

$$A_r = \mu_r \times \sigma_r \tag{1}$$

$$A_r = \mu_f \times \sigma_f \tag{2}$$

where the left side, i.e., the sum of A_r and A_f corresponds to the bone remodeling space, while the right side to the new site activation frequency (μ) and sites turnover period (σ). The latter together determine the ongoing remodeling dynamics. On the endosteal and periosteal envelope the following equations obtain:

$$R = \mu_r \times \sigma_r \tag{3}$$

$$F = \mu_f \times \sigma_f \tag{4}$$

In this case, the number of resorption (R) and formation (F) sites does not accurately define the RSp; neither does the fraction or absolute perimeter undergoing resorption and formation. Even the osteoid volume (in mm² or as a percent of the total bone volume) will not equal the true contribution of bone forming sites to the RSp, since a certain number of them will be still in the process of closing and have not reached the virtual bone surface (see Figure 2). The temporary deficit which the resorption sites constitute, particularly on the periosteal and endosteal surfaces, is even more difficult to assess because it is difficult to determine the virtual bone surface from the shape and length of the resorption perimeter (Figure 2). The following reasoning, however, may be applied to estimate the contribution of bone resorption sites to the RSp using the available data in the literature.[13]*

The fraction of surface perimeter covered by osteoid is about 10.5% and that covered by resorption about 3.5%, a ratio of 3:1. Osteoid volume is about 3% of trabecular bone. If the perimeter ratio holds true for volume, one may expect the formation portion of the RSp to be about 3% of trabecular volume and resorption portion about 1%. In other words, the RSp under normal circumstances should constitute approximately 3 to 4% of trabecular bone. This volume, which is perhaps underestimated, may increase considerably in pathological conditions. Thus, as data of Meunier[13] show in primary hyperparathyroidism, the reduction of mineralized trabecular bone volume to 11% (normal 23% ± 4%) corresponds largely to reversible expansion of the RSp.

---

* This is why in order to determine a change in the total bone volume (BSU-mediated balance) based on the serial bone biopsies in a patient or a group of patients with a similar condition, total bone volumes including mineralized plus remodeling bone space have to be compared. If one measures the mineralized bone volume only, or the mineralized bone volume plus the osteoid volume, one may be assessing not the difference in the total bone volume, i.e., BSU-mediated balance, but the change in the mineralized bone volume in relation to the bone remodeling space resulting from altered bone remodeling dynamics.

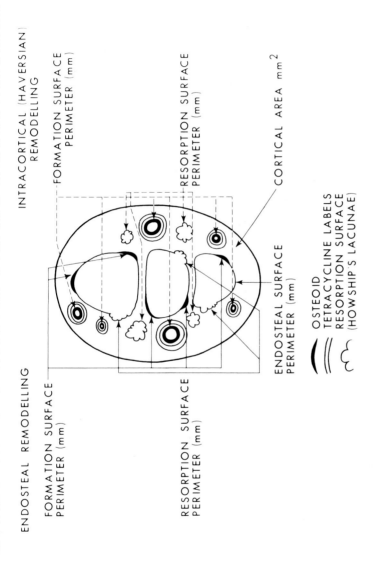

FIGURE 2. Primary data in bone histomorphometry. Schematic representation of the cortical and trabecular bone remodeling units (BSU) identified for counting and measuring to compute static and dynamic parameters of bone remodeling. On the right are the intracortical and on the left the endosteal remodeling primary data. Note that on a given surface the resorption (R) or formation (F) surface perimeter lengths are the sums of the circumferences or perimeters of individual sites (S_r and S_f); this transforms the static remodeling features (individual units of remodeling in various stages of evolution and variously distributed) into static turnover parameters.

In osteomalacia and renal osteodystrophy it may reach 50% and more. The decrease of mineralized bone volume in such instances has to be clearly distinguished from the loss of total bone volume due to BSU-mediated negative total bone balance.

While the left side of Equations 1 and 2 indicates the actual remodeling space, the right side lists the factors which determine it. The task of histomorphometric analysis is to discover to what extent each of them operates in a given case, the cell kinetics (μ factors) and/or cell function (σ factors). For BSU balance, neither the past nor ongoing one can be inferred from data from a single biopsy; the former because ongoing remodeling may bear no relation to past processes responsible for the total bone volume (mineralized plus RSp) at the time of the biopsy, and the latter because at the time of biopsy the ongoing cycles of local remodeling are not completed and hence remain indeterminate.[4]

C. Computation of Dynamic Parameters

Chapters 5 to 7 outline the computation of various measured and calculated values in bone histomorphometry. The mean distance between the two tetracycline labels from which one calculates the appositional rate ($\overline{M}$) is the one which allows calculation of all other dynamic parameters of bone remodeling and turnover.

Thus, the appositional rate allows one to calculate directly the BSU bone formation period (sigma f), and tissue level bone formation rate and to compute the linear resorption rate, the new site activation frequency (μ), the tissue level bone resorption rate, the BMU bone resorption period (sigma r), and eventually the bone tissue turnover rate.

III. PATHOGENESIS

A. Clinical Material

To study the pathogenesis of metabolic bone disease by means of biopsy one begins by tentatively grouping patients with similar manifestations and presumably similar disease. Next one attempts to establish the basic cause of the skeletal mechanical failure, whether it is a decrease in the total bone volume (mineralized plus remodeling space) or expansion of the bone remodeling space. Finally one tries to determine which cell(s) are involved and the type of underlying disturbance affecting them (kinetic or functional).

The advantage of serial studies is that the measurements of static parameters of bone histology at various times during the disease provide rates, such as the change in the true bone volume, mineralized bone volume, or remodeling space. But tetracycline labeling adds a new dimension to such studies because dynamic parameters of bone histology are compared between individual biopsies. Such changes in rates necessarily imply changes in either cell kinetics or cell function, the basic cellular phenomena underlying the diseased bone remodeling process. Thus, a picture of the bone disease emerges from analysis of bones sampled from a group of symptomatic patients who individually are at various stages of the disease.

It may be found that in some cases the morphologic features and dynamic disturbances not only change in relation to the duration and severity of the disease but differ significantly from the group. Such "odd" cases may constitute a subgroup and hint towards an unrecognized etiological factor causing the disease or a variant of the disease studied. For example, the histomorphometric analysis of biopsies from patients with apparent postmenopausal osteoporosis unexpectedly may show features of mild hyperparathyroidism or osteomalacia. The question then arises whether to label them as osteoporotics or as cases of hyperparathyroidism grafted onto age related bone loss.

B. Structural Causes of Skeletal Mechanical Failure

The major structural causes of skeletal mechanical failure are listed in Table 1. The first (decreased total bone volume) and the second (decreased mineralized bone volume due to an expanded remodeling space) appear characteristically on radiographs as osteopenias or "osteoporotic" skeletons. However, morphologically and dynamically the osteopenias are of two kinds: reversible and irreversible.

Reversible osteopenias are characterized in the compacta of the cortex by an increase in the number of resorption and formation sites (cortical porosity) and on the periosteal and endosteal (cortical and trabecular) surfaces by an increase in the bone resorption and formation surface perimeter. In other words, the mineralized bone volume is decreased because the remodeling space is expanded. This is a reversible form of osteopenia because an appropriate treatment may reduce the mineralized bone deficit by either decreasing the rate of new site activation or by accelerating osteoid formation and mineralization in existing bone-forming sites.

The two main mechanisms of reversible osteopenias are conditions associated with high rates of bone turnover (increased activation frequency of new sites) or prolonged lifespan of turnover sites either in their resorption or formation phases. The latter results from impairment of osteoclast or osteoblast function, respectively (sigma factors of Frost).[3]

Among the transient osteopenias, therefore, one can list the early phase of disuse osteoporosis,[3,16] acute reaction to injury,[3] and metabolic bone diseases such as hyperparathyroidism (primary or secondary), osteomalacia, or thyrotoxicosis. Irreversible osteopenias, on the other hand, are characterized by thinning of the cortex from within and thinning and loss of the trabeculae. They are irreversible since in such instances the loss of bone tissue implies also the disappearance of the structural component. They result from an accumulation of negative bone balances in the individual bone turnover sites occurring mainly on the endosteal surface. The mechanism here is either a disproportion between the osteoclasts and osteoblasts recruited in such sites and/or impairment of the osteoblast function, i.e., the amount of bone produced per osteoblast.

The list of irreversible osteopenias includes age-related bone loss (although it doesn't necessarily imply mechanical failure of the skeleton) and postmenopausal and senile osteoporosis.

Total bone loss, i.e., BMU mediated characterizing the irreversible osteoporoses, may be aggravated by the intervention of factors which produce reversible osteopenias. Thus, the reaction to immobilization, injury, hyperparathyroidism, thyrotoxicosis, etc., which are conditions associated with the high bone turnover, may accentuate the bone loss from the endosteal bone surface by increasing the number of the individual turnover sites each producing negative bone balance.

Furthermore, the morphologic features of irreversible and reversible osteopenias may be combined with an additive deleterious effect on the mechanical strength of the skeleton. Expansion of the remodeling space may reduce sufficiently the volume of mineralized bone tissue to the point of fractures (mild hyperparathyroidism, vitamin D deficiency, transient immobilization, etc.).

C. Total Bone Volume and Balance

As shown in Table 2 trabecular bone volume may be measured by a variety of methods. However, except for the Singh Index,[17] histomorphometry is the only one which allows us to measure it directly in vivo.

Any of the methods listed in Table 2 applied serially will allow us to determine bone balance (B), which is the difference in bone volume (Vol) measured on two occasions,

$$B = Vol_1 - Vol_2 \qquad\qquad (5)$$

To be exact, however, this equation should be rewritten as follows:

$$B = (Vol_1 + RSp_1) - (Vol_2 + RSp_2) \qquad\qquad (6)$$

in order to account for the remodeling space, RSp. Methods which measure bone volume indirectly (by its mineral content) or fail to consider the remodeling space in assessing total bone volume may be misleading as to the nature of a disease or the effect of therapy because the mineralized bone volume may transiently change in relation to the bone remodeling space without affecting the total bone volume. Similarly, measurement of external mineral balance cannot distinguish between a change in total bone volume and the mineralized bone volume due to a change in the remodeling space.

On the other hand, bone histomorphometry or radiocalcium kinetics measure bone balance as the difference between the rates of bone resorption and formation during a short time period. Neither of these methods can distinguish between the total bone balance accomplished at the BSU level and mineralized bone volume or mineral balance resulting from transients due to changes in new site activation frequency or BMU turnover time. The difference between bone resorption and formation rates reflects total bone balance only during steady-state remodeling, a situation which is difficult to verify.[4]

By performing two biopsies in succession one may combine measurements of total bone volume in different stages of a disease and measurements of bone resorption and formation rates during the time interval between the two total bone volume measurements. If the total bone volumes on two occasions are measured as the mineralized bone plus the remodeling space (usually only mineralized bone volume is measured), the mean of the bone formation rate obtained on those two occasions can be used to measure the bone resorption rate.[2]

Since the total bone balance and bone turnover rates (Figure 3) may evolve independently, the measurement of both may be useful. The only method which allows us to measure them conjointly are histomorphometry and radiokinetics, and to some extent urinary hydroxyproline and serum alkaline phosphatase.

The mechanism which sets the bone balance at individual (usually endosteal) turnover sites can be assessed only by histomorphometry in serial long range studies. It is at this site that total bone balance is determined in normal circumstances and in disease, and it is at this level that treatment aimed at the prevention or cure of involutional osteoporosis must operate. As shown by serial bone biopsies, the age-related bone loss from the endosteal surface occurs over the years accompanied by a near normal bone turnover rate.

Consequently one must conclude that with advancing age the difference between the number of osteoclasts and osteoblasts recruited in such sites changes in favour of osteoclasts. While of crucial importance in the production of age-related bone loss and postmenopausal osteoporosis, this is the least studied aspect of bone remodeling.

D. Pathology of Bone and Etiologic Factors

In a metabolic bone disease with an apparently clearcut etiology one may be tempted to ascribe all the morphologic changes to a given etiological agent and thus infer from them its mode of action on the remodeling system. Although important for diagnosis and the study of the pathogenesis of a bone disorder, the specific disturbance in the bone remodeling process (in bone cell recruitment or function) may be due secondary to disturbances produced indirectly by the etiological factor, for instance, in plasma calcium or phosphorus homeostasis. As shown in Figure 4 an agent may act directly

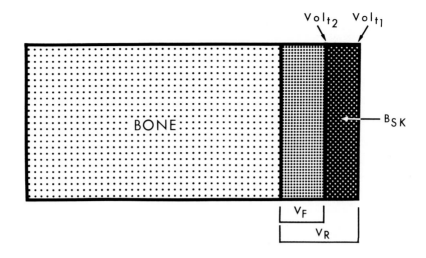

FIGURE 3.   Bone balance, turnover, and volume. Total bone balance (B_{sk}) during the time interval t_1 to t_2 can be assessed in three ways: (1) as the difference between the total bone volume (mineralized + RSp) measured directly by morphometry at t_1 and t_2 (note that with only one biopsy, total bone volume at t_1 is unknown); (2) by external mineral metabolic balance; and (3) as the difference between the amount of bone resorbed (V_r) and formed (V_f) during the time interval t_1 to t_2. Total bone balance thus can be estimated directly by means of tetracycline-based histomorphometry in a biopsy sample or indirectly by means of whole skeletal measurements such as radiocalcium kinetics. The determination of bone balance by the indirect approaches (or tetracycline-based histomorphometry if limited to mineralized bone volume) cannot distinguish between the two possible mechanisms causing it, irreversible change in BSU balance or reversible expansion of the remodeling space due to changes in μ or σ factors.

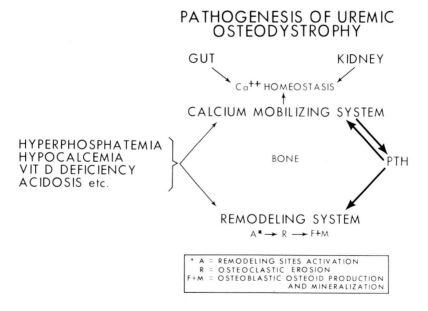

FIGURE 4.   Calcium homeostasis and the remodeling system in the pathogenesis of renal osteodystrophy.

on calcium and phosphorus homeostasis in body fluids as well as on the remodeling system; or it may directly affect the remodeling system first and produce secondary changes in calcium and phosphorus homeostasis which in turn affect the bone remodeling system; or the agent may directly affect first calcium and phosphorus homeostasis and through compensatory changes in them affect the bone remodeling system. Renal osteodystrophy is a case in point (Figure 4). An attempt to correlate the endocrine and biochemical parameters in advanced chronic renal failure with the specific disturbances in bone remodeling as revealed by histomorphometry becomes very difficult. Consequently, the study of the interaction of normal or pathogenic factors and the remodeling system requires special models or isolated systems in vivo or in vitro where the bone cells can be studied undisturbed. Such knowledge is needed not only to reconstruct the pathogenesis of bone disease but also to elaborate therapeutic strategies.

E. Sample-Based vs. Whole Skeleton Measurements

We dealt with the problem of converting static to dynamic measures of bone remodeling and turnover based on histomorphometric analysis in Section II. (Histomorphometry) leaving untouched the problem of converting two-dimensional values to three-dimensional ones (see Chapter 5) and extrapolation from the sample fragment to the whole skeleton.

In order to evaluate the need for such conversions, two questions have to be answered: (1) to what extent does a given sample represent the skeleton as a whole? and (2) what is the relation between measurements of turnover integrated over the entire skeleton (radiokinetics) and turnover in various discrete regions of the skeleton sampled by bone biopsy?

With regard to bone balance, "whole skeletal" measurements average ongoing negative BSU balance occurring on the endosteal envelope and positive balance on the periosteal envelope in various regions of the skeleton.

It has been shown by Amprino and Marotti[5] that bone turnover varies in different regions of the skeleton according to the local new site activation frequency; changes in turnover, however, occur in parallel in all regions with age and probably with systemic diseases as well. Consequently, the change in bone turnover rate found in a bone biopsy sample is probably representative of the change in the skeleton as a whole and vice versa, i.e., while the absolute values differ the magnitude of change in bone turnover as assessed by whole skeletal parameters will allow us to predict sample values.

On the other hand cell function, which is the linear bone resorption and new bone appositional rates, can only be assessed by means of histomorphometry based on tetracycline labeling. It can be validly extrapolated from a biopsy sample to the skeleton as a whole, taking into consideration the anatomy of the skeleton and the map of bone remodeling.[3] A decreased appositional rate occurring with a metabolic bone disease will likely be found on all envelopes and regions of the skeleton.

Neither bone biopsy or radiocalcium kinetics can provide information regarding the status of bone cell kinetics within the individual bone remodeling sites and whether negative balance results from increased recruitment of osteoclasts or impairment of osteoblast recruitment and function.

F. Histomorphometry and Other Methods

From the foregoing analysis, the potential and the limitations of bone histomorphometry emerge (Table 2). The importance of the information sought in a given context has to be balanced against its inconvenience (invasive vs. noninvasive methods) and reliability. Bone biopsy with histomorphometric analysis may be used in bone diseases to study: (1) diagnosis, (2) pathogenesis, (3) natural history (allowing the detection of subjects at risk), and (4) the response to therapy.

The informational value of tetracycline-based histomorphometry is very high. It offers the possibility to assess in a single sample of the rib or iliac bone almost all the parameters which otherwise have to be studied (Table 2) by different methods simultaneously; some parameters, such as the bone remodeling space or mean bone tissue age can be assessed only in nondecalcified properly stained bone sections. The drawback of histomorphometry is that because of the inconvenience of minor surgery, it cannot be applied repeatedly in the same subject or used in mass surveys.

The role of bone biopsy in the diagnosis of metabolic bone disease remains controversial. It does not need to be a routine diagnostic tool since in many instances the diagnosis of a metabolic bone disease may be made using routine clinical tools and judgment.

There is, however, a widespread agreement that bone histomorphometry is indispensible in discovering the fundamental mechanisms of human bone diseases. Further, it is vitally necessary in testing the mechanism of action of new treatments. For example, bone diseases such as involutional osteoporosis evolve very slowly and any proposed treatment will require very long periods of observation in order to document efficacy. Bone biopsy with histomorphometry may allow one to predict the likelihood of efficacy quite accurately long before any other method.

IV. DISEASES

A. Irreversible Osteopenias
1. Aging

Loss of bone with age is a universal phenomenon in humans;[6,15,18,19] the bone marrow cavity becomes larger, the cortex becomes thinner, and the trabeculae become decreased in number and size. Thus, bone is lost mostly from the endosteal surface.[6,7,20] While its onset is a matter of dispute it is easily detectable in males after the age of 50 and somewhat sooner in women who show an acceleration of bone loss around the time of the menopause.[15,21,22] The subsequent loss per decade amounts to some 10% of the value at maturity.[3]

The number of remodeling sites on the corticoendosteal or trabecular bone surface do not increase with age nor does the ratio of resorption to formation site. The appositional rate and the tissue level bone formation rate decrease; so necessarily does the bone turnover rate.

From these observations it has been concluded that age-related bone loss results from the summation of minute negative bone balances occurring within each individual microscopic site of bone turnover on the endostial surfaces[3,6] implying a preferential recruitment of osteoclasts over osteoblasts or an impairment of osteoblast function.

A progressive loss of bone from the endosteal surface, however, leads to a second mechanism which accelerates or amplifies the first and renders the bone loss irreversible. As the plates and trabeculae become thinner than the depth of the resorption thrust (some 60 to 70 μm), circular defects in the trabeculae occur which are too extensive to refill again with new bone.[7] In time, they add to a substantial net loss of trabeculae bone and to erosion or trabeculation of the cortex.[5,23]

Since the volume and microscopic organization of bone tissue within the skeleton evolve together during growth and development, the loss of bone volume in adults must be associated with increasingly irreversible loss of the structural component. The intimate mechanism of the negative bone loss within the endosteal remodeling units, however, is poorly understood and is difficult to study by any method. It must be due to a discrepancy between the number of osteoclasts and osteoblasts respectively recruited within such units or from some combination of increased osteoclast recruitment with decreased recruitment or function of osteoblasts. It is at this level of the bone remodeling system that the etiological factor of age-related bone loss in both sexes and estrogen withdrawal in women must operate.

2. Involutional Osteoporosis

The incidence of fractures increases with age and parallels the decrease in bone volume.[24-28] While bone loss is universal and all women at the time of the menopause show its acceleration, only some women develop fractures. But as a group, symptomatic osteoporotics have reduced bone volume when compared to a population of the same age and sex without fractures.[26,29,30]

The symptomatic osteoporotics have less bone either because they enter their mature life with an underdeveloped skeleton, (subsequent rate of bone loss being normal) or they lose it faster.[29] There is no reliable data allowing a choice between these two hypotheses and it is likely that both may combine.

The predominance of endosteal bone loss explains why fractures of the vertebrae and distal radius occur earlier than the femoral neck which has a strong cortical support.[26,27]

The static and dynamic histomorphometric data collected from rib and iliac bone biopsies do not differ greatly between osteoporotics and the rest of the aging population. Some consider osteoporosis an extension of ''physiological'' bone loss to the point of fracture, a quantitative but not qualitative difference from aging.[29]

The etiology of involutional osteoporosis is unclear since age and estrogen withdrawal along do not explain it. Repeated periods of increased bone turnover related to changing levels of physical activity or other intercurrent conditions may intermittently accentuate the bone loss. Decreased bone turnover and consequently increased mean age of bone tissue may account for the increased fragility of bone due to the accumulation of microdamage.[3]

More recently in patients with osteoporosis a variety of bone remodeling disturbances has been found such as increased bone turnover or increased amounts of osteoid tissue.[31] However, while these factors may be responsible for a further increase in fragility of bone by further reducing the mineralized bone volume, they are not the basic underlying cause of the bone loss.

B. Reversible Osteopenias

1. Primary Hyperparathyroidism

In advanced primary hyperparathyroidism there is an increase in the number of microscopic areas of bone resorption on the endosteal surface, increased numbers of osteoclasts and fibrosis of the marrow, along with increased numbers of resorption sites on the periosteal surface and within the cortex. Since the number of bone-forming sites is also increased,[32-35] the volume of mineralized bone is decreased because of expansion of the bone remodeling space.[35]

Surprisingly, the appositional rate is moderately but consistently decreased[32,34,36] and there are indications that osteoclast function is also impaired.[3,37] The bone turnover rate varies.[3]

The excess of parathyroid hormone appears to activate osteoclast precursors[38,39] resulting in an increased number of remodeling sites and an increased number of osteoclasts within the individual remodeling sites. It is likely that the function and recruitment of osteoblasts[38] in severe forms of hyperparathyroidism are simultaneously inhibited as reflected in the decreased appositional rate. With time, not only osteoclasts but also fibroblasts and capillaries increase in number. Finally, these lesions may fuse producing cysts of various sizes containing osteoclasts, capillaries, and fibrous tissue.

The question arises whether the low serum phosphorus associated with this disorder affects the bone apposition rate rather than parathyroid hormone itself. Bingham et al.,[38] however, observed an early impairment of osteoblasts under the influence of parathyroid hormone.

The effect of hyperparathyroidism on bone volume depends largely on the age of the patient.[34] Mineralized bone volume is reduced because of expansion of the bone remodeling space rather than the true negative (BSU) bone balance. Thus in older people who already have age-related bone loss, the mineralized bone tissue may be reduced to the point of fractures.[40]

The restoration of normal parathyroid function results in decreased osteoclast recruitment, reappearance of osteoblasts, and the return of bone morphology to normal. The cysts, however, heal very slowly if ever.

2. Osteomalacias
a. Vitamin D Deficiency

In advanced symptomatic osteomalacia, the osteoid volume is significantly increased both on the periosteal and endosteal surfaces as well as within the cortex. It may represent 20 to 40% or more of the total bone volume.[30,41-43] The mean thickness of the osteoid seams is increased.[13,30,41,44] The number of resorption sites and the resorption surface perimeter are also moderately increased. Consequently, the bone remodeling space is considerably expanded. The appositional rate and percentage of seams labeled decrease early.[4] In the advanced cases there is a marked mineralization defect accounting for the wide osteoid seams reflecting the impairment of osteoblast function.

It would appear then that vitamin D deficiency results in a tendency toward hypocalcemia inducing compensatory hyperparathyroidism (Stage 1). Initially, normal serum calcium with or without hypophosphatemia may prevail (Stage 2). At the bone level the excess of parathyroid hormone stimulates new remodeling site activation, (osteoclast precursor proliferation and osteoclast recruitment) which accounts for the increased number of resorption sites and increased bone tissue turnover. This is reflected by the increase in plasma alkaline phosphatase and urinary hydroxyproline excretion. It is difficult to know in this early stage to what extent the osteoblasts are also impaired but in primary hyperparathyroidism the appositional rate has been found moderately but consistently decreased.[32,34,36] The specific mineralization defect supervenes probably quite early. As it worsens the mean thickness of osteoid seams increases, while at the same time the signs of secondary hyperparathyroidism at the bone level appear to subside. The ensuing hypocalcemia and hypophosphatemia characterize the advanced stages of vitamin D deficiency (Stage 3). The progressive replacement of functional mineralized bone tissue by osteoid affects the mechanical competence of the skeleton accounting for pain, deformities, and fractures.[30]

The institution of appropriate therapy induces a rapid healing of osteomalacia. The mineralization front recovers early and apparently before a rise in serum calcium and phosphorus levels occur.[41]

b. Hypophosphatemic Osteomalacia

The main causes of hypophosphatemia are renal tubular defects and depletion by antacid phosphate binders.[30] The gross and microscopic bone lesions resemble that produced by vitamin D deficiency. There is a marked increase in the osteoid volume due both to an increased number of bone-forming sites as well as thickening of the osteoid seams on trabecular surfaces and within the cortex. Tetracycline uptake is markedly decreased so that many osteoid borders remain unlabeled[45] indicating that mineralization of the osteoid is markedly impaired. Resorption sites and surfaces appear normal in number and appearance. In contrast to vitamin D deficiency, there is no evidence of hyperparathyroidism.[44,46]

The following sequence of events appears to explain the pathogenesis of hypophosphatemic osteomalacia. Hypophosphatemia affects mineralization of the osteoid. Since there is no hyperparathyroidism, the activation frequency of new remodeling sites remains normal and bone turnover is not increased. But with time a marked accumulation of osteoid occurs at the expense of mineralized bone tissue in the compacta of the cortex and in the spongiosa.

Correction of hypophosphatemia induces rapid mineralization of the existing osteoid which may remain imperfect for a long period of time.

3. Uremic Osteodystrophy
a. Acute Renal Failure

There is only one reported histomorphometric study of bone (ten patients) in acute renal failure.[47] The rapid increase in circulating parathyroid hormone which follows the onset of renal failure is paralleled by an increase in the resorption surface perimeter and the size of the osteocytic lacunae, and by a similar increase in the osteoid surfaces without an impairment of the appositional rate or increase in osteoid thickness.

b. Chronic Renal Failure and Dialysis

With the advent of renal dialysis and transplantation, bone disorders previously seen in few cases of chronic renal insufficiency are now observed in large numbers following institution of maintenance dialysis in patients surviving long enough to develop them[48,49] and after successful renal transplant.[50,51]

Typical renal osteodystrophy shows lesions of hyperparathyroidism and osteomalacia mixed in various proportions.[47,48,50-57] There is a reduction of mineralized bone volume due to expansion of the remodeling space.

The disease starts with decreasing glomerular filtration accompanied by a tendency for phosphate retention in the body fluids. This creates a tendency for hypocalcemia (mechanism unclear) which secondary hyperparathyroidism tends to overcome.[49,58] For a period of time, depending on the progress of the renal disease, serum calcium remains in the normal range and high normal phosphatemia will prevail.[54] At the bone level, changes in remodeling dynamics characteristic of primary hyperparathyroidism are produced. Meunier[47] found that only 24% of patients with renal osteodystrophy on dialysis showed a disturbance in osteoid mineralization (decreased appositional rate and increased osteoid thickness). Huffer,[51] on the other hand, found increased osteoid thickness in patients nondialyzed and on dialysis which approached the findings of Frost in rib biopsies.[37] Sooner or later the disturbance in vitamin D metabolism appears to supervene, affecting the bone-forming phase of the remodeling process. Thus, defective mineralization of osteoid appears with widening of the osteoid seams and decreasing tetracycline uptake. In time this leads to the accumulation of osteoid and the mixed picture of osteitis fibrosa and osteomalacia. Various abnormalities within the collagen and preosseous matrix may be also present.[50] Correction of the hyperparathyroid component of renal osteodystrophy without eliminating vitamin D deficiency may result in transformation of the lesion into almost pure osteomalacia.

A successful renal transplant usually removes all the abnormalities leading to the production of osteitis fibrosa and osteomalacia and results in healing, provided that hyperplasia of the parathyroid glands is not excessive and there is no parathyroid adenoma. Healing is much faster and more complete than occurs following institution of dialysis and treatment with calcium, vitamin D, and phosphate binders.

c. Posttransplant Osteodystrophy

Melsen and Mosekilde studied 17 patients[42,55] up to 52.5 months after successful kidney transplantation maintained on immunosuppressive therapy. Osteoid surfaces were above normal while the appositional rate as well as the fraction of osteoid seams taking tetracycline label were moderately decreased. This would account for the decreased rate of bone formation at the BMU and tissue level and slightly thinner osteoid borders. In these patients decreased serum phosphorus may be a factor in the altered bone remodeling dynamics.[55] The commonest and most crippling complication of immunosuppressive therapy is aseptic necrosis of the bone.[55]

4. Thyrotoxic Osteopathy

Symptomatic metabolic bone disease may occur in hyperthyroidism.[59-61] There is rarefaction of the skeleton and characteristic tunneling of the cortex.[62] In older people, fractures and vertebral collapse can occur.[63]

Microscopically there is an increased number of remodeling sites[63,64] which in the cortex are unusually large.[62] The appositional and bone formation rates are markedly increased.[63,64] High bone turnover therefore characterizes thyrotoxic osteopathy.

The osteopenia is reversible and subsides following treatment of thyrotoxicosis, especially in younger patients.[62] In older people, on the other hand, the incomplete recovery of lost bone and excessive fractures may be due to the fact that thyrotoxic bone disease is grafted onto physiological bone loss.[63]

The increased bone rate of turnover in this condition is due to the effect of thyroid hormone on osteoclast and osteoblast recruitment and probably their function as well.[65] The plasma PTH level is not elevated.[63]

5. Corticosteroid-Induced Osteopathy

Fragility of the skeleton in Cushing's syndrome[53,66] or in patients treated with corticosteroids for a variety of reasons is frequently observed.[8,13] Grossly, there is a marked decrease in bone density on radiographs with vertebral collapse in older patients.[3,18,13]

Histomorphometry in advanced stages shows thinning of both the cortex and trabeculae.[13,67,68] There is an increased number of resorption sites while the number of formation sites, the osteoid volume, and osteoid thickness are all decreased.[13,68] The appositional and bone formation rates are decreased and appear to be more depressed than the bone resorption rate.[67,68]

Clinical observation[13,68] and experimental data[13,67,68] suggest that glucocorticoids inhibit the recruitment and function of osteoblasts, and stimulate first and then depress the recruitment of osteoclasts. The first phase may be due to secondary hyperparathyroidism.[69] With prolonged use of corticosteroid drugs, however, the recruitment of osteoclasts becomes also decreased[3] and bone balance becomes less negative than in early stages of the disease. The low bone turnover rate in the later stages and the consequent increase in the mean age of the bone tissue may be an added factor in the excessive bone fragility.

While the osteopenia induced by corticosteroids has been so far considered permanent, in younger patients remarkable recovery of bone volume has been observed.[13,70] This suggests that there is a fair degree of reversible expansion of the bone remodeling space (mostly resorption). In the older age group, however, the degree of the recovery of lost bone will depend on the preexisting age-related bone loss and on the primary disease which required the use of glucocorticoids. Often these patients (asthmatics, arthritics, etc.) are older and are partially immobilized from their original disease.

6. Regional Acceleratory Phenomenon and Disuse Osteoporosis

A sudden increase in bone turnover due to the activation of osteoclast precursors following trauma or injury is a well-known phenomenon.[3,8] The bone loss occurring during the first (resorption) phase of any state of increased bone turnover is reversible.[3] This reversible reaction is observed also during the first stage of immobilization leading to disuse osteoporosis of the immobilized limb or in the whole skeleton following an injury.[3,16]

The second stage is characterized by a more protracted and permanent bone loss probably due to the continued increase in osteoclast recruitment and depressed recruitment of osteoblasts.[16] If immobilization persists the individual is left with a markedly decreased total bone volume, while bone turnover returns to normal (i.e., irreversible disuse osteoporosis). It is important to distinguish the "spent stage" from the preceding one during which remobilization may result in the recovery of lost bone.[16]

C. Desorganized Structure

Since patients with Paget's disease of bone are usually first seen in their advanced stages, the earliest stages of this local bone disease have not been investigated. The disease may begin locally in the cortex as a large area of bone resorption seen on radiographs as osteolysis circumscripta. In the pelvis and short bones areas of increased bone density may be seen.[71] Numerous areas of bone resorption with characteristic wide osteoclastic fronts and contiguous areas of bone formation occupy most trabecular bone surfaces. In bone-forming sites woven bone is produced.[71,72] Markedly increased local bone turnover associated with positive bone balance is therefore characteristic of this disorder.[71,72] Since the osteoclastic fronts do not follow the usual patterns, new units of bone tend to have bizarre shapes.

It appears that the disease is due to an inordinate focal proliferation of osteoclast precursors and increased osteoclast recruitment. The excessive osteoblast recruitment and deposition of woven bone appear as a secondary repair reaction.

Effective therapy results in the normalization of osteoclast recruitment and repair by deposition of normal lamellar bone.[72,73]

REFERENCES

1. **Frost, H. M.,** *Mathematical Elements of Lamellar Bone Remodeling,* Charles C Thomas, Springfield, Ill., 1964.
2. **Frost, H. M.,** Tetracycline-based analysis of bone remodeling, *Calcif. Tissue Res.,* 3, 211, 1969.
3. **Frost, H. M.,** Bone remodeling and its relationship to metabolic bone diseases, *Orthopaedic Lecture Series,* Vol. 3, Charles C Thomas, Springfield, Ill., 1973.
4. **Jaworski, Z. F. G.,** *Parameters and Indices of Bone Resorption in Bone Histomorphometry,* Meunier, P., Ed., Armour Montegu, Paris, 1977, 193.
5. **Amprino, R. and Marotti, G.,** A topographic quantitative study of bone formation and reconstruction, in *Bone and Tooth,* Blackwood, H. J. J., Ed., Pergamon Press, New York, 1964, 21.
6. **Epker, B. N., Kelin, M., and Frost, H. M.,** Magnitude and location of bone loss with aging, *Clin. Orthoped.,* 41, 198, 1965.
7. **Jaworski, Z. F. G.,** Some morphologic and dynamic aspects of remodeling on the endosteal cortical and trabecular surface, *Isr. J. Med. Sci.,* 7, 491, 1971.
8. **Albright, T. and Reifenstein, E. C.,** *The Parathyroid Glands and Metabolic Bone Disease,* Williams & Wilkins, Baltimore, 1948.
9. **Heaney, R. P.,** Evaluation and interpretation of calcium kinetic data in man, *Clin. Orthoped.,* 31, 153, 1963.
10. **Beck, J. S. and Nordin, B. E. C.,** Histological assessment of osteoporosis by iliac crest biopsy, *J. Pathol. Bacteriol.,* 80, 391, 1960.

11. **Bordier, Ph., Matrajt, H., Miravet, L., and Hioco, D.,** Mesure histologique de la masse et de la resorption des travees osseuses, *Pathol. Biol.,* 12, 1238, 1964.

12. **Sedlin, E. D., Frost, H. M., and Villaneuva, A. R.,** The eleventh rib biopsy in the study of metabolic bone disease, *Henry Ford Hosp. Med. Bull.,* 11, 217, 1963.

13. **Bressot, C., Courpron, P., Edouard, C., and Meunier, P.,** Histomorphometrie des Osteopathies Endocriniennes, Lyon, 1976, 117.

14. **Meunier, P., Vignon, G., and Vauzelle, J. L.,** Methodes histologiques quantitatives en pathologie osseuse, *Rev. Lyon Med.,* 18, 133, 1969.

15. **Garn, S. M., Rohmann, C. G., and Wagner, B.,** Bone loss as a general phenomenon in man, *Fed. Proc.,* 26, 1729, 1967.

16. **Uhthoff, H. K. and Jaworski, Z. F. G.,** Bone loss in response to long term immobilization, *J. Bone Jt. Surg.,* 60B, 420, 1978.

17. **Singh, M., Riggs, B. L., Beabout, J. W., and Jowsey, J.,** Femoral trabecular pattern index for evaluation of spinal osteoporosis, *Ann. Intern Med.,* 77, 63, 1972.

18. **Exton-Smith, A. N., Millard, P. H., Payne, P. R., and Wheeler, E.,** Pattern of development and loss of bone with age, *Lancet,* 2, 1154, 1969.

19. **Morgan, D. B. and Newton-John, H. F.,** Bone loss and senescence, *Gerontologia,* 15, 140, 1969.

20. **Arnold, J. S. and Wei, C. T.,** Quantitative morphology of vertebral trabecular bone, in *Radiobiology of Plutonium,* Stover, B. J. and Jee, W. S. S., Eds., University of Utah, Salt Lake City, 1972, 333.

21. **Meema, H. E., Bunker, M. L., and Meema, S.,** Loss of compact bone due to menopause, *Obstet. Gynecol.,* 26, 333, 1965.

22. **Nordin, B. E. C., MacGregor, J., and Smith, D. A.,** The incidence of osteoporosis in normal women: its relation to age and the menopause, *Q. J. Med.,* 35, 25, 1966.

23. **Bromley, R. G., Dockum, N. L., Arnold, J. S., and Jee, W. S. S.,** Quantitative histological study of human lumbar vertebrae, *J. Gerontol.,* 21, 537, 1966.

24. **Alfram, P.,** An epidemiologic study of cervical and trochanteric fractures of the femur in an urban population, *Acta Orthopaed. Scand.,* 65, 1, 1964.

25. **Arnold, J. S.,** Amount and quality of trabecular bone in osteoporotic vertebral fractures, *Clin. Endocrinol. Metab.,* 2, 221, 1973.

26. **Bauer, G. C. H.,** Epidemiology of fractures, in *Osteoporosis,* Brazell, U. S., Ed., Grune & Stratton, New York, 1970, 153.

27. **Stewart, I. M.,** Fracture of neck of femur, *Br. Med. J.,* 2, 922, 1957.

28. **Stevens, J. and Abrani, G.,** Osteoporosis in patients with femoral neck fractures, *J. Bone Jt. Surg.,* 468, 24, 1964.

29. **Doyle, F.,** Involutional osteoporosis, *Clin. Endocrinol. Metab.,* 1, 143, 1972.

30. **Meunier, P., Courpron, P., Edouard, C., Bernard, J., Bringuier, S., and Vignon, G.,** Physiological senile involution and pathological rarefaction of bone, *Clin. Endocrinol. Metab.,* 2, 239, 1973.

31. **Teitelbaum, J. L., Rosenberg, E. M., Richardson, C. A., and Avioli, L. V.,** Histological study of bone from normocalcemic post menopausal osteoporotic patients with increased circulated parathyroid hormone, *J. Clin. Endocrinol. Metab.,* 42, 537, 1976.

32. **Bordier, P. J.,** Osteoid mineralization defect in primary hyperparathyroidism, *Clin. Endocrinol.,* 2, 377, 1973.

33. **Byers, P. D. and Smith, R.,** Quantitative histology of bone in hyperparathyroidism: its relation to clinical features, x-ray, and biochemistry, *Q. J. Med.,* 40, 471, 1971.

34. **Meunier, P., Vignon, G., Bernard, J., Edouard, C., Courpron, P., and Porte, J.,** La lecture quantitative de la biopsie osseuse, moyen de diagnostic et d'etude de 106 hyperparathyroidies primitives, secondaires et paraneoplastiques, *Rev. Rhoum.,* 39, 635, 1972.

35. **Merz, W. A., Olah, A., Schenk, R. K., Dambacher, M. A., Guncava, J., and Haas, H. G.,** Bone remodeling in primary hyperparathyroidism, *Isr. J. Med. Sci.,* 7, 494, 1971.

36. **Wilde, C. D., Jaworski, Z. F. G., Villaneuva, A. R., and Frost, H. M.,** Quantitative histological measurements of bone turnover in primary hyperparathyroidism, *Calcif. Tissue Res.,* 12, 137, 1973.

37. **Villanueva, A. R., Jaworski, Z. F. G., Hitt, O., Sarnsethsiri, P., and Frost, H. M.,** Cellular-level bone resorption in chronic renal failure and primary hyperparathyroidism, *Calcif. Tissue Res.,* 5, 288, 1970.

38. **Bingham, P. J., Brazell, I. A., and Owen, M.,** The effect of parathyroid extract on cellular activity and plasma calcium levels in vivo, *J. Endocrinol.,* 45, 387, 1969.

39. **James, L., McGuire, C., Sandy, C., and Marks, J. R.,** The effects of parathyroid hormone on bone cell, structure and function, *Clin. Orthoped. Rel. Res.,* 100, 392, 1974.

40. **Dauphine, R. T.,** Back pain and vertebral crush fractures. An unemphasized mode of presentation for primary hyperparathyroidism, *Ann. Int. Med.,* 83, 365, 1975.

41. **Bordier, P., Matrajt, H., Hioco, D., Hepner, G. W., Thomson, G. R., and Booth, C. C.,** Subclinical vitamin D deficiency following gastric surgery, *Lancet,* 1, 437, 1968.

42. Melsen, F. and Mosekilde, L., Dynamic studies of trabecular bone formation and osteoid maturation in normal and certain pathological conditions, *Metab. Bone Dis. Rel. Res.*, 1, 45, 1978.

43. Mosekilde, L. and Melsen, F., Anticonvulsant osteomalacia determined by quantitative analyses of bone changes. Population study and possible risk factors, *Acta Med. Scand.*, 199, 349, 1976.

44. Jaworski, Z. F. G., Pathophysiology, diagnosis and treatment of osteomalacia, *Orthoped. Clin. of N. Am.*, 3, 623, 1972.

45. Frame, B., Arnstein, A. R., Frost, H. M., and Smith, W. R., Resistant osteomalacia, *Am. J. Med.*, 38, 134, 1965.

46. Fraser, D., Hooh, S. W., Kind, H. P., Holick, M. F., Tanaka, Y., and DeLuca, H. F., Pathogenesis of hereditary vitamin D-dependent rickets, *N. Engl. J. Med.*, 289, 817, 1973.

47. Meunier, P., Edouard, C., Bressot, C., Valat, J. N., Courpron, Ph., and Zech, P., Histomorphometrie osseuse dans l'insuffisance renale aigue et chronique, *J. d'Urol. Nephrol.*, 12, 931, 1975.

48. Garner, A. and Ball, J., Quantitative observations on mineralised and unmineralised bones in chronic renal azotaemia and intestinal malabsorption syndrome, *J. Pathol. Bacteriol.*, 91, 545, 1966.

49. Stanbury, S. W., Azotemic renal osteodystrophy, *Clin. Endocrinol. Metab.*, 1, 267, 1972.

50. Avioli, L. V. and Teitelbaum, S. L., *The Renal Osteodystrophies in the Kidney*, Brenner, B. M. and Rector, F. C., Eds., W. B. Saunders, Philadelphia, 1976, 1562.

51. Huffer, W. E., Kuzeka, D., and Popovtzer, M. M., Metabolic bone disease in chronic renal failure, *Am. J. Pathol.*, 3, 365, 1975.

52. Duursma, S. A., Visser, W. J., and Nuo, L., A quantitative histological study of bone in thirty patients with renal insufficiency, *Calcif. Tissue Res.*, 9, 216, 1972.

53. Ellis, H. A. and Peart, K. M., Azotaemic renal osteodystrophy a quantitative study of iliac bone, *J. Clin. Pathol.*, 26, 83, 1973.

54. Jaworski, Z. F. G., Morphology and dynamics of bone remodeling in chronic renal failure, in *Proc. 5th Int. Congr. Nephrol. Mex.*, Vol. 3, Villareal, H., Ed., S. Karger, Basel, 1974, 149.

55. Melsen, F. and Nielsen, H. E., Osteonecrosis following renal allotransplantation, *Acta Pathol. Microbiol. Scand (A)*, 85, 99, 1977.

56. Sherrard, D. J., Baylink, D. J., Wergedal, J. E., and Maloney, N. A., Quantitative histological studies on the pathogenesis of uremic bone disease, *J. Clin. Endocrinol. Metab.*, 39, 119, 1974.

57. Teitelbaum, S. L., Hruska, K. A., Sheiber, W., Debnam, J. W., and Nichols, S. H., Tetracycline fluorescence in uremic and primary hyperparathyroid bone, *Kidney Int.*, 12, 366, 1977.

58. Reiss, E. and Canterbury, J., Genesis of hyperparathyroidism, *Ann. J. Med.*, 50, 679, 1971.

59. Follis, R. H., Jr., Skeletal changes associated with hyperthyroidism, *Johns Hopkins Med. J.*, 92, 405, 1953.

60. Laake, H., Osteoporosis in association with thyrotoxicosis, *Ann. Med. Scand.*, 151, 229, 1955.

61. Nielsen, H., The bone system in hyperthyroidism. A clinical and experimental study, *Acta Med. Scand.*, 142 (Suppl. 266), 783, 1952.

62. Meema, H. E. and Schatz, D. I., Simple radiologic demonstration of cortical bone loss in thyrotoxicosis, *Radiology*, 97, 9, 1970.

63. Meunier, P. J., Bianchi, G. G. S., Edouard, C. M., Bernard, J. C., Courpron, Ph. and Vignon, G. E., Bony manifestations of thyrotoxicosis, *Orthoped. Clin. of N. Am.*, 3, 745, 1972.

64. Melsen, F. and Mosekilde, L., Morphometric and dynamic studies of bone changes in hyperthyroidism, *Acta Pathol. Microbiol. Scand. (A)*, 85, 141, 1977.

65. Jowsey, J. and Detenbeck, L. C., The importance of thyroid hormones to bone metabolism and calcium homeostasis, *Endocrinology*, 85, 87, 1969.

66. Riggs, B. L., Kelly, P. J., Jowsey, J., and Keating, F. R., Skeletal alterations in hyperparathyroidism: determination of bone formation, resorption and morphologic changes by microradiography, *J. Clin. Endocrinol.*, 25, 777, 1965.

67. Duncan, H., Hanson, C. A., and Curtiss, A., The different effect of soluble and crystalline hydrocortisone on bone, *Calcif. Tissue Res.*, 12, 159, 1973.

68. Klein, M., Villaneuva, A. R., and Frost, H. M., A quantitative histological study of rib from 18 patients treated with adrenal cortical steroids, *Acta Orthopaed. Scand.*, 35, 171, 1965.

69. Jee, W. S. S., Roberts, W. E., Park, H. Z., Julian, G., and Kramer, M., Interrelated effects of glucocorticoid and parathyroid hormone upon bone remodeling, in *Calcium, Parathyroid Hormone and the Calcitonins*, Talmage, R. V. and Munson, P. L., Eds., Excerpta Medica, Amsterdam, 1971, 430.

70. Skeels, R. F., Reversibility of osteoporosis in Cushing's disease: case report, *J. Clin. Endocrinol. Metab.*, 18, 61, 1958.

71. Harris, E. D., Jr. and Krane, S. M., Paget's disease of bone, *Bull. Rheum. Dis.*, 18, 506, 1968.

72. Meunier, P., La maladie osseuse de Paget. Histologie quantitative, histopathogenie et perspectives therapeutiques, *Lyon Med.*, 233, 839, 1975.

73. Khairi, M., Rashid, A., Johnston, C. C., Jr., Altman, R. D., Wellman, H. N., Serafini, A. N., and Sankey, R. R., Treatment of Paget's disease of bone (osteitis deformans). Results of a one-year study with sodium etidronate, *JAMA,* 230, 562, 1974.

74. Dunnill, M. S., Anderson, J. A., and Whitehead, R., Quantitative histological studies on age changes in bone, *J. Pathol. Bacteriol.,* 94, 275, 1967.

75. Merz, W. A. and Schenk, R. K., Quantitative structural analysis of human cancellous bone, *Acta Anat. (Basel),* 75, 54, 1970a.

76. Merz, W. A. and Schenk, R. K., A quantitative histological study of bone formation in human cancellous bone, *Acta Anat. (Basel),* 76, 1, 1970b.

77. Wakamatsu, E. and Sisson, H. A., The cancellous bone of the iliac crest, *Calcif. Tissue Res.,* 4, 147, 1969.

78. Becker, F. O., Eisenstein, R., Schwartz, T. B., and Economon, S. G., Needle bone biopsy in primary hyperparathyroidism, *Arch. Intern. Med.,* 131, 650, 1973.

79. Bordier, P., Arnaud, C., Hawker, C., Tun Chot, S., and Hioco, D., Relationship between serum immunoreactive parathyroid hormone, osteoclastic and osteocytic bone resorptions and serum calcium in primary hyperparathyroidism and osteomalacia, *Excerpta Medica Int. Congr. — Ser. 1973,* 270, 222, 1973.

80. Jowsey, J., Bone histology and hyperparathyroidism, *Clin. Endocrinol. Metab.,* 3, 267, 1974.

81. Jung Kuei, W. and Arnold, J. S., Staining osteoid seams in thin slabs of undecalcified trabecular bone, *Stain Technol.,* 45, 193, 1970.

82. Mazess, R. B., Judy, P. F., Wilson, C. R., and Cameron, J. R., Progress in clinical use of photon absorptiometry, in *Clinical Aspects of Metabolic Bone Disease,* Frame, B., Parfitt, A. M., and Duncan, H., Eds., Excerpta Medica, Amsterdam, 1973, 37.

83. Meema, H. E., The combined use of morphometric and microradioscopic methods in the diagnosis of metabolic bone diseases, *Radiologe Berl.,* 1313, 111, 1973.

84. Meema, H. E. and Meema, S., Cortical bone mineral density versus cortical thickness in the diagnosis of osteoporosis. A roengenologic-densitometric study, *J. Am. Geriatr. Soc.,* 17, 120, 1969.

85. Nordin, B. E. C., International patterns of osteoporosis, *Clin. Orthoped.,* 45, 17, 1966.

86. Sorenson, J. A. and Cameron, J. R., A reliable in vivo measurement of bone mineral content, *J. Bone Jt. Surg.,* 49, 481, 1967.

87. Wu, K., Schubeck, K. E., Frost, H. M., and Villaneuva, A. R., Haversian bone formation rates determined by a new method in a Mastodon, and in human diabetes mellitus and osteoporosis, *Calcif. Tissue Res.,* 6, 204, 1970.

88. Jaworski, Z. F. G., Lok, E., and Wellington, J. L., Impaired osteoclastic function and linear bone erosion rate in secondary hyperparathyroidism associated with chronic renal failure, *Clin. Orthoped.,* 107, 298, 1975.

89. Benoit, F. L., Theil, G., and Watren, R. H., Hydroxyproline excretion in endocrine disease, *Metabolism,* 12, 1072, 1963.

Chapter 12

METABOLIC BONE DISEASES AS EVALUATED BY BONE HISTOMORPHOMETRY

F. Melsen, L. Mosekilde, and J. Kragstrup

TABLE OF CONTENTS

I. INTRODUCTION

Histomorphometrically metabolic bone diseases can be characterized by the amount and structure of cortical and trabecular bone; the amount, extent, and width of osteoid seams; and the extent of osteoclastic resorption surfaces. These *static* parameters, however, do not give any information on the pathogenic alterations in bone remodeling leading to the observed condition. An increase in the surface extent of bone resorption or bone formation, for instance, may be caused by an increase in the formative rate of new remodeling cycles (activation frequency) and/or by a prolongation of the resorptive and formative phases (life span), respectively, of the remodeling cycle. Furthermore, as seen in Figure 1, variations in the amount of osteoid may be caused by alterations in the extent and/or in the mean width of the osteoid seams. The surface extent of osteoid is, as mentioned above, proportional to the activation frequency and the life span of bone-forming sites. The mean osteoid seam width is determined by the product of the appositional rate of osteoid and the mineralization lag time (osteoid maturation period), where the mineralization lag time denotes the time interval between the formation of osteoid and its subsequent mineralization. Information on these *dynamic* variables of bone remodeling cannot be obtained from histomorphometric analyses of bone samples without a bone marker introducing the dimension time^{-1} in the measurements. Several markers exist but tetracycline[8] is the only bone marker suitable for human studies (see Chapter 2). Following *intravital* double labeling with tetracycline, bone formation rates and indirectly resorption rates can be estimated at different levels of organization (cellular level, BMU level, and tissue level) and the mineralization lag time can be calculated.

Bone remodeling can be described by the formation rate of new remodeling cycles and the duration of the resorptive and formative phases of the cycles. In order to obtain information on these variables it is necessary to determine the amount and the thickness of bone turned over per remodeling cycle. These structural bone units are well known in cortical bone as osteons. In trabecular bone a less well-defined profile, the packet, has been identified as an often crescent-shaped image outlined by the trabecular bone surface and a cement line in the depth of the trabecular bone.

The following contains a description of the stereological background, details of the tetracycline labeling schedule, biopsy procedure, preparation procedure, sampling procedure and measuring principles used at our institutes, and a description of the results obtained in various metabolic diseases affecting bone.

II. TECHNICAL BACKGROUND

A. Stereological Background

Stereology (see Chapter 5) is a body of mathematical methods relating three-dimensional parameters defining the structure to two-dimensional measurements obtainable on sections of the structure.[44] Theoretically, the main limitations are transformations

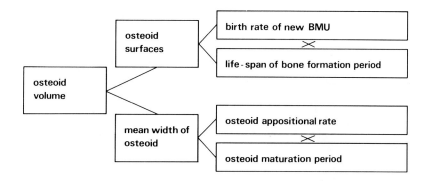

FIGURE 1. Schematic diagram showing the variables, which cause changes in the amount of osteoid in trabecular bone. BMU = basic multicellular unit.

of number and degree of connectedness from two-dimensional observations. Unbiased sampling procedures and measuring methods are the main preconditions for reliable stereological transformations. Ideally representative specimens should be sectioned randomly and the measurements performed by uniform or random sampling procedures.[15,20]

B. Tetracycline Labeling

Tetracycline antibiotics given intravitally are incorporated in actively mineralizing bone surfaces.[8] In undecalcified bone sections this labeling can be identified as bright fluorescent lines in the fluorescent microscope. Different labeling procedures have been proposed. We use 600 mg of demethylchlortetracycline given orally day 14, 13, 4, and 3 before biopsy. All double-labeling schedules will give some single-labeled surfaces (see Chapter 8) due to the nature of the remodeling system and to what has been termed the ON-OFF phenomenon. If one of the labelings "hits" a remodeling site in its formative period the other label may have been given before the formation period starts or after it has stopped producing a single-labeled surface. The occurrence of single-labeled surfaces due to this mechanism depends on the labeling interval and the life span of the formative sites. Furthermore, it has been shown that the functional state of bone-forming (i.e., osteoid covered) surfaces may vary, some being active and some inactive (ON-OFF phenomenon). If one of two labels hits an ON period and the other an OFF period a single-labeled surface is produced. Besides these mechanisms a short-labeling interval relative to the bone appositional rate may lead to no true or detectable separation of the markers.[8,9] The 10-day labeling interval suggested by Frost[8,9] has in our hands given double labels that could easily be discriminated in most metabolic states. Whatever the mechanism is for single labeling we feel that some bone formation occurs at these sites in the labeling period and we have therefore chosen to include all labeled surfaces (single- and double-labeled) in the determination of the surface extent of active mineralization.[26]

C. Biopsy Procedure

The iliac crest is now used by most investigators as a standard biopsy site because of the easy approach to the region, the possibility of repeated biopsy, and the fact that the specimen includes both cortical and trabecular bone. Because of a spatial variability in bone mass in the vertical direction[4] we prefer a horizontal biopsy 2 cm below the summit and 2 cm behind the anterior superior spine of the iliac bone. Since no difference in bone mass or remodeling has been found between the right and left iliac crests,[3,20] both sides can be used for biopsies. This is essential for evaluation of the

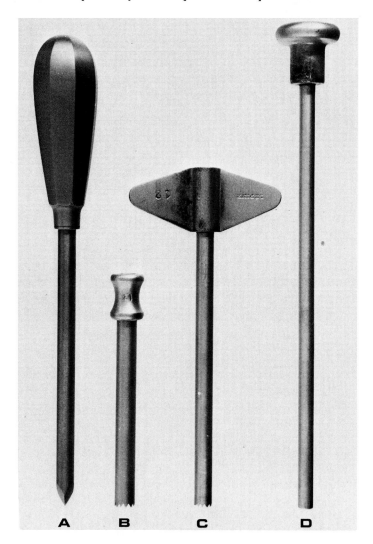

FIGURE 2.   Bordier's trephine for iliac crest biopsies. See text for further explanation.

result of a given therapy or of the natural cause of a disease. A biopsy taken at one site may affect the bone remodeling in that region (RAP phenomenon), therefore it is recommended that repeat biopsy should be obtained from the contralateral side.

In each individual two specimens are removed from the standard site. The patient is sedated with meperidine and diazepam intravenously, and a local anaesthetic is used to infiltrate the skin, muscle, and periosteum. The biopsy instrument (the Bordier trephine[1] modified by P. Meunier) is shown in Figure 2. Skin, muscle, and periosteum are incised. The trochar (A) and the external cylinder with a serrated edge (B) are introduced perpendicular to the bone surface. After removal of the trochar, the internal cylinder with a serrated edge (C) is passed down inside B and a transiliac bone specimen with a diameter of 8 mm is obtained by rotating the internal cylinder (C) through the iliac crest. The pin (D) is used to expel the biopsy from the instrument. The wound is sutured and hemostasis is obtained by use of a compress. Very few complications have been reported after this technique.[5]

D. Histological Preparation Technique

To prevent shrinkage and distortion and to maintain the differentiation between mineralized and unmineralized bone, one of the specimens is embedded undecalcified in methylmethacrylate. Using the Jung® Microtome (Model K) seven 7-μm-thick sections are cut and stained with Goldner trichrome for resorption parameters and with Masson trichrome for osteoid. Further 20-μm-thick sections are cut and mounted unstained for fluorescent microscopy.

The second bone specimen is decalcified in nitric acid or EDTA, paraffin embedded, cut, and stained conventionally with hematoxylin-eosine for determination of the periosteocytic lacunar size.

E. Sampling Design

Since biopsies are obtained for diagnostic purposes they should optimally comprise the two cortical layers and interjacent trabecular bone, and are, therefore, cut in a plane perpendicular to the cortical surface but with a random rotation. The sectioning is, thus, not performed completely at random. This deviation from the preconditions of unbiased stereology does not, however, influence the histomorphometric estimates significantly.[16,20] Sections spaced by an average of 100 μm are selected and measurements performed in microscopic fields sampled by a random[20] or uniform[16] design.

F. Measuring Methods

The following five types of measurements are generally used in our laboratory:

1. Simple counting of one specific structure as a fraction of the total number of structures (number of canals in cortical bone with resorption, osteoid, or tetracycline).
2. Simple measurements of width or length, or both, of a given structure (osteoid seam width, appositional rate, mean wall thickness, and periosteocytic lacunae).
3. Relative surface measurements (surface extent of osteoid, tetracycline, osteoblasts, Howship's lacunae, and osteoclasts) are carried out using an integrating grid with a number of lines. The surface structure to be measured as a fraction of the total surface extent is determined as the ratio of the number of intersections of the grid lines with the specific surface structure over the total number of surface intersections. This determination is more easily performed manually than by automatic devices.
4. Relative area measurements (amount of bone or osteoid and cortical porosity) are performed using a reticle having a certain number of points. The area to be measured as a fraction of the test area is the number of points projected over the area as a fraction of the total number of test points.
5. Determination of surface to volume ratio or surface density (trabecular bone surface per total trabecular bone volume) can be carried out by various integration grids with known constants or by automatic image analyzing devices. This variable expresses a property of the structure that is important for conversion of surface-related parameters to volume-related parameters.

G. Data Transformation

Two-dimensional information obtained by simple counting and point counting can, if the estimates are expressed in fractions, be transformed directly to the third dimension. Simple measurements on the other hand cannot directly be transformed since the sectional plane may influence the image of the three-dimensional structure. However, based on the geometric probability density function, measurements of the mean apparent width of uncurved plates can be transformed to the three-dimensional thickness

Table 1
CONVERSION OF SYMBOLS AND TERMINOLOGY

Aarhus	Changed to	Terminology	Dimensions
$t_{vfract(b)}$	V	Trabecular bone volume	$\mu m^3/\mu m^3$
$t_{vfract(o)}$	V_{osf}	Relative osteoid volume	$\mu m^3/\mu m^3$
AVO	V_{os}	Osteoid volume	$\mu m^3/\mu m^3$
$S_{fract(f)}$	S_f	Fractional formation surface	$\mu m^2/\mu m^2$
$u_{wf}{}^a$	$\overline{OSW}$	Osteoid seam width	μ
$u_m/{}_t{}^a$	$\overline{M}$	Appositional rate	$\mu m/day$
$S_{fract(lab)}{}^b$	S_{fa}	Fractional active formation surface	$\mu m^2/\mu m^2$
$S_{vf}{}^{a,b}$	F_f	Bone formation rate, formation surface referent	$\mu m^2/\mu m^2$
$S_{vf(BMU)}{}^{a,b}$	F_b	Bone formation rate, BMU surface referent	$\mu m^3/\mu m^2$
$t_m{}^b$	MLT	Mineralization lag time	days
$S_{fract(r)}$	S_r	Fractional resorption surface	$\mu m^{2'}/\mu m^2$
$S_{vr(BMU)}{}^{a,b}$	R_b	Bone resorption rate, BMU surface referent	$\mu m^3/\mu m^2$
$\mu_{mwt}{}^a$	MWT	Mean wall thickness	μm
Sigma-f^b	Sigma-f	Bone formation period	Days
Sigma-r^b	Sigma-r	Bone resorption period	Days
Sigma-t^b	Sigma	BMU Life span	Days

Cortical Bone

$C_{vfract(b)}$	V_c	Cortical bone volume	$\mu m^3/\mu m^3$
$C_{cfract(r)}$	C_r	Fraction of canals with resorption	—
C_{APol}	A_L	Mean area, periosteocytic lacunae	μm^2

[a]   Measurement affected by the absence of the orientation correction factor.
[b]   Measurement affected by the use of the entire labeled surface, single plus double, in determining the active bone-forming surface.

by multiplication with $\pi/4$. Deviations from this model (uncurved plates) can be corrected by a data transformation[14,16] which furthermore reconstructs the true three-dimensional thickness distribution from the measured two-dimensional widths.

H. Histomorphometric Parameters

The derivation of histomorphometric parameters used in this chapter along with their physiologic meaning are discussed in detail in Chapters 4, 5, 7, and 9 of this volume. This laboratory has used a different group of symbols in prior publications and there are some minor differences in the calculations between our laboratory and those published elsewhere in this volume. Table 1 is constructed so that the reader may interpret the work reported here using the terminology introduced in this volume and may easily refer to earlier publications using our symbols and calculations.

There are two important differences in our calculations that should be specially noted: (1) the orientation correction factor, $\pi/4$, is not used here to correct any of the width measurements and (2) the active bone-forming surface is taken as the entire surface covered by tetracycline labels, whether singly or doubly labeled. The items affected by these two differences are noted in Table 1.

Table 2
HISTOMORPHOMETRIC VALUES FROM NORMAL MALES AND THEIR RELATION TO AGE

				Age relation	
	x	SE	N	R	P
Bone mass					
V, $\mu m^3/\mu m^3$	0.203	0.008	72	−0.56	0.001
V_c, $\mu m^3/\mu m^3$	0.934	0.004	40	−0.09	n.s.
Bone turnover					
F_f, $\mu m^3/\mu m^2/day$	0.128	0.02	12	−0.15	n.s.
Cellular activity					
M, $\mu m/day$	0.68	0.04	12	−0.33	n.s.
F_b, $\mu m^3/\mu m^2/day$	0.58	0.06	12	−0.37	n.s.
R_b, $\mu m^3/\mu m^2/day$	2.56	0.31	12	−0.02	n.s.
Formation and mineralization					
S_f, $\mu m^2/\mu m^2$	0.177	0.009	71	−0.04	n.s.
S_{fa}, $\mu m^2/\mu m^2$	0.183	0.021	12	0.13	n.s.
OSW, μm	10.15	0.38	43	−0.24	n.s.
MLT, days	21.2	2.3	12	0.39	n.s.
Resorption					
S_r, $\mu m^2/\mu m^2$	0.042	0.002	70	0.30	
C_r	0.024	0.004	40	−0.37	0.05
A_L, μm^{2a}	51.2	0.4	25	0.04	n.s.

[a]   Pooled values for males and females.

Table 3
HISTOMORPHOMETRIC VALUES FROM NORMAL FEMALES AND THEIR RELATION TO AGE

				Age relation	
		SE	N	R	P
Bone mass					
V, $\mu m^3/\mu m^3$	0.228	0.010	63	−0.58	0.001
V_c, $\mu m^3/\mu m^3$	0.946	0.004	35	−0.67	0.001
Bone turnover					
F_f, $\mu m^3/\mu m^2/day$	0.074	0.006	29	−0.43	0.05
Cellular activity					
M, $\mu m/day$	0.64	0.02	29	−0.14	n.s.
F_b, $\mu m^3/\mu m^2/day$	0.48	0.03	29	−0.49	0.01
R_b, $\mu m^3/\mu m^2/day$	2.00	0.18	29	−0.46	0.02
Formation and mineralization					
S_f, $\mu m^2/\mu m^2$	0.165	0.01	63	−0.11	n.s.
S_{fa}, $\mu m^2/\mu m^2$	0.115	0.008	29	−0.38	0.05
OSW, μm	9.14	0.37	56	−0.20	n.s.
MLT, days	24.7	2.7	28	0.28	n.s.
Resorption					
S_r, $\mu m^2/\mu m^2$	0.039	0.002	62	0.07	n.s.
C_r	0.023	0.003	39	0.18	n.s.
A_L, μm^{2a}	51.2	0.4	25	0.04	n.s.

[a]   Pooled values for males and females.

Table 4
MEAN WALL
THICKNESS (MWT),
BONE RESORPTION
PERIOD (SIGMA$_R$),
FORMATION PERIOD
(SIGMA$_F$), AND
TOTAL REMODELING
PERIOD (SIGMA) (±
SE) FOR 20 YOUNG
NORMAL
INDIVIDUALS

	$\overline{x}$	SE
MWT, μm	72.6	1.5
Sigma$_r$, days	39	7
Sigma$_f$, days	137	12
Sigma, days	176	16

III. BONE REMODELING IN NORMAL AND CERTAIN PATHOLOGICAL STATES

A. Normal Individuals[16,18,21,23,25]

The bone specimens included necropsy and biopsy material. The necropsies were obtained from individuals who fulfilled the following criteria: sudden unexpected death, no immobilization before death, no congenital diseases, no information of endocrine or metabolic disorders, no autopsy diagnosis apart from the cause of death, a normal degree of arteriosclerosis, and signs of senile atrophi. The biopsy material was obtained from normal volunteers who showed no biochemical evidence of disturbed thyroid, parathyroid, kidney, or liver function. Prior to the biopsy double labeling was performed.

The histomorphometric values for normal males and females are shown in Tables 2 and 3.

Among the males, reduction in the trabecular bone mass and in the fraction of Haversian canals with resorption was found with increasing age. All other static and dynamic parameters were unrelated to age.

Among the females the trabecular and cortical bone masses were inversely correlated to age. All the other static parameters were unrelated to age. Bone turnover at tissue level and BMU level was inversely correlated to age due to a normal appositional rate (bone turnover at cellular level) and a significant decrease in labeled surfaces with increasing age.

Sex differences were observed only in the fractional-labeled surfaces and in the bone formation rate at tissue level.

Table 4 gives the mean wall thickness (MWT) of trabecular bone packets in 20 young normal individuals and the calculated values of sigma$_r$, sigma$_f$, and sigma.

B. Hypothyroidism[17,25,34,38]

The diagnosis was based on clinical symptoms and signs, reduced serum levels of thyroxine and serum triiodothyronine uptake tests, and elevated serum levels of TSH.

1. Histomorphometry

The histomorphometric values for untreated hypothyroid patients are shown in Table 5. The amount of trabecular and cortical bone was normal. The bone turnover was

Table 5
HISTOMORPHOMETRIC VALUES FOR PATIENTS WITH HYPOTHYROIDISM AND AGE- AND SEX-MATCHED NORMAL CONTROLS

	Hypothyroid patients			Normal controls				
	$\bar{x}$	SE	N	$\bar{x}$	SE	N	P	Deviation
Bone mass								
V, $\mu m^3/\mu m^3$	19.6	1.5	13	20.2	1.4	27	n.s.	
V_c, $\mu m^3/\mu m^3$	94.2	0.7	14	94.3	0.4	24	n.s.	
Bone turnover								
F_f, $\mu m^3/\mu m^2$/day	0.017	0.007	10	0.077	0.014	10	<0.01	↓
Cellular activity								
M, μm/day	0.36	0.08	10	0.60	0.03	10	<0.05	↓
F_b, $\mu m^3/\mu m^2$/day	0.12	0.05	10	0.38	0.06	10	<0.01	↓
R_b, $\mu m^3/\mu m^2$/day	0.5	0.1	10	1.7	0.2	10	<0.01	↓
Formation and mineralization								
S_f, $\mu m^2/\mu m^2$	0.140	0.018	13	0.167	0.014	27	n.s.	
S_{fa}, $\mu m^2/\mu m^2$	0.042	0.014	12	0.143	0.011	32	<0.01	↓
OSW, μm	6.4	0.7	14	8.7	0.4	27	<0.01	↓
MLT, days	92.1	26.9	10	29.9	3.0	10	<0.05	↑
MWT, μm	58.9	1.6	14	54.0	1.4	28	<0.05	↑
Resorption								
S_r, $\mu m^2/\mu m^2$	0.037	0.005	13	0.640	0.003	27	n.s.	
C_r	0.010	0.003	14	0.023	0.004	26	<0.05	↓
A_L, μm^2	54.0	1.3	8	52.0	0.7	20	n.s.	

Table 6
BONE RESORPTION PERIOD (SIGMA$_R$), FORMATION PERIOD (SIGMA$_F$), AND TOTAL REMODELING PERIOD (SIGMA) (± SE) IN ILIAC TRABECULAR BONE FROM PATIENTS WITH HYPO- AND HYPERTHYROIDISM

	Hypothyroidism			Hyperthyroidism			
	$\bar{x}$	SE	N	$\bar{x}$	SE	N	P
Sigma$_r$, days	208	57	8	29	7	10	<0.01
Sigma$_f$, days	865	275	8	88	12	10	<0.01
Sigma, days	1067	323	8	117	16	10	<0.01

reduced in trabecular bone. The osteoblasts were less active than normal at both cellular and BMU levels. The bone resorption rate was decreased at BMU level. The surface extent of osteoid in trabecular bone was normal, whereas the mean width of the osteoid seams was reduced. The extent of tetracycline-labeled surfaces was reduced. The fractional resorption surface in trabecular bone was normal. The mineralization lag time was markedly increased. In cortical bone the number of canals with resorption was reduced, whereas the mean size of the periosteocytic lacunes was normal. The MWT of trabecular bone packets was increased which, in combination with the decreased cellular activity, resulted in a prolongation of the bone remodeling periods (Table 6).

2. Biochemistry

Serum levels of calcium, phosphorus, and iPTH were normal. Serum alkaline phosphatase was reduced. The renal excretion of calcium was increased.

Table 8

HISTOMORPHOMETRIC VALUES FOR PATIENTS WITH PRIMARY HYPERPARATHYROIDISM AND AGE- AND SEX-MATCHED NORMAL CONTROLS

	Hyperparathyroid patients			Normal controls				
	$\bar{x}$	SE	N	$\bar{x}$	SE	N	P	Deviation
Bone mass								
V, $\mu m^3/\mu m^3$	0.179	0.011	23	0.185	0.014	25	n.s.	
V_c, $\mu m^3/\mu m^3$	0.916	0.010	22	0.938	0.005	22	n.s.	
Bone turnover								
F_f, $\mu m^3/\mu m^2/day$	0.113	0.011	19	0.099	0.010	32	<0.05	↑
Cellular activity								
M, $\mu m/day$	0.56	0.03	21	0.67	0.02	32	<0.05	↓
F_b, $\mu m^3/\mu m^2/day$	0.40	0.03	18	0.55	0.04	31	<0.01	↓
R_b, $\mu m^3/\mu m^2/day$	1.59	0.16	18	2.24	0.19	29	<0.01	↓
Formation and mineralization								
S_f, $\mu m^2/\mu m^2$	0.285	0.023	23	0.213	0.017	25	<0.01	↑
S_{fa}, $\mu m^2/\mu m^2$	0.204	0.019	19	0.143	0.014	32	<0.02	↑
OSW, μm	10.0	0.6	24	9.9	0.5	25	n.s.	
MLT, days	28.6	2.8	18	21.6	2.1	29	n.s.	
MWT, μm	64.3	2.0	12	71.1	2.2	12	<0.05	↓
Resorption								
S_r, $\mu m^2/\mu m^2$	0.076	0.004	24	0.040	0.003	24	<0.01	↑
C_r	0.039	0.005	24	0.026	0.005	24	<0.05	↑
A_L, μm^2	63.0	1.5	17	51.2	0.4	25	<0.01	↑
Remodeling periods								
Sigma$_r$	44	8	12	46	9	12	n.s.	
Sigma$_f$	171	13	12	147	15	12	n.s.	
Sigma	217	18	12	191	20	12	n.s.	

3. Interpretation

Excess parathyroid hormone increases bone turnover which explains the increased surface extent of bone formation and bone resorption. The elevated serum alkaline phosphatase levels may be explained by an increased number of osteoblasts in accordance with an observed positive relation between serum alkaline phosphatase and fractional-labeled surfaces. The activity of the osteoblasts and osteoclasts was reduced which may be due to the combined effect of parathyroid hormone and hypercalcemia or hypophosphatemia since the acute effect of parathyroid hormone is to stimulate the cells. The osteoid seam width was normal in spite of a slightly reduced appositional rate of osteoid. This may be explained by the high normal mineralization lag time.

The amount of trabecular and cortical bone was normal showing that loss of bone mineral is not a major factor in the pathogenesis of hypercalcemia. The increased osteocytic osteolysis, however, may together with an enhanced function of the lining cells be of importance for maintaining the hypercalcemia by creating a functional barrier between extraosseous and intraosseous fluid.[43]

E. Intestinal Bypass Patients[25,39,41]

All the patients had undergone an intestinal bypass operation for morbid obesity 3 to 8 years before bone biopsy. The patients were not selected for their symptoms. The average weight loss after operation was 38 kg or 31% of their initial weight.

Table 9
HISTOMORPHOMETRIC VALUES FOR INTESTINAL BYPASS PATIENTS AND AGE- AND SEX-MATCHED NORMAL CONTROLS

	Intestinal bypass patients			Normal controls				
	$\bar{x}$	SE	N	$\bar{x}$	SE	N	P	Deviation
Bone mass								
V, $\mu m^2/\mu m^3$	0.22	0.01	25	0.22	0.02	17	n.s.	
Bone turnover								
F_f, $\mu m^3/\mu m^2/day$	0.053	0.008	17	0.072	0.008	17	<0.05	↓
Cellular activity								
M, $\mu m/day$	0.46	0.03	17	0.63	0.03	17	<0.01	↓
F_b, $\mu m^3/\mu m^2/day$	0.25	0.03	17	0.45	0.04	17	<0.01	↓
R_b, $\mu m^3/\mu m^2/day$	1.	0.24	17	1.86	0.26	17	<0.01	↓
Formation and mineralization								
S_f, $\mu m^2/\mu m^2$	0.234	0.032	25	0.173	0.014	17	<0.05	↑
S_{fa}, $\mu m^2/\mu m^2$	0.117	0.014	20	0.119	0.011	17	n.s.	
OSW, μm	12.3	1.4	25	9.0	0.6	17	<0.01	↑
MLT, days	42.7	6.4	17	22.5	2.2	17	<0.01	↑
Resorption								
S_r, $\mu m^2/\mu m^2$	0.052	0.005	25	0.041	0.003	17	<0.05	↑

1. Histomorphometry

The histomorphometric values for intestinal bypass patients and normal controls are shown in Table 9. The amount of trabecular bone was normal. The bone turnover was reduced and the osteoblasts were less active than normal at both cellular and BMU levels. The bone resorption rate at BMU level was reduced. The extent of osteoid surface and resorption surface was increased whereas the extent of tetracycline-labeled surface was normal giving a reduced proportion of labeled to osteoid covered surfaces. The mean osteoid seam width was increased and the mineralization lag time markedly prolonged.

2. Biochemistry

Serum levels and renal excretion rates of calcium, magnesium, and phosphorus were reduced. Serum iPTH concentrations was normal or slightly increased whereas the serum levels of 25-OHD and 1,25-$(OH)_2D$ were reduced. Serum levels of alkaline phosphatase were increased.

3. Interpretation

The low serum levels of vitamin D metabolites indicate a vitamin D deficient state probably due to reduced intestinal absorption of vitamin D and an increased loss of vitamin D during the enterohepatic circulation. The absorption of calcium, phosphorus, and magnesium is reduced partly because of a reduced absorptive intestinal surface and partly because of vitamin D deficiency. The malabsorption leads to hypocalcemia, hypomagnesiemia, parathyroid stimulation, hypophosphatemia, and reduced renal excretion of these electrolytes. The vitamin D deficiency is so pronounced in these patients that a clear osteomalacic state with a mineralization defect is found. The bone turnover is reduced and the increased surface extent of bone formation and bone resorption may therefore be explained by longer life spans of bone formation and bone resorption in accordance with the reduced cellular activity. The osteoid seam width is increased because of a marked prolongation of the mineralization lag time and in spite of a reduced osteoid appositional rate. The enlarged osteocytic lacunae may be explained by parathyroid stimulation and hypocalcemia.

Table 10

HISTOMORPHOMETRIC VALUES FOR EPILEPTIC PATIENTS RECEIVING ANTICONVULSANT TREATMENT AND FOR AGE- AND SEX-MATCHED NORMAL CONTROLS

	x	SE	N	x	SE	N	P	Deviation
Bone mass								
F, $\mu m^3/\mu m^3$	0.217	0.011	20	0.239	0.011	24	n.s.	
Bone turnover								
F_f, $\mu m^2/\mu m^2$/day	0.171	0.007	20	0.096	0.013	24	<0.01	↑
Cellular activity								
M, m/day	0.66	0.02	20	0.64	0.03	24	n.s.	
F_b, $\mu m^3/\mu m^2$/day	0.58	0.02	20	0.50	0.04	24	n.s.	
R_b, $\mu m^3/\mu m^2$/day	2.71	0.19	20	2.34	0.23	30	n.s.	
Formation and mineralization								
S_f, $\mu m^2/\mu m^2$	0.281	0.014	20	0.189	0.016	24	<0.01	↑
S_{fa}, $\mu m^2/\mu m^2$	0.247	0.012	20	0.145	0.015	24	<0.01	↑
OSW, μm	12.2	0.4	20	9.7	0.5	24	<0.01	↑
MLT, days	21.5	1.1	20	22.4	2.1	24	n.s.	
MWT, μm	67.1	1.0	20	72.6	1.5	20	<0.05	
Resorption								
S_r, $\mu m^2/\mu m^2$	0.063	0.003	20	0.038	0.002	24	<0.01	↑
A_L, μm^2	71.9	1.9	14	51.9	0.7	20	<0.01	↑
Remodeling periods								
Sigma$_r$	27	2	20	39	7	20	n.s.	
Sigma$_f$	114	7	20	137	12	20	n.s.	
Sigma	141	8	20	176	16	20	n.s.	

F. Epileptic Patients Receiving Anticonvulsants[2,18,25,35,36,39]

All the patients had been treated for more than 10 years with diphenylhydantoin in combination with other anticonvulsant drugs. They were selected from a group of 60 adult epileptic outpatients in whom bone biopsy was performed. The selection was based on an increased amount of osteoid and/or an increase in the mean size of the periosteocytic lacunes, which were the two most frequent histomorphometric findings among the epileptics.

1. Histomorphometry

The histomorphometric values for epileptic patients receiving anticonvulsants and normal controls are shown in Table 10. The amount of trabecular bone was normal. The bone turnover was increased in trabecular bone. The activity of the osteoblasts was normal at both cellular level and BMU level. The bone resorption rate at BMU level was normal. The extent of osteoid surfaces, labeled surfaces, and resorption surfaces was increased. The osteoid seam width and the mineralization lag time were normal. The MWT of remodeling sites in trabecular bone was reduced, which in combination with the normal cellular activity, resulted in small but statistically insignificant reductions in bone formation and bone resorption periods.

In cortical bone the mean size of the periosteocytic lacunes was increased.

2. Biochemistry

Serum levels and renal excretion of calcium were reduced. Serum phosphorus levels were normal. Serum iPTH levels were normal or slightly increased whereas serum concentrations of the three vitamin D metabolites 25-OHD, 1,25-$(OH)_2$D, and 24,25-$(OH)_2$D were reduced. Serum levels of alkaline phosphatase and renal hydroxyproline excretion were increased. The fractional intestinal calcium absorption was reduced but responded normally to exogenous 1,25-(OH)₂D.

3. Interpretation

The low serum levels of vitamin D metabolites indicate a vitamin D deficient state probably because of a drug-induced enhanced hepatic catabolism and excretion of these compounds. The resulting reduced intestinal calcium absorption is followed by slight hypocalcemia, parathyroid stimulation, and reduced renal excretion of calcium. The bone changes with enhanced bone turnover and enlarged periosteocytic lacunae are characteristic for secondary hyperparathyroidism and may be explained by the slight parathyroid stimulation possibly combined with an increased target cell sensitivity to parathyroid hormone due to hypocalcemia. The increased serum levels of alkaline phosphatase and renal excretion rates of hydroxyproline reflect the high bone turnover. The mineralization of osteoid was normal in the present population as measured by the osteoblastic activity and the mineralization lag time. The histomorphometric values differ in this aspect from those found in intestinal bypass patients with malabsorption. The epileptic patients may reflect a slight vitamin D deficiency with bone changes mainly characterized by secondary hyperparathyroidism, whereas the intestinal bypass patients have a more pronounced vitamin D deficiency with a mineralization defect.

G. Medullary Thyroid Carcinoma[6]

The diagnosis was based on elevated basal- and/or pentagastrin-stimulated serum immunoreactive calcitonin levels and histological evidence of medullary thyroid carcinoma.

1. Histomorphometry

The histomorphometric values for patients and normal controls are shown in Table 11. The amount of trabecular bone was normal. The bone turnover in trabecular bone was slightly but insignificantly increased. The osteoblasts were less active than normal at cellular level. The same tendency was found at BMU level; the difference from normal, however, was not significant. The bone resorption rate at BMU level was normal. The extent of osteoid surfaces, labeled surfaces, and resorption surfaces was increased. The osteoid seam width and mineralization lag time were normal.

In cortical bone the mean size of the periosteocytic lacunes was increased.

2. Biochemistry

Serum levels of calcium, phosphorus, and iPTH were normal. Serum alkaline phosphatase levels were increased. Serum concentrations of 1,25-(OH)₂D were increased in spite of reduced serum levels of 25-OHD. Serum 24,25-(OH)₂D levels were normal.

3. Interpretation

Excess calcitonin increases renal excretion of calcium and phosphorus. The high serum levels of 1,25-(OH)₂D in spite of low serum 25-OHD indicates an enhanced activity of the renal 1-alpha-hydroxylase. This may be caused by a direct effect of calcitonin on the renal enzyme or may represent adaptive changes secondary to a relative lack of calcium and phosphorus. The dynamic bone changes are similar to those found in primary and secondary hyperparathyroidism and may be caused by an enhanced sensitivity to circulating PTH induced by the increased 1,25-(OH)₂D. The increased extent of formative and resorptive surfaces may be explained partly by an enhanced bone turnover and partly by a slight prolongation of the formative period due to subnormal cellular activity. The increased serum levels of alkaline phosphatase

Table 11
HISTOMORPHOMETRIC VALUES FOR PATIENTS WITH MEDULLARY THYROID CARCINOMA AND FOR AGE- AND SEX-MATCHED NORMAL CONTROLS

	Hypercalcitonemic patients			Normal controls				
	$\overline{x}$	SE	N	$\overline{x}$	SE	N	P	Deviation
Bone mass								
V, $\mu m^3/\mu m^3$	0.251	0.024	10	0.206	0.016	20	n.s.	
V_c, $\mu m^3/\mu m^3$	0.918	0.018	9	0.950	0.004	10	<0.02	↓
Bone turnover								
F_f, $\mu m^3/\mu m^2/day$	0.117	0.007	9	0.092	0.011	10	n.s.	
Cellular activity								
M, $\mu m/day$	0.53	0.03	9	0.64	0.03	10	<0.05	↓
F_b, $\mu m^3/\mu m^2/day$	0.39	0.01	9	0.48	0.06	10	n.s.	
R_b, $\mu m^3/\mu m^2/day$	1.6	0.2	9	2.2	0.2	10	n.s.	
Formation and mineralization								
S_f, $\mu m^2/\mu m^2$	0.295	0.035	10	0.202	0.022	10	<0.05	↑
S_{fa}, $\mu m^2/\mu m^2$	0.230	0.021	9	0.146	0.018	10	<0.05	↑
OSW, μm	10.6	0.8	10	9.8	0.7	10	n.s.	
MLT, days	29.7	3.5	9	21.7	1.9	10	n.s.	
Resorption								
S_r, $\mu m^2/\mu m^2$	0.076	0.008	10	0.043	0.005	10	<0.005	↑
C_r	0.048	0.012	9	0.013	0.005	10	<0.02	↑
A_L, μm^2	73.9	4.3	9	51.2	0.7	20	<0.001	w[a]

may reflect the high bone turnover although alterations due to metastatic disease could not be ruled out.

H. Immunosuppressive Treatment After Renal Transplantation[25,29]

All the patients had undergone a renal allotransplantation at an average of 50 months before biopsy and were receiving an average dose of 3.1 mg prednisone and 125 mg azathioprine. Of the 17 patients, 10 showed radiologic signs of osteonecrosis at various sites of the skeleton.

1. Histomorphometry

The histomorphometric values for the renal transplant patients and the normal controls are shown in Table 12. The amount of trabecular bone was reduced. The turnover was reduced in trabecular bone and the osteoblasts were less active than normal at both cellular and BMU levels. The resorption rate was reduced at BMU level. The surface extent of osteoid in trabecular bone was increased, whereas the mean width of the osteoid seams was reduced. The extent of tetracycline-labeled surfaces was normal as was the fractional extent of resorption surfaces. The mineralization lag time was markedly increased.

In cortical bone the size of periosteocytic lacunae was found normal.

2. Biochemistry

All the patients had normal serum values of creatinine and a creatinine clearance of above 50 mℓ/min. The mean serum concentration of calcium was normal, whereas the mean concentration of phosphorus was reduced. The mean serum level of iTPH was increased.

Table 12

HISTOMORPHOMETRIC VALUES FOR PATIENTS RECEIVING IMMUNOSUPPRESSIVE TREATMENT AFTER RENAL TRANSPLANTATION AND FOR AGE- AND SEX-MATCHED NORMAL CONTROLS

	Renal transplant patients			Normal controls				
	$\bar{x}$	SE	N	$\bar{x}$	SE	N	P	Deviation
Bone mass								
V, m³/m³	0.145	0.012	17	0.222	0.012	34	<0.01	↓
V_c, m³/m³	—	—		—	—		—	
			—			—		
Bone turnover								
F_t, m³/m²/day	0.060	0.012	17	0.084	0.011	12	<0.05	↓
Cellular activity								
M, μm/day	0.52	0.02	17	0.61	0.03	12	=0.02	↓
F_b, μm³/μm²/day	0.21	0.03	17	0.46	0.05	12	<0.01	↓
R_b, μm³/μm²/day	1.49	0.25	17	2.28	0.26	12	<0.05	↓
Formation and mineralization								
S_f, μm²/μm²	0.289	0.026	17	0.189	0.020	12	<0.02	↑
S_{fa}, μm²/μm²	0.114	0.020	17	0.136	0.017	12	n.s.	
OSW, μm	8.1	0.5	17	10.1	0.6	12	<0.02	↓
MLT, days	54.5	9.7	17	24.1	2.7	12	<0.01	↑
Resorption								
S_r, μm²/μm²	0.043	0.004	17	0.039	0.002	34	n.s.	
C_r	—	—	—	—	—	—	—	
A_L, μm²	51.9	1.3	17	51.2	0.4	25	n.s.	

3. Interpretation

In renal allotransplanted patients the bone turnover was low and the activity of the osteoblasts and osteoclasts was reduced. The static histomorphometric data, i.e., the normal fraction of resorptive surfaces and the increased fraction of formative surfaces, might suggest a "decoupling phenomenon" but from a dynamic point of view the proportion of resorptive to formative surfaces was normal. The osteoid seam thickness was reduced because of a reduced appositional rate of osteoid in spite of a prolongation of the mineralization lag time. The reduced amount of trabecular bone may be due to the altered bone formation activity. An initial effect of the immunosuppressive treatment on bone, not present at the time of biopsy, and the presence of renal osteodystrophy before transplantation may, however, play important roles in this reduction.

I. Acromegaly[7,10-12]

The diagnosis was based upon clinical symptoms and signs and raised levels of serum growth hormone (radioimmunoassay technique).

1. Histomorphometry

The morphometric values for patients and normal controls are shown in Table 13. The fractional volume of trabecular bone was increased. Bone turnover in trabecular bone was increased as was the formation rate at cellular level. A small increase in formation rate at the BMU level was statistically insignificant. The fractional extents of osteoid covered, tetracycline-labeled, and resorptive surfaces were increased.

2. Biochemistry

Elevated serum levels of iPTH and 1,25(OH)₂D were found. Fasting serum concen-

Table 13

HISTOMORPHOMETRIC VALUES FOR PATIENTS WITH ACROMEGALY AND FOR AGE- AND SEX-MATCHED NORMAL CONTROLS

	Acromegally patients			Normal controls				
	$\overline{x}$	SE	N	$\overline{x}$	SE	N	P	Deviation
Bone mass								
V, $\mu m^3/\mu m^3$	0.264	0.012	18	0.203	0.016	17	<0.01	↑
Bone turnover								
F_f, $\mu m^3/\mu m^2/$ day	0.198	0.038	9	0.067	0.011	9	<0.01	↑
Cellular activity								
M, μm/day	0.87	0.09	11	0.60	0.04	9	<0.01	↑
F_b, $\mu m^3/\mu m^2/$ day	0.66	0.13	9	0.36	0.06	9	n.s.	(↑)
Formation and mineralization								
S_f, $\mu m^2/\mu m^2$	0.278	0.029	18	0.191	0.017	17	<0.01	↑
S_{fa}, $\mu m^2/\mu m^2$	0.231	0.026	9	0.112	0.019	9	<0.01	↑
Resorption								
S_r, $\mu m^2/\mu m^2$	0.067	0.005	18	0.043	0.004	17	<0.05	↑

trations of calcium and phosphorus were increased. The urinary excretion of hydroxyproline was raised, as were excretions of calcium and phosphorus.

3. Interpretation

No significant correlations were found between the histomorphometric data and values of calcium-phosphorus metabolism. The appositional rate, however, was positively related to the fasting serum growth hormone levels. Since the appositional rate represents the metabolic output of the active osteoblasts and since osteoblasts have been shown to be enlarged in acromegaly, growth hormone might have a direct effect on the cells. To what extent secondary changes in PTH and vitamin D affect bone remodeling in acromegaly is unknown.

REFERENCES

1. Bordier, P., Matrajt, H., Miravet, L., and Hioco, D., Mesure histologique de la masse et de la resorption des travees osseuses, *Pathol. Microbiol.*, 12, 1238, 1964.
2. Christensen, C. K., Lund, Bi, Lund, Bj, Sorensen, O. H., Nielsen, H. E., and Mosekilde, L., Reduced 1,25-dihydroxyvitamin D and 24,25-dihydroxyvitamin D in epileptic patients receiving chronic combined anticonvulsant therapy, *Metab. Bone Dis. Rel. Res.*, 3, 17, 1981.
3. Courpron, P., Meunier, P., Bressot, C., and Giroux, J. M., Amount of bone in iliac crest biopsy. Significance of the trabecular bone volume. Its values in normal and in pathological conditions, in *Bone Histomorphometry*, Meunier, P. J., Ed., Societe de la Nouvelle Imprimerie Fournie, Toulouse, France, 1977, 39.
4. Delling, G., Age-related bone changes, *Curr. Top. Pathol.*, 58, 117, 1973.
5. Duncan, H., Sudhakher, D. R., and Parfitt, A. M., Complications of bone biopsy, *Metab. Bone Dis. Rel. Res.*, 2S, 483, 1980.
6. Emmertsen, K., Melsen, F., Mosekilde, L., Lund, Bi, Lund, Bj, Sorensen, O. H., Nielsen, H. E., Solling, H., and Hansen, H. H., Altered vitamin D metabolism and bone remodeling in patients with medullary thyroid carcinoma and hypercalcitoninemia, *Metab. Bone Dis. Rel. Res.*, in press.

7. **Eskildsen, P. C., Lund, Bj, Sorensen, O. H., Lund, Bi, Bishop, J. E., and Normal, A. W.,** Acromegaly and vitamin D metabolism: Effect of bromocriptine treatment, *J. Clin. Endocrinol. Metab.,* 49, 484, 1979.

8. **Frost, H. M.,** Tetracycline-based histological analysis of bone remodeling, *Calcif. Tissue Res.,* 3, 211, 1969.

9. **Frost, H. M.,** A method of analysis of trabecular bone dynamics, in *Bone Histomorphometry,* Meunier, P. J., Ed., Societe de la Nourvelle Imprimerie Fournie, Toulouse, France, 1977, 445.

10. **Halse, J. and Gordeladze, J.,** Total and non-dialyzable urinary hydroxyproline in acromegalics and control subjects, *Acta Endocrinol.,* 96, 451, 1981.

11. **Halse, J. and Haugen, H. N.,** Calcium and phosphate metabolism in acromegaly, *Acta Endocrinol.,* 94, 459, 1980.

12. **Halse, J., Melsen, F., and Mosekilde, L.,** Iliac crest bone mass and remodeling in acromegaly, *Acta Endocrinol.,* 97, 18, 1981.

13. **Jastrup, B., Mosekilde, L., Melsen, F., Lund, Bi, Lund, Bj, and Sorenson, O. H.,** Serum levels of vitamin D metabolites and bone remodeling in hyperthyroidism, *Metabolism,* 31, 126, 1982.

14. **Jensen, E. B., Gundersen, H. J. G., and Osterby, R.,** Determination of membrane thickness distribution from orthogonal intercepts, *J. Microsc.,* 115, 19, 1979.

15. **Kragstrup, J., Eriksen, E., Melsen, F., and Mosekilde, L.,** An improved method for determining mean wall thickness (mwt) of trabecular bone packets, *Calcif. Tissue Int.,* Suppl. 33, 56, 1981.

16. **Kragstrup, J., Gundersen, H. J. G., Melsen, F., and Mosekilde, L.,** Estimation of the 3-dimensional wall thickness of completed remodeling sites in iliac trabecular bone, *Metab. Bone Dis. Rel. Res.,* in press.

17. **Kragstrup, J., Melsen, F., and Mosekilde, L.,** Effects of thyroid hormone(s) on mean wall thickness of packets in iliac trabecular bone, *Metab. Bone Dis. Rel. Res.,* 3, 181, 1981.

18. **Kragstrup, J., Melsen, F., Mosekilde, L., and Bergmann, S.,** Reduced trabecular bone mean wall thickness (mwt) in patients receiving chronic anticonvulsant therapy, *Calcif. Tissue Int.,* Suppl. 33, 55, 1981.

19. **Lips, P., Courpron, P., and Meunier, P.,** Mean wall thickness of trabecular bone packets in human iliac crest: changes with age, *Calcif. Tissue Res.,* 26, 13, 1978.

20. **Melsen, F., Melsen, B., and Mosekilde, L.,** An evaluation of the quantitative parameters applied in bone histology, *Acta Pathol. Microbiol. Scand. (A),* 86, 63, 1978.

21. **Melsen, F., Melsen, B., Mosekilde, L., and Bergmann, S.,** Histomorphometric analysis of normal bone from the iliac crest, *Acta Pathol. Microbiol. Scand. (A),* 86, 70, 1978.

22. **Melsen, F. and Mosekilde, L.,** Morphometric and dynamic studies of bone changes in hyperthyroidism, *Acta Pathol. Microbiol. Scand. (A),* 85, 141, 1977.

23. **Melsen, F. and Mosekilde, L.,** Tetracycline double-labeling of iliac trabecular bone in 41 normal adults, *Calcif. Tissue Res.,* 26, 99, 1978.

24. **Melsen, F. and Mosekilde, L.,** Dynamic studies of trabecular bone formation and osteoid maturation in normal and certain pathological conditions, *Metab. Bone Dis. Rel. Res.,* 1, 45, 1978.

25. **Melsen, F. and Mosekilde, L.,** Trabecular bone mineralization lag time determined by tetracycline double-labeling in normal and certain pathological conditions, *Acta Pathol. Microbiol. Scand. (A),* 88, 83, 1980.

26. **Melsen, F. and Mosekilde, L.,** Interpretation of single labels after in vivo double labeling, *Metab. Bone Dis. Rel. Res.,* 2S, 171, 1980.

27. **Melsen, F. and Mosekilde, L.,** Tetracycline based bone morphometry in primary hyperparathyroidism and hyperthyroidism, *Metab. Bone Dis. Rel. Res.,* 2S, 249, 1980.

28. **Melsen, F., Mosekilde, L. and Christensen, M. S.,** Interrelationships between bone histomorphometry, S-iPTH and calcium-phosphorus metabolism in primary hyperparathyroidism, *Calcif. Tiss. Res.,* 24S, 16, 1977.

29. **Melsen, F. and Nielsen, H. E.,** Osteonecrosis following renal allotransplantation, *Acta Pathol. Microbiol. Scand. (A),* 85, 99, 1977.

30. **Melsen, F., Viidik, A., Melsen, B., and Mosekilde, L.,** Some relations between bone strength, ash weight and histomorphometry, in *Bone Histomorphometry,* Meunier, P. J., Ed., Societe de la Nouvelle Imprimerie Fournie, Toulouse, France, 1977, 89.

31. **Meunier, P.,** La Dynamique de Remaniement Osseux Etudiee par Lecture Quantitative de la Biopsie Osseuse, Thesis, Lyon, 1968.

32. **Meunier, P.,** Tetracycline Dynamics and Bone Histomorphometry in Normal and Osteoporotic Man, Sun Valley Workshop on Mineralized Tissues, Sun Valley Idaho, 1977.

33. **Meunier, P., Bianchi, G. G. S., Edouard, C. M., Bernard, J., Courpron, P., and Vignon, G.,** Bone manifestations of thyrotoxicosis, *Orthoped. Clin. N. Am.,* 3, 745, 1972.

34. **Mosekilde, L.,** Effects of Thyroid Hormone(s) on Bone Remodeling, Bone Mass and Calcium-Phosphorus Homeostasis in Man, Thesis, Aarhus, 1979.

35. Mosekilde, L., Christensen, M. S., Hansen, H. H., Melsen, F., and Norman, A. W., Effect of 1,25-dihydroxycholecalciferol and 25-hydrocholecalciferol on fractional intestinal calcium absorption in anticonvulsant osteomalacia, *Calcif. Tissue Res.,* 24S, 18, 1977.

36. Mosekilde, L., Christensen, M. S., Lund, B., Sorensen, O. H., and Melsen, F., The interrelationships between serum 25-hydroxycholecalciferol, serum parathyroid hormone and bone changes in anticonvulsant osteomalacia, *Acta Endocrinol.,* 84, 559, 1977.

37. Mosekilde, L. and Melsen, F., A tetracycline based histomorphometric evaluation of bone resorption and bone turnover in hyperthyroidism and hyperparathyroidism, *Acta Med. Scand.,* 204, 97, 1978.

38. Mosekilde, L. and Melsen, F., Morphometric and dynamic studies of bone changes in hypothyroidism, *Acta Pathol. Microbiol. Scand. (A),* 86, 56, 1978.

39. Mosekilde, L. and Melsen, F., Dynamic differences in trabecular bone remodeling between patients after jejuno-ileal bypass for obesity and epileptic patients receiving anticonvulsant therapy, *Metab. Bone Dis. Rel. Res.,* 2, 77, 1980.

40. Mosekilde, L., Melsen, F., Bagger, J. P., Myhre-Jensen, O., and Sorensen, H. S., Bone changes in hyperthyroidism: interrelationships between bone morphometry, thyroid function and calcium-phosphorus metabolism, *Acta. Endocrinol.,* 85, 515, 1977.

41. Mosekilde, L., Melsen, F., Hessov, I., Christensen, M. S., Lund, Bj, Lund, Bi, and Sorensen, O. H., Low serum levels of 1,25-dihydroxyvitamin D and histomorphometric evidence of osteomalacia after jejuno-ileal bypass for obesity, *Gut.,* 21, 624, 1980.

42. Rasmussen, H. and Bordier, P., *The Physiological and Cellular Basis of Metabolic Bone Disease,* Williams & Wilkins, Baltimore, 1974.

43. Talmage, R. V., Morphological and physiological considerations in a new concept of calcium transport in bone, *Am. J. Anat.,* 129, 467, 1970.

44. Weibel, E. R., *Sterological Methods,* Vol. 1, Academic Press, New York, 1979, 1.

Chapter 13

AUTOMATED SKELETAL HISTOMORPHOMETRY

James M. Smith and Webster S. S. Jee

TABLE OF CONTENTS

I. INTRODUCTION

A major problem in the field of bone histomorphometry, a problem which it shares with all fields of quantitative science, is the limited "capacity (of the human observer) to assimilate and evaluate information in a given time span".[1] There have been recent developments in instrumentation which have allowed significant improvements in the reduction of time required to perform routine measurements of bone parameters, although these advanced instrumentation techniques are not without shortcomings. Any attempt to duplicate the eye/brain interface of the histologist by machine is still a goal of instrumentational development which will not be realized in the present generation of histologists. It is the purpose of this chapter to review some of the quantitative image analyzers which, nevertheless, offer significant aid to the bone histomorphometrist.

II. INSTRUMENTATION

Quantitative image analyzers fall into two broad categories: automatic and manual scanners.

The manual scanner (often referred to as a "digitizer") requires the operator to perform pattern recognition by manually tracing with a stylus those features to be quantified. The automatic scanner, on the other hand, performs pattern recognition of specific features automatically by optoelectronic signal processing. Both instruments digitize the features of interest and subsequently perform measurements on the digitized image which is stored in memory. These instruments, therefore, are generally interfaced with dedicated computers of varying levels of complexity.

A. Manual Scanner

The basic principle of operation of a manual digitizer is illustrated in Figure 1. The image of the specimen (or a photograph of the specimen) to be evaluated is projected onto a "digitizing tablet". Beneath the tablet surface is an encoding matrix which senses the position of the stylus. Placing the stylus at a particular point on the surface of the tablet results in assignment of grid coordinates (x,y) in memory. As the stylus traces an image, the coordinates of points along the line are continuously stored in memory. Instruments vary in their degree of resolution of the coordinate system; a resolution on the order 0.1 mm is typical of commercial instruments.

Subsequent to the above image digitization process, various parameters may be calculated (e.g., area, perimeter, projection, count, etc.) with the computer operating on the image stored in memory.

Sherrard et al.[2] used a Model 1010A Grafacon® digitizing tablet* interfaced to a computer for a number of measurements on 7-μm-thick human bone biopsy sections, stained by Goldner's method.[3] (With this staining method, mineralized bone appears green and osteoid red.) They used a series of mirrors to project the microscopic image onto the digitizing tablet where the stylus tracing was performed. Their measurements included (1) total bone area (osteoid and mineralized bone), (2) mineralized bone area, (3) forming surface perimeter (identified by the presence of osteoid), (4) fibrous tissue area in the marrow space, and (5) resorbing surface perimeter (identified by surface scalloping and absence of osteoid). They report a precision of 1 to 8% in their measurements of all histological parameters.

Baylink and Wergedal[4] used the Model 1010A Grafacon® tablet for area measurements on 30-μm sections (stained with toluidine blue) of rat bone. These measurements

---

* Compunetics Inc., Monroeville, Pa.

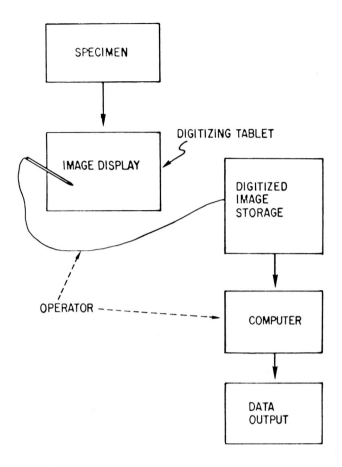

FIGURE 1.   Block diagram of a manual digitizer for image analysis.

were made by tracing the microscopic images found on the tablet with a camera lucida. Their measurements included endosteal and periosteal tetracycline-labeled areas as well as bone and marrow areas. They report a precision of ± 5%.

Holtrup[5] used a digitizing system to determine the ultrastructural morphometry of osteoclasts. Electron micrographs of bone sections from a rat were projected onto a digitizing tablet interfaced to a computer. The circumference of each area of interest (ruffled border, clear zone, cytoplasmic, and nuclear areas) within the osteoclast image was traced and the areas were calculated by the computer program. A precision of 1 to 4% for one operator was reported.

B. Automatic Scanner

The basic principle of an automatic scanner takes advantage of feature contrast within an image as the primary mode of pattern recognition. Feature detection may be determined by the shade of gray of the features of interest. However, further pattern recognition may also be achieved by the shape of the features, using stereological shape or form factors which will not be discussed here. Bone is particularly amenable to automatic image analysis because of its high contrast both optically (with appropriate staining) and radiographically.

A block diagram of a general automatic scanner (similar to commercial instruments such as Cambridge-Imanco's Quantimet® or Bausch and Lomb's Omnicon®) is shown in Figure 2. The specimen (or a photograph or radiograph of the specimen) is optically coupled to a television-type scanner by way of a microscope or a macroim-

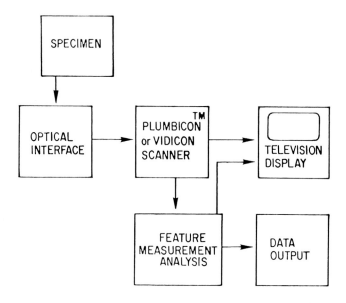

FIGURE 2.   Block diagram of a general automatic image analyzer.

aging system. As the image is electronically scanned, measurements are made instantaneously on all features which are detected based upon their shading of gray. The operator generally chooses the appropriate gray level discrimination, although the instrument in some cases may be able to determine automatically the proper detection of the features of interest.

The television display conveniently allows both the field of view and the feature measurements to be monitored. A variety of measurements may be made, as in the case of the manual scanners. These will be discussed in Section III.

With the automatic scanners a considerable degree of operator interaction is possible by means of a light pen to select individual features for measurement from the full field of view on the video screen. In addition, an image editing capability allows significant modification of the image. In this mode the operator, using a stylus with a specially designed photocell, may modify or construct images on the screen which will likewise change the measurements taken in the field of view. For example, if a bone image has a poorly defined endosteal surface boundary due to faulty polishing of the bone section, the operator may "fill in" a distinct surface boundary and this modified image can be measured.

Robb and Jowsey[6] describe a "digital scanning videodensitometer" which operates on the same basic principles as the automatic scanner illustrated in Figure 2. They used it for measurements of fractional bone volumes from microradiographs of bone biopsy sections.

Lloyd and Hodges[7] used an automatic scanner (CHLOE) to make measurements of bone surfaces and volume, trabecular thickness, and marrow cavity size from microradiographs of 100-μm bone sections. In this scanner a 25-μm diameter light beam generated by a cathode ray tube is focused onto the microradiograph and the light which is transmitted is detected by a photomultiplier tube. In a single raster scan or frame, the system records the coordinates of all points where the transmitted light changes abruptly (from black to white or vice versa).

The group at Cookridge Hospital in Leeds[8-10] use a unique scanning device for chord (path length) distributions in bone. Instead of electronically scanning the bone specimen as in the instruments described above, they mount the section on a mechanical

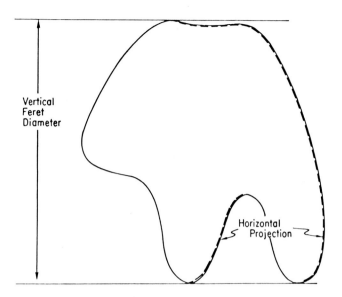

FIGURE 3.   Illustration of the vertical Feret diameter and the horizontal projection of an arbitrary feature.

stage which is traversed beneath a conventional projection microscope and the magnified image is viewed through an aperture by a photomultiplier.

III. MEASUREMENTS

A. Stereological Parameters

With the image of specific features of the field stored in memory of the automatic or manual scanner, a number of measurements may be made, depending upon the versatility of the instrument, for example: (1) area, (2) perimeter and length, (3) projection, (4) Feret diameter, (5) shape factors, (6) feature count, and (7) optical density. The raw data are often in units of picture points or pixels which are readily calibrated into the appropriate metric units using a stage micrometer or a standard metric scale. A picture point or pixel is the smallest unit of resolution of the measuring field.

The Feret diameter and projection of a feature are illustrated in Figure 3. The Feret diameter is used as a measure of "size" of a feature and is defined as the maximum dimension in a given direction. This is sometimes referred to as a "caliper" measurement. The projection of a feature is a boundary length of the feature proportional to the number of scanning line segments (in a given direction) which overlies the feature.

With these basic measurements a number of important stereological parameters of interest to the bone histomorphologist may be derived. Some of these are discussed below.

1. Specific Surface-to-Volume Ratio

The ratio of the total perimeter (P) to the total area (A) of bone in the section is proportional to the specific surface-to-volume ratio (S/V). Expressing the constant of proportionality as k, we may write

$$(S/V)_{bone} = k \left(\frac{P}{A} \right) \tag{1}$$

In the case of cortical bone, in which the thin sections have been cut or sawed perpendicular to the shaft of the bone, k is unity. For trabecular bone, in which the sections are cut such that the orientation of the trabeculae is isotropic, k is $4/\pi$.[7,11]

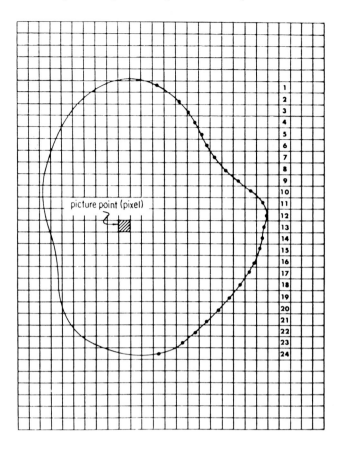

FIGURE 4.   Plane section of a single bone trabeculum. A grid array
of picture points is superimposed on the image.

2. Bone and Marrow Chord Distributions

The mean width of trabeculae or marrow cavities (and the distribution about the mean) in a thin section is a common parameter in bone histomorphometry.

Consider a plane section of a single, highly magnified bone trabeculum as illustrated in Figure 4. A square of the grid array represents a single picture point or pixel. The size of the picture points (pp) are exaggerated for illustration. The number of picture points (multiplied by the appropriate calibration factor) overlying the bone spicule image represents a close approximation to the area of the trabeculum. Each occurrence of a row of picture points which intersects the right hand boundary of the image is marked with a closed circle. The number of occurrences in this case is 24 and the horizontal projection (also referred to as horizontal intercept) of the image is, therefore, 24 pp in length. (The left-hand boundary may be chosen also; it is arbitrary.)

The number of pp in a given row intersecting the image represents the chord length of the feature for that particular scan of the image. Note that in the third row the corresponding chord length is 10 pp. The mean chord length (c) for the feature is, therefore, given by

$$\bar{c} = \frac{1}{n} \sum_{i=1}^{n} L_i \qquad\qquad (2)$$

where L_i is the chord length in the ith row and n is the number of rows intersecting the feature. Therefore, n is the horizontal projection (H). However, note that the total area of the feature (A) is the sum of all chord lengths

$$\sum_{i=1}^{n} L_i$$

Since A and H for a single feature or all features detected in the field of view are readily measured by the image analyzers, the mean chord is readily obtained by

$$\bar{c} = \frac{A}{H} \tag{3}$$

For anisotropic sturctures the mean chord will obviously vary with the orientation of the structure. This measurement is, indeed, a test for anisotropy.

3. Trabecular Bone Volume

Trabecular bone volume (V) may be readily obtained from measurements of area of bone and marrow on a thin section. It is given by

$$V = \frac{A_b}{A_T} \tag{4}$$

where A_T is the total area of tissue (including bone and marrow) in the measurement frame and A_b is the area of bone only.[12]

Boyce et al.[13] define trabecular bone volume as a measure of the amount of trabecular bone and express it as a percentage of trabecular space between the cortices. These authors use a measurement frame which is bounded by and approximately parallel to the inner surfaces of the cortices. However, they separate the frame boundaries from the cortices by a distance roughly equal to 1.5 times the width of those trabeculae emanating from the endosteal surface of the cortex. This criterion is used to avoid the problem of determining the cortical/trabecular boundary. The reader should refer to the discussion by Parfitt[14] and Chapter 5. They define trabecular bone volume as the percentage of the frame area which is trabecular bone (trabecular bone area/test frame area × 100).

These authors have also defined the parameter total bone density (TBD) as a percentage of the total amount of bone (cortical and trabecular) expressed as a percentage of the total area between the periosteal surfaces. These and related parameters are included in the study by Courpron et al.[15] and by Meunier and Courpron.[16]

B. Other Measurements and Applications

1. Mineral Density

Microradiographs of bone sections are amenable to measurements of relative mineral density and its variation in microscopic volumes. For these measurements, one may use a microdensitometer which may be scanned manually or automatically across the microradiograph. Commercial image analyzers usually have available a microdensitometer as an optional module.

Rowland et al.[17] performed measurements of microscopic mineral density using microradiographs of cortical bone from man, dog, and a few other species, concluding that species-specific differences in calcium metabolism do exist.

Recently Phillips et al.[18] reported on a computerized method of mapping relative mineral density in bone from microradiographs using a Joyce-Loebl double beam recording microdensitometer with a scanning attachment.

2. Lacunar Size

A number of authors in the field of bone histomorphometry have used morphological methods to quantify the metabolic activity of osteocytes. Boyd[19] and Meunier et al.[20,21] have used an automatic scanner (the Quantimet®) for measuring cross-sectional lacunar areas in human iliac bone sections. The latter group points out the advantage of using the system to directly measure the projected areas (which appear ellipsoidal) without taking into consideration their shapes as was required in their manual technique. They demonstrated a highly significant correlation between the two methods. Boyd discusses in depth the pitfalls of various methods of determining lacunar size. This author had measured lacunar areas directly from the scanning electron microscope image by coupling the instrument to a Quantimet® system.

3. Skeletal Remodeling Rates by Fluorescent Labeling

Determination of skeletal remodeling rates is conventionally done by administration of fluorescent compounds (e.g., tetracycline) which are incorporated at bone formation surfaces. Fluorescence microscopy of bone sections is used to quantify the distance (or bone area) between fluorescent labels, and this information is used in the calculation of bone formation rates.

Sontag[22] has developed an automated scanning method for determination of fluorescent labels in bone in order to measure directly bone remodeling rates. Groups of rats were given intraperitoneal injections of a calcein-NaHCO$_3$ solution followed by sacrifice at weekly intervals. Femurs and vertebrae were removed, embedded in methylmethacrylate, and sectioned at 60 μm with a sawing microtome. The fluorescence from the calcein complexes was measured with a Leitz® scanning microphotometer equipped with a fluorescence illuminator. The data analysis was performed by interfacing the system to a PDP 8E computer (which also controlled the stage movement). Each scan field was 20 μm square. Following scanning of the section for fluorescence intensity, the bone section was stained with alizarin red S, the section was again scanned, and the bone structure digitized by processing the transmitted light signal. By superposition of the data from the two scans the amount of bone growth between the time of label (calcein) uptake and death was determined using a mathematical algorithm.

C. Magnification

With automatic scanners it is advantageous to make measurements at as low a magnification as possible since a measurement of a single field of view can be performed rapidly. However, there is a lower limit to the magnification consistent with precision. Since the dimensions of features appearing on the measurement screen can be measured with an accuracy at best ± 1 pp, features (e.g., trabeculae) which have dimensions of only a few picture points may not be measured accurately.

On the other hand, one must be careful not to operate at too high a magnification. When measuring surface-to-volume ratios in sections of primate or canine bone, optical magnifications greater than about ×50 may resolve lacunae and include their perimeter, thus overestimating endosteal surface area.

In the authors' laboratory, measurements on bone sections are generally done at optical magnifications of ×10 to ×40.

IV. SPECIMEN PREPARATION

In preparing specimens for analysis using manual scanners, the conventional histological care and precautions should be taken. However, in preparing specimens for image analysis by automatic scanners, extra precautions are necessary. The reason for this is that the latter will readily include measurements of artifacts which have the same gray level contrast as the bone specimens unless criteria are used which automatically discriminate artifact from bone or marrow images. For example, if the bone features are large compared with artifactitious features, then the device may be programmed to measure only features with an area greater than a preselected threshold. Depending upon the versatility of the instrument, a wide range of criteria may be selected which the features of interest must meet in order to qualify for measurement (it is here that shape factors may play a role).

In preparing sections of bone for microradiography, special care must be used to insure that the sections are of uniform thickness. The thinning out of edges during polishing is a particular problem which results in poorly defined boundaries. In addition the gray level will not appear uniform throughout the section, so that feature detection varies throughout the field of view of the section. Although such conditions can usually be corrected by operator interaction (especially with those instruments having image editor capabilities), the time which is consumed in so doing may become comparable to the time which it would take to measure the specimens manually.

In sawing, polishing, or cutting with a microtome, particular care must be taken to avoid cracks and other artifacts in the section. The boundaries of cracks, for example, may be detected as additional perimeter of the specimen and thus artificially increase surface-to-volume ratios determined from the section (see Equation 1).

It is important to prepare specimens under constant conditions. Microradiography of bone sections should be done with appropriate standardization. As conditions of development (time, temperature, etc.) and X-ray exposure (voltage, current, etc.) change, the measurement parameters may change significantly (by as much as a factor of 2 in extreme cases) for the same bone sections. In order to monitor that these conditions remain constant, it is advisable to use several thicknesses of aluminum foil (or an aluminum step wedge) to be X-rayed and developed along with the batch of bone sections. The optical density of the image of the foil should remain approximately constant for all batches of microradiographs.

With bone section images produced by transmitted light or by radiation (such as microradiographs), the sections must be sufficiently thin to avoid systematic error introduced by section thickness.[23,24] This presents a fundamental problem in microradiography, since obtaining sections of trabecular bone over large areas much thinner than $\simeq 100$ μm is difficult to achieve without destroying or distorting the section. The systematic error introduced using 100-μm sections compared to infinitesimally thin sections has not been quantified experimentally.

In contrast to the parameter dependence on variability in processing microradiographs, the dependence of parameters on section thickness is insignificant in the range in which bone sections usually are sawed and polished ($\simeq 100$ μm). Lloyd and Hodges[7] found no difference in fractional bone volume measurements for sections which were 30 or 100 μm thick using their automatic scanning device (CHLOE). Srisukonth[25] measured perimeter, area, and horizontal projection of microradiographs made from a cross section of a beagle lumbar vertebra as it was successively ground from 120 to 70 μm. He used the Cambridge-Imanco QTM 720® automatic scanner for all measurements. The measurements did not change within the precision of his technique (<10%). A similar study was done by Albright et al.[26] For their measurements they also used the QTM 720® and contact microradiographs of a single rib section from an adult male dog. The section thickness varied from 124 to 50 μm and they found that the coefficient of variation for total bone area and fractional bone volume was $\leqslant$ 5%.

It should be kept in mind that although microradiographs are often used for automatic image analysis (because of their high contrast), only mineralized bone is included in the measurements and osteoid is not included. The bone histodynamics group in Lyon (under the direction of P. Meunier) stain their 8-μm plastic-embedded bone section (cut with a Jung® K microtome) with Solochrome Cyanin R in order to include the osteoid as bone tissue in their automatic image analyses.[13,15]

In normal, adult human bones the specific surface-to-volume ratio one obtains from microradiographs would be a few percent higher than that determined by including osteoid.[27] In cases involving osteomalacia or rickets where an abnormal amount of osteoid formation occurs, the discrepancy could be significant.

V. CONCLUSION

A need for rapid measurements of a variety of parameters from many sites of numerous skeletons is obvious to those in the field of bone histomorphometry. Various methodologies of automated image analysis are commercially available and under development. Their principal advantages are their speed and reduction of observer bias and fatigue and thus they provide a realistic potential for large-scale, precise characterization of skeletal sites for analysis of the processes of skeletal growth and disease.

Although a completely automatic scanner (with on-line computer processing of the data generated) is the goal of automated image analysis, presently its use is restricted to the analysis of the highest quality specimens. Even in these cases, man/instrument interaction is often required to define or exclude areas of measurement on the specimen. Nevertheless, measurements may be made both rapidly and with precision. Presently, most automatic scanners use a television system for image production which allows feature detection based upon gray-level contrast. The introduction of imaging in color, which certainly will be available in the near future, will be a significant advantage for measurements on stained sections.

The manual scanner (or digitizer) is a compromise between the automatic scanner and the conventional manual methods using a planimeter or integrating eyepiece. This method is tedious and time consuming, but the main advantage is that the observer may direct his attention to delineating the regions of features on the specimen for measurement (which is automatically performed by the instrument) without having to perform directly any counting or measuring.

REFERENCES

1. **Ream, A. K.**, Advances in medical instrumentation, *Science,* 200, 959, 1978.
2. **Sherrard, D. J., Baylink, D. J., Wergedal, J. E., and Maloney, N. A.,** Quantitative histological studies on the pathogenesis of uremic bone disease, *J. Clin. Endocrinol. Metab.,* 39, 119, 1974.
3. **Goldner, J.,** A modification of the Masson trichrome technique for routine laboratory purposes, *Am. J. Pathol.,* 15, 237, 1938.
4. **Baylink, D. J. and Wergedal, J. E.,** Bone formation by osteocytes, *Am. J. Physiol.,* 221, 669, 1971.
5. **Holtrop, M. E.,** Quantitation of the ultrastructure of the osteoclast for the evaluation of cell function, in *Bone Histomorphometry,* 2nd Int. Workshop, Meunier, P. J., Ed., Societe de la Nouvelle Imprimerie Fournie, Toulouse, France, 1977, 103.
6. **Robb, R. A. and Jowsey, J.,** Quantitative measurement of fractional bone volume using digital scanning videodensitometry, *Calcif. Tissue Res.,* 25, 265, 1978.
7. **Lloyd, E. and Hodges, D.,** Quantitative characterization of bone: a computer analysis of microradiographs, *Clin. Orthoped. Rel. Res.,* 78, 230, 1971.
8. **Darley, P. J.,** Measurement of linear path length distributions in bone and bone marrow using a scanning technique, in *Proceedings of the Symposium on Microdosimetry,* Ispra, EUR d-f-e, 1967, 509.
9. **Beddoe, A. H., Darley, P. J., and Spiers, F. W.,** Measurements of trabecular bone structure in man, *Phys. Med. Biol.,* 21, 589, 1976.
10. **Spiers, F. W. and Beddoe, A. H.,** Radial scanning of trabecular bone: consideration of the probability distribution of path lengths through cavities and trabeculae, *Phys. Med. Biol.,* 22, 670, 1977.
11. **Whitehouse, W. J.,** A stereological method for calculating internal surface areas in structures which have become anisotropic as the result of linear expansions or contractions, *J. Microsc.,* 101, 169, 1974.
12. **Jee, W. S. S., Kimmel, D. B., Hashimoto, E. F., Dell, R. B., and Woodbury, L. A.,** Quantitative studies of beagle lumbar vertebral bodies, in *Proceedings of the First Workshop on Bone Morphometry,* Jaworski, Z. F. G., Ed., University of Ottawa Press, Ottawa, Canada, 1976, 110.
13. **Boyce, B. F., Courpron, P., and Meunier, P. J.,** Amount of bone in osteoporosis and physiological senile osteopenia: comparison of two histomorphometric parameters, *Metab. Bone Dis. Rel. Res.,* 1, 35, 1978.
14. **Parfitt, A. M.,** Some problems in measuring the amount of bone by histologic techniques, in *Bone Histomorphometry,* 2nd Int. Workshop, Meunier, P. J., Ed., Societe de la Nouvelle Imprimerie Fournie, Toulouse, France, 1977, 103.
15. **Courpron, P., Meunier, P., Bressot, C., and Giroux, J. M.,** Amount of bone in iliac crest biopsy: significance of the trabecular bone volume. Its values in normal and in pathological conditions, in *Bone Histomorphometry,* 2nd Int. Workshop, Meunier, P. J., Ed., Societe de la Nouvelle Imprimerie Fournie, Toulouse, France, 1977, 39.
16. **Meunier, P. and Courpron, P.,** Iliac trabecular bone volume in 236 controls — representativeness of iliac samples, in *Proceedings of the First Workshop on Bone Morphometry,* Jaworski, Z. F. G., Ed., University of Ottawa Press, Ottawa, 1976, 100.
17. **Rowland, R. E., Jowsey, J., and Marshall, J. H.,** Microscopic metabolism of calcium in bone. III. Microradiographic measurements of mineral density, *Radiat. Res.,* 10, 234, 1959.
18. **Phillips, H. B., Owen-Jones, S., and Chandler, B.,** Quantitative histology of bone: a computerized method of measuring the total mineral content of bone, *Calcif. Tissue Res.,* 26, 85, 1978.
19. **Boyd, A.,** Resolution, sampling and the determination of lacunar size, in *Bone Histomorphometry,* 2nd Int. Workshop, Meunier, P. J., Ed., Societe de la Nouvelle Imprimerie Fournie, Toulouse, France, 1977, 399.
20. **Meunier, P., Courpron, P., Edouard, C., Bernard, J., Bringuier, J., and Vignon, G.,** Physiological senile involution and pathological rarefaction of bone, *Clin. Endocrinol. Metab.,* 2, 239, 1973.
21. **Meunier, P. and Bernard, J.,** Morphometric analysis of periosteocytic osteolysis, in *Proceedings of the First Workshop on Bone Morphometry,* Jaworski, Z. F. G., Ed., University of Ottawa Press, Ottawa, 1976, 279.
22. **Sontag, W.,** An automatic microspectrophotometric scanning method for the measurement of bone remodeling rates *in vivo, Calcif. Tissue Int.,* 32, 63, 1980.
23. **Holmes, H. H.,** *Petrographic Methods and Calculations,* Thomas Murby, London, 1927.
24. **Whitehouse, W. J.,** Errors in area measurement in thick sections, with special reference to trabecular bone, *J. Microsc.,* 107, 183, 1976.
25. **Srisukonth, W.,** Quantitative Morphometry of the Vertebrae and Femur of the Beagle as a Function of Age and Sex, Ph.D. thesis, University of Utah, Salt Lake City, 1978.
26. **Albright, J., Martin, R., and Flohr, R.,** Automatic image analysis of the bone biopsy: variations in rib architecture, *Microscope,* 26, 1, 1978.
27. **Smith, J. M. and Jee, W. S. S.,** unpublished data.

W

X